# Systems Analysis Approach for Complex Global Challenges

Priscilla Mensah · David Katerere
Sepo Hachigonta · Andreas Roodt
Editors

# Systems Analysis Approach for Complex Global Challenges

Springer

*Editors*
Priscilla Mensah
Human and Infrastructure Capacity Development
National Research Foundation
Pretoria
South Africa

David Katerere
Department of Pharmaceutical Sciences
Tshwane University of Technology
Pretoria
South Africa

Sepo Hachigonta
Strategic Partnerships
National Research Foundation
Pretoria
South Africa

Andreas Roodt
Department of Chemistry
University of the Free State
Bloemfontein
South Africa

ISBN 978-3-030-10065-0 ISBN 978-3-319-71486-8 (eBook)
https://doi.org/10.1007/978-3-319-71486-8



Printed on acid-free paper

This Springer imprint is published by the registered company Springer International Publishing AG part of Springer Nature
The registered company address is: Gewerbestrasse 11, 6330 Cham, Switzerland

# Preface

There is no doubt that the challenges facing humanity and our planet are complex and interconnected. Tackling them requires holistic approaches and mechanisms that transcend the traditional barriers between disciplines and sectors and takes into account the interactions within a complex system. Systems analysis in this context, is a problem-solving process that involves multiple sectors, countries and experts to develop integrated solutions for multidimensional challenges such as poverty, climate change, energy and rapid population growth in developing countries. For example, climate change driven by rising levels of anthropogenic emissions of greenhouse gases has a progressively increasing impact on the environment. Using systems analysis to integrate these facets has provided new insights for developing nexus solutions, as well as evidence-based policy formulation, implementation and monitoring. Furthermore, systems analysis and integrated approaches will be key to supporting a sustainable policy implementation process for *inter alia* the African Union's Agenda 2063, the Sustainable Development Goals (SDGs) and the Paris Agreement, by enhancing policy cohesion.

The use of applied systems analysis as a research approach in the context of emerging economies is not common. Recognising the role that systems analysis plays in addressing complex social and policy challenges, South Africa identified the strengthening of capacity in systems analysis as a strategic area for increased investment. South Africa's membership as a National Member Organisation (NMO) to the International Institute for Applied Systems Analysis (IIASA) in 2007 was a move towards cultivating high level human resource capacity and promoting this methodological approach in the country and region. While it is reassuring that emerging economies such as BRICS (Brazil, Russia, India, China and South Africa) are member countries of IIASA, only South Africa and Egypt represent the African continent. This further underscores the need for an expanded training programme targeted at future systems analysts on the African continent. Inspired by the success of IIASA's Young Scientists Summer Programme (YSSP), the *Southern African* Young Scientists Summer Programme (SA-YSSP) was launched in November 2012. For a period of 3 years, 80 young scientists from 30 countries, including 35 from South Africa, worked under the supervision of senior scientists from IIASA and South Africa to advance their research in areas related to the environment, energy, risk, population, food, water, and climate change, linked to policy and socio-economic studies.

This peer-reviewed book contains a collection of review articles by those promising young scientists and their academic mentors that showcase the unique ability of systems science to provide practical solutions to almost overwhelmingly complex societal challenges and demonstrates how systems approaches can be incorporated into the science-policy-society interface. As editors, we are confident that this book will not only be an important resource for future scholars, but will also serve as an introductory reference text to researchers and policy makers.

Pretoria and Bloemfontein, South Africa

Priscilla Mensah
David Katerere
Sepo Hachigonta
Andreas Roodt

# Acknowledgement

We thank and acknowledge all authors for their key contributions to this book and expert reviewers for their critical feedback on the initial drafts of the chapters. Without their unwavering support, the book would not have been possible. Our special thanks to Dr. Gansen Pillay, the Deputy CEO for Research and Innovation Support and Advancement at the National Research Foundation (NRF), South Africa and Prof. Pavel Kabat, the Former Director General and Chief Executive Officer of the International Institute for Applied Systems Analysis (IIASA), Austria; Chief Scientist/Director of Research World Meteorological Organization, Geneva, Switzerland, who have been instrumental in initiating joint programmes between South Africa and IIASA. In addition, we would like to express our gratitude to the University of the Free State for hosting the Southern African Young Scientists Summer Programme (SA-YSSP). Furthermore, we wish to thank Dr. Aldo Stroebel (NRF) and Dr. Ulf Dieckmann (IIASA) for their exceptional leadership and support for the SA-YSSP. Our special thanks to South Africa's Department of Science and Technology (DST) for the continuous investment and support in growing the systems approaches knowledge base in Africa.

Finally, we thank the following reviewers who provided us with vital feedback on the book chapters:

Rabidyuti Biswas (School of Planning and Architecture, Indian Institute of Technology, New Delhi, India), Aiyetan Olatunji (Department of Construction Management and Quantity Surveying, Durban University of Technology, Durban, South Africa), Everisto Benyera (Department of Political Science, University of South Africa, Pretoria, South Africa), Tariro Kamuti (Centre for Africa Studies, University of the Free State, Bloemfontein, South Africa), Olusola Ogunnubi (Department of Political Science and Public Administration, University of Zululand, KwaZulu-Natal, South Africa), Gideon Wolfaardt (Stellenbosch University Water Institute, Stellenbosch University, South Africa), Kevin Winter (Environmental and Geographical Science, University of Cape Town), Kevin

Harding (Industrial and Mining Water Research Unit, University of the Witwatersrand, Johannesburg, South Africa), Claudious Chikozho (Africa Institute of South Africa Programme, Human Sciences Research Council, Pretoria, South Africa), Katye Altieri (Energy Research Centre, University of Cape Town, South Africa), Caradee Wright (Medical Research Council, Pretoria, South Africa), Fadiel Ahjum (Energy Research Centre, University of Cape Town, South Africa), Ben Sebitosi (Mechanical and Mechatronic Engineering Department, University of Stellenbosch, South Africa), Olivier Crespo (Environmental and Geographical Science Department, University of Cape Town, South Africa), Voster Muchenje (Faculty of Science and Agriculture, University of Fort Hare, Alice, South Africa), Martin de Wit (School of Public Leadership, University of Stellenbosch, South Africa), Ali Kharrazi (Global Education for Innovation and Leadership Program, Graduate School of Public Policy, University of Tokyo, Japan), Dan Baird (Department of Botany and Zoology, University of Stellenbosch, South Africa), Kenny Uren (School of Electrical, Electronic and Computer Engineering, North-West University Potchefstroom, South Africa), Colleen Seymour (South African National Biodiversity Institute, Pretoria, South Africa), Mirowslaw Lachowicz (Institute of Applied Mathematics and Mechanics, University of Warsaw, Poland), Matthias Langer (Department of Mathematics and Statistics, University of Strathclyde, Glasgow, United Kingdom), Corine de Wet (University of the Free State, Bloemfontein, South Africa), Jan Heystek (Department of Educational Policy, Educational Management, Educational Leadership, North West University, Potchefstroom, South Africa), David Katerere (Department of Pharmaceutical Science, Tshwane University of Technology, Tshwane, South Africa), Alfred Maroyi (Department of Botany, University of Fort Hare, Alice, South Africa), Takwanisa Machemedze (DataFirst Department, University of Cape Town, South Africa), Candice Morkel (School of Governance, University of the Witwatersrand, Johannesburg, South Africa), David Mphuthi (Department of Health Studies, University of South Africa, Pretoria, South Africa), Victor Naidu (Department of Performance Monitoring and Evaluation, University of Fort Hare, Alice, South Africa), Andreas Roodt (Department of Chemistry, University of the Free State, Bloemfontein, South Africa), Roy Tapera (Department of Environmental Health, University of Botswana, Gaborone, Botswana), Peter Tshisikhawe (Department of Botany, University of Venda, Thohoyandou, South Africa).

# Contents

# Editors and Contributors

## About the Editors

**Dr. Priscilla Mensah** is Director in the Human and Infrastructure Capacity Development Directorate of the National Research Foundation (NRF) of South Africa. Prior to joining the NRF, She held a number of positions at the University of the Free State, Bloemfontein, South Africa, including Deputy-Director: International Academic Programs and Deputy Director of the Postgraduate School. She served as Director for the three consecutive summer schools of the Southern African Young Scientist Summer Program (SA-YSSP) in Applied Systems Analysis hosted at the University of the Free State from 2012–2015. She has extensive knowledge of South Africa's Higher Education and Training landscape and the National System of Innovation. She is a specialist in systems analysis with specific emphasis on science and policy within South Africa and beyond. In addition, she is a committee member of South Africa's National Member Organization (NMO) of the International Institute of Applied Systems (IIASA). The work of this committee contributes to the national strategy for developing systems analysis capacity and establishing systems analysis as a research approach in the country. Priscilla holds Masters and Doctoral degrees in Chemistry from the University of Cape Town.

**Prof. David Katerere** is a full Professor of Pharmaceutical Science at Tshwane University of Technology (TUT) where he teaches, conducts research and consults. He has worked for the University of Pretoria, Farmovs-Parexel CRO, SA Medical Research Council and Hwesa Consulting. His areas of interest and expertise are pharmaceutical and analytical chemistry, regulatory science and IKS pharmacognosy. He was involved in the South African Young Scientists Summer Programme (SA-YSSP) as a trainer and resource person on Scientific and Proposal Writing, which is something that he is passionate about. He is an expert advisor/trainer to the South African Medical Research Council (MRC) Grants and Scholarships Division (GASD) and reviewer of grant proposals for the National Research Foundation (NRF) and Medical Research Council (MRC). David has been an official trainer/mentor for the SA Medical Research Council's Research Capacity Programme, including the Emerging Scientists programmes and the Institutional Strengthening Initiatives. He was recently involved in writing and delivering e-learning modules for the Association of Commonwealth Universities (ACU) Structured Training for African Researchers (STARS). This course was initially delivered to over 20 universities in 10 African countries via youtube and webcasting. David holds a BPharm (Hons) degree from the University of Zimbabwe and a Doctoral degree in Pharmaceutical Science from the University of Strathclyde, Scotland.

**Dr. Sepo Hachigonta** is Director of Strategic Partnerships at the NRF of South Africa. He has extensive work experience in trans-disciplinary fields and the regions' science, technology and policy landscape. Prior to joining the NRF, he was a programme manager at FANRPAN, a regional policy analysis network on food security and agriculture based in Pretoria. He has published widely in environmental and agriculture systems, with a specific focus on research to policy interface. Dr. Hachigonta has played a significant role in developing systems analysis expertise in the region through various programmes including the Southern African Systems Analysis Centre (SASAC), a multi-year initiative that take cognisance of multi-level systems analysis capacity interventions, and a comprehensive approach to policy related activities in southern Africa. Having participated in the 2007 Young Scientist Summer Programme, at IIASA, Dr. Hachigonta has remained an active member of the IIASA alumni serving as a lead national coordinator for IIASA engagements in South Africa. Dr. Hachigonta holds a Masters and a Doctoral degree in Environmental Science from the University of Cape Town.

**Prof. Andreas Roodt** is a distinguished professor at the University of the Free State (UFS) and leads the Inorganic Chemistry research group consisting of four senior colleagues and 20 Ph.D. and M.Sc. students and post-doctoral fellows. His prime research focus is on complete reaction mechanisms in Coordination Chemistry systems and chemical processes with application in homogeneous catalysis, radiopharmaceuticals and separation technology. Therein he utilises X-ray crystallography, reaction kinetics, equilibrium and time resolved spectroscopy at ambient, low and high temperature and pressure, and Computational Chemistry. He actively collaborates with South African industries such as SASOL Technology, the Nuclear Energy Corporation of South Africa and Pet Labs Pharmaceuticals, and international research groups in Switzerland, Sweden, Germany, Russia, Croatia, India, Tunisia and the USA. As immediate past president of the European Crystallographic Association (35 member countries across Europe, the Middle East and Africa) he has a passion for the expansion of science in Africa via crystallography and also using a systems approach. He is a Fellow of the Royal Society of Chemistry, international elected member of the Swedish Physiographic Society as well as the Ruđer Bošković Institute in Zagreb, Croatia. He was the winner of the South African Department of Trade and Industry Award for the best THRIP research programme, focusing on Small and Medium Enterprise

Development in 2010 and served as chairperson of the Organising Committees of more than fifteen international conferences in South Africa and abroad. He presented more than 90 lectures at international venues and has made more than 25 international research visits as visiting professor to universities in Switzerland and Sweden. He regularly reviews papers for more than ten international chemistry journals, served on the editorial boards of three international scientific journals and has published more than 300 scientific papers and chapters in books (Scopus H-index of 32). He served as dean for the three consecutive summer schools of the *Southern African* Young Scientist Summer Programme (SA-YSSP) in Applied Systems Analysis hosted at the University of the Free State from 2012–2015. Andreas holds a Doctoral degree in Chemistry from the University of the Free State.

## Contributors

**Mao Amis** African Centre for a Green Economy, Cape Town, South Africa; Graduate School of Business, University of Cape Town, Cape Town, South Africa

**Harold J. Annegarn** Department of Geography and Environmental Sciences, North-West University, Potchefstroom, South Africa

**Jacek Banasiak** Department of Mathematics and Applied Mathematics, University of Pretoria, Pretoria, South Africa; Institute of Mathematics, Łódź University of Technology, Łódź, Poland

**Arnab Banerjee** Ecological Modelling Laboratory, Department of Zoology, Visva-Bharati University, Bolpur, India

**M. Bruce Beck** Department of Civil & Environmental Engineering, Imperial College London, London, UK

**Åke Brännström** Evolution and Ecology Program, International Institute for Applied Systems Analysis, Laxenburg, Austria; Department of Mathematics and Mathematical Statistics, Umeå University, Umeå, Sweden

**Everardt Burger** Department of Civil Engineering, Central University of Technology, Westdene, Bloemfontein, South Africa

**Wisemen Chingombe** School of Biology and Environmental Sciences, University of Mpumalanga, Mpumalanga, South Africa

**Innocent Chirisa** Department of Urban and Regional Planning, University of Zimbabwe, Harare, Zimbabwe

**Dillip K. Das** Department of Civil Engineering, Central University of Technology, Westdene, Free State, South Africa

**Ulf Dieckmann** Evolution and Ecology Program, International Institute for Applied Systems Analysis, Laxenburg, Austria

**Stephen Eromobor** Department of Civil Engineering, Central University of Technology, Westdene, Free State, South Africa

**Aleksandra Falkiewicz** Institute of Mathematics, ŁódźUniversity of Technology, Łódź, Poland

**Delin Fang** State Key Laboratory of Water Environment Simulation, School of Environment, Beijing Normal University, Beijing, China

**Brian D. Fath** Department of Biological Sciences, Towson University, Towson, MD, USA; Advanced Systems Analysis Program, International Institute for Applied Systems Analysis, Laxenburg, Austria

**Sumbul Fatima** Aligarh Muslim University, Aligarh, India

**Le Feng** State Key Laboratory of Water Environment Simulation, School of Environment, Beijing Normal University, Beijing, China

**Adam French** International Institute for Applied Systems Analysis, Laxenburg, Austria

**Jacqueline Goldin** Environmental and Water Sciences Program, Department of Earth Sciences, University of the Western Cape, Bellville, South Africa

**Lucas R. F. Henneman** Department of Biostatistics, Harvard University T.H. Chan School of Public Health, Boston, MA, USA

**Cang Hui** Department of Mathematical Sciences, Stellenbosch University, Stellenbosch, South Africa; Mathematical and Physical Biosciences, African Institute for Mathematical Sciences, Muizenberg, South Africa

**Vain D. B. Jarbandhan** Department of Public Management and Governance, University of Johannesburg, Johannesburg, South Africa

**Pavel Kabat** Former Director General and Chief Executive Officer, International Institute for Applied Systems Analysis, Laxenburg, Austria; Chief Scientist/Director of Research World Meteorological Organization, Geneva, Switzerland

**Thokozani Kanyerere** Environmental and Water Sciences Program, Department of Earth Sciences, University of the Western Cape, Bellville, South Africa

**Carmen Klausbruckner** University of Natural Resources and Life Sciences, Vienna, Austria

**Nadejda Komendantova** Risk and Resilience Program, International Institute for Applied Systems Analysis, Laxenburg, Austria

**Serge Kubanza** School of Geography, Archaeology, and Environmental Studies, University of the Witwatersrand, Johannesburg, South Africa

**Pietro Landi** Department of Mathematical Sciences, Stellenbosch University, Stellenbosch, South Africa; Evolution and Ecology Program, International Institute for Applied Systems Analysis, Laxenburg, Austria

**Johanna Ledwaba** WITS Graduate School of Governance, University of the Witwatersrand (WITS), Johannesburg, South Africa

**Audrey D. Levine** Environmental and Water Sciences Program, Department of Earth Sciences, University of the Western Cape, Bellville, South Africa; University of California, Santa Cruz, CA, USA

**Wolfgang Lutz** Wittgenstein Centre for Demography and Global Human Capital (IIASA, ÖAW and WU), Vienna, Austria

**Maria Marcello** Department of Environmental Engineering, Louisiana State University, Baton Rouge, USA

**Paul Mensah** Institute for Water Research, Rhodes University, Grahamstown, South Africa

**Henintsoa O. Minoarivelo** Department of Mathematical Sciences, Stellenbosch University, Stellenbosch, South Africa; Centre of Excellence in Mathematical and Statistical Sciences, Wits University, Johannesburg, South Africa

**Portia Mokoena** Department of Water and Sanitation, Surface and Groundwater Information, Pretoria, South Africa

**Shakespear Mudombi** Trade and Industrial Policy Strategies (TIPS), Pretoria, South Africa

**Joyita Mukherjee** Ecological Modelling Laboratory, Department of Zoology, Visva-Bharati University, Bolpur, India

**Walter Mupangwa** International Maize and Wheat Improvement Center (CIMMYT), ILRI Sholla Campus, Addis Ababa, Ethiopia

**Bongani Ncube** Centre for Water and Sanitation Research, Cape Peninsula University of Technology, Bellville, South Africa

**Cheryl V. Nikodem** School of Health Sciences, University of Fort Hare, East London, South Africa

**Elvis Nkoana** University of South Africa, Pretoria, South Africa

**Shonali Pachauri** Energy Program, International Institute for Applied Systems Analysis, Laxenburg, Austria

**Dorsamy G. Pillay** Deputy Chief Executive Officer (DCEO): Research and Innovation Support and Advancement, National Research Foundation, Pretoria, South Africa

**Mayank Prakash** International Institute for Population Science, Mumbai, India

**Peter Rafaj** Air Quality and Greenhouse Gases Program, International Institute for Applied Systems Analysis, Laxenburg, Austria

**Tejas Rewal** Department of Architecture and Planning, Indian Institute of Technology, Roorkee, India

**Ursula M. Scharler** School of Life Sciences, University of KwaZulu-Natal, Durban, South Africa

**Mercy Shoko** Statistics South Africa and School of Public Health, University of the Witwatersrand, Johannesburg, South Africa

**Blessing Silaigwana** School of Applied Human Sciences, University of KwaZulu-Natal, Scottsville, Pietermaritzburg, South Africa

**Milaine S. S. Tchamga** School of Mathematics, Statistics and Computer Science, University of KwaZulu-Natal, Durban, South Africa

**Michael Thompson** Risk and Resilience Program, International Institute for Applied Systems Analysis (IIASA), Laxenburg, Austria

**Sylvia Tramberend** Water Program, International Institute for Applied Systems Analysis, Laxenburg, Austria

**Mapula Tshangela** National Sustainable Development, Department of Environmental Affairs, Pretoria, South Africa

**Ikechukwu Umejesi** Department of Sociology, University of Fort Hare, East London, South Africa

**Emmanuel Vellemu** Institute for Water Research, Rhodes University, Grahamstown, South Africa

**Kambidima Wotela** WITS Graduate School of Governance, University of the Witwatersrand (WITS), Johannesburg, South Africa

**Romao Xavier** University of Witwatersrand, Johannesburg, South Africa

**Linlin Xia** State Key Laboratory of Water Environment Simulation, School of Environment, Beijing Normal University, Beijing, China

# Introduction

**Dorsamy (Gansen) Pillay and Pavel Kabat**

**Keywords** Systems analysis · Integrated approaches · Climate change Sustainability · Environment

*"Systems Analysis Approach for Complex Global Challenges"* represents a culmination of collaborative research that directly addresses socioeconomic development challenges. As noted in the Preface, systems analysis plays a pivotal role in solving global challenges, contributing to development and impacting on policy. Applied systems analysis takes into account the interconnectedness of multiple development goals and offers the best chance of overcoming the substantial barriers to sustainability, now and for future generations [1].

The papers in this book represent a microcosm of the larger world of research in systems analysis and demonstrate that complex global challenges have no geographical boundaries. An integrated approach to bringing together the world's best thinkers is critical to finding smarter solutions to these complex global challenges. Systems analysis at IIASA is a problem-solving process in which scientists of various disciplines, stakeholders and decision makers participate. Research in systems analysis lends itself to translation and application for the benefit of society. This book comprises four sections which elaborate on conceptual and practical aspects of systems analysis.

Understanding risk and governance issues in a complex global context requires the inclusion of multiple sources of knowledge with vertical and horizontal links. The opening section, "Risk and Governance", presents different depth and breadth of knowledge relating to the environment (both natural and built), climate change

D. (Gansen) Pillay
Deputy Chief Executive Officer: Research and Innovation Support and Advancement, National Research Foundation, Pretoria, South Africa

P. Kabat
Former Director General and Chief Executive Officer,
International Institute for Applied Systems Analysis, Laxenburg, Austria
Chief Scientist/Director of Research World Meteorological Organization, Geneva, Switzerland

risk, resilience and adaptation and governance, through the development and application of innovative methods and models. From an evidence-informed policy making perspective, this section provides valuable insights into risk and governance which are applicable in local and regional implementation of global policy commitments relating to urban development, climate changes and natural resource management.

The need for integrative and participatory frameworks to better understand interactions between and to promote efficient, inclusive and sustainable management of "Water-Energy-Food" (WEF) resources is the focus of the second section. Adopting a systems perspective to specific WEF resources permits the interrogation of questions such as the following: "How will future environmental and socioeconomic change affect the supply of WEF services, and how can these systems be made resilient to such change?" "How can policy coherence in WEF resource management be used to maximize synergies and avoid tradeoffs?" Thus, this section sets its focus on the intersection of water, energy and food security in sub-Saharan Africa in order to develop nexus solutions under different development and climate pathways.

The functions and services of "Ecosystems" which brings together the concepts of resilience, complexity and advanced analytical methods is investigated in the penultimate section. Ecosystems provide vital goods and services to humanity and must be protected from the adverse effects of global change. However, progress can only be made when a systems approach is used to develop understanding of the inherent complexity of socio-ecological systems. Policy responses also need to be designed in a manner that recognises this complexity, for example by ensuring that in policy development both top down and bottom up strategies are mainstreamed.

Research findings related to people-based systems analysis of health, education and institutions, are presented in the final section. The role of education and health related interventions for improving the quality of life of Africans is addressed from a systems analytical perspective to help broaden the picture and provide deeper insights.

This book is a result of direct investments in systems approaches supported by the National Research Foundation (NRF) of South Africa, in collaboration with the Department of Science and Technology (DST) and the International Institute for Applied Systems Analysis (IIASA) based in Laxenberg, Austria.

## *About the NRF*

*The National Research Foundation (NRF) is an independent government agency that promotes and supports research in all fields of knowledge in order to help improve the quality of life of all the people of South Africa through funding research, the development of high-end Human Capacity and the provision of critical research infrastructure. The NRF promotes South African research interests across the country and internationally working together with communities, research institutions, business, industry and international partners. The NRF is the National Member Organization, representing South Africa's membership of IIASA.*

## *About IIASA*

*Founded in 1972, the International Institute for Applied Systems Analysis (IIASA) conducts policy-oriented research into problems of a global nature that are too large or too complex to be solved by a single country or academic discipline. IIASA's nine research areas are Advanced Systems Analysis; Ecosystems Services and Management; Energy; Evolution and Ecology; Air Quality and Greenhouse Gases; Risk and Resilience; Transitions to New Technologies; World Population; and Water. About 350 researchers from more than 50 countries currently work at IIASA. IIASA is at the centre of a global research network of over 3000 scholars and has active and formalised collaboration with over 720 institutions worldwide. IIASA has 23 National Member Organizations (NMOs). It is sponsored by its NMOs in Africa, Asia, Europe, and the Americas. Its research is independent and completely unconstrained by political or national self-interest.*

## Reference

1. Retrieved November 2, 2017, from http://www.iiasa.ac.at/web/home/about/whatisiiasa/research/what-is-systems-analysis.html.

# Part I
# Risk and Governance

# Chapter 1
# Introduction to Part I: The Role of System Dynamics, Systems Thinking and Systems Perspective

**Mapula Tshangela**

**Abstract** The concept of risk is not new, however, focus on risk is gaining more attention as it relates to addressing increasing environmental risks and their impacts on business and society.

The concept of risk is not new. However, focus on risk is gaining more attention as it relates to addressing increasing environmental risks and their impacts on business and society. The 2015 global policy commitments such as the Sendai framework for disaster risk reduction, the Paris Agreement on climate change and the Sustainable Development Goals are useful in providing the necessary policy signals to address climate related risks. It is argued that environmental risks should be systematically assessed, including the overall vulnerability (Hsu et al. 2017: 3). Proactively managing and reducing the relevant risks once they are identified, is a key response measure towards a sustainable development path (Bennett and Smyth 2016: 252; Janowiak et al. 2017: 223). Chapters 2, 3 and 4 of this book contribute different depth and breadth of knowledge relating to climate and environmental risk assessment and management through systems approaches.

Understanding integrated natural resources risks is not to be approached in isolation. Effective governance arrangements are deemed necessary to speed transition to a just greener economy (UNEP 2011: 148). Improved governance is even argued as more important for the African continent in its responses to climate change (UNECA 2016: 155). Governance in this context includes the governance relations between the multiple actors in natural resources (Wamsler 2015: 2). The multiple actors encompass not only government, but the private sector and civil society organisations' arrangements. Such non-state actors are important contributors to broader climate change governance activities including raising awareness, implementation action and representing marginalised voices (Nasiritousi et al. 2016: 113). Chapters 2, 3 and 4 of this book contribute knowledge on multiple

M. Tshangela (✉)
National Sustainable Development, Department of Environmental Affairs,
Pretoria, South Africa
e-mail: mapula.tshangela@gmail.com; MTshangela@environment.gov.za



P. Mensah et al. (eds.), *Systems Analysis Approach for Complex Global Challenges*,
https://doi.org/10.1007/978-3-319-71486-8_1

pronged systems of governance including local government, institutional settings and natural resources.

The 2011 United Nations Environment Programme (UNEP) Green Economy Report has set the tone for the use of system dynamics models to understand the green economy and sustainability transition. Such a model also enables the understanding of risks, impacts and costs associated with climate change (UNECA 2015: 73). In Ethiopia, for example, the model demonstrated the linkages between the risks of climate change and patterns of development across different economic sectors (UNECA 2015: 21). The model also enabled development of the South African Green Economy Modelling (SAGEM) report that looked at the management of natural resources, investments and economic growth (Musango et al. 2014: 258). Various global, regional and national initiatives further contributed to, for example, the cascading of system dynamics at provincial level. As the case was with the Western Cape Green Economy Model (WeCaGEM). At a local government level, modeling urban systems such as the one conducted for the Australian Logan City, complements the urban planning and the interlinkage with environment, economy and society systems (Brits et al. 2014: 76).

The contribution by Beck et al. (2018) in Chapter 2 is therefore timely as it discusses the use of system dynamics at cities level. The authors are concerned with how a systems approach can be employed to meet the challenges of Cities as Forces for Good in the Environment. They use four cases from Kanyakumari city-region in India, cities of Harare in Zimbabwe, Kinshasa in Democratic Republic of Congo and Bloemfontein in South Africa. The authors draw on cities level experiences, particularly on transport systems, waste water infrastructure, governance in the provision of solid waste management and sustainable energy use. Importantly, Beck et al. (2018) also reflect on their real life experiences with the Southern Africa Young Scientists Summer Programmes. The authors' understanding of Applied Systems Analysis (ASA) were drawn from multi-disciplinary and cross-disciplinary research. The authors report on aspects of ASA prominent in each of the cases in term of six parts. These are on initial crafting of the "problem", identification of the system, its environment, and the system's sub-systems, preliminary analysis of logic of visual-analog representation, computational analysis with a model, screening and analysis of computational results and lastly implementing the decision and handling uncertainty. From a risk point of view, the authors note that care should be taken to avoid the risk of sole reliance on quantitative analysis. Also, from a governance perspective, a key issue is that the structures of participatory governance need to be fit for purpose. While using computational modelling in understanding various interventions and infrastructure at cities level is important, the authors conclude in their key lessons, reflections that engaging with humanities, governance and resilience is also necessary in meeting the challenges of cities as Forces for Good in the Environment.

Climate change is a grand challenge not only identified in policy documents but also as a business risk identified by the World Economic Forum (2016:11). Over the years climate change adaptation specifically is increasingly receiving due attention.

Chapter 3 by Mudombi contributes to the promotion of systems thinking in climate change adaptation, particularly in understanding global change, risks and responses. The systems thinking approach, the author argues, is a necessary and relevant approach to utilise because climate change is complex and involves multiple stakeholders. It is especially important where risks and required governance mechanisms are associated with multiple stressors and multiple scales that interact with each other globally. Through a literature review, the chapter reveals what efforts are underway in understanding climate change risks, resilience and adaptation. The magnitude of global risks already experienced over the years and the required resilience and adaptation cannot be over emphasised. It is in that context, and with a systems thinking approach, that Mudombi draws attention to the linkages between risks, vulnerability and stressors. A systems thinking approach is used to analyse key enabling responses relevant to global change risks and governance. These areas are on taking a "glocal" approach, information generation and dissemination, relevant and responsive institutions, flexibility and learning, building the asset base of households and communities, ownership, participation, and stewardship, enhancing ecological infrastructure and lastly forging partnerships and collaborations. From a risk point of view, the chapter reveals that as a holistic approach that is focused on the whole system, resilience is better placed to address global change issues rather than managing individual risks. Such resilience building includes consideration for risk landscape at local and global contexts, while including environmental, social and economic risks underpinned by relevant research, innovation, information and knowledge. From a governance perspective, key issues revealed include the need for relevant institutions to embrace risks in a holistic manner, since sector-based governance systems have a limited ability to be reactive. The local "good" governance is argued as contributing to resilience building, and that governance systems need to be appropriate to local settings. The author concludes that more work is required, and that such work should include strengthening multi-disciplinary partnerships, raising awareness and applying the correct assessment methodologies in order to facilitate appropriate adaptation responses.

Development has different meanings for different actors and often with competing interests, hence, it is expected that governance will be a challenge (Perrot 2015: 37). Furthermore, the complex interplay between natural resources, energy- and carbon-intensive industries remain a challenge in addressing development imperatives (Montmasson-Clair and Ryan 2014: 44).

In Chapter 4, Umejesi et al. (2018) focus on understanding the struggles between four kinds of social solidarity through historical survey and three cases on natural resource governance. With a focus on Africa, the authors base their argument on the role of policy and governance as it relates to individualism and hierarchy dominance, thereby excluding the other two social solidarities: egalitarianism and fatalism in resource related conflicts. The chapter reveals that individualism and hierarchy are dominating within the cases of Reducing Emissions from Deforestation and Degradation plus (REDD+) in the Democratic Republic of Congo, acid mine drainage in South Africa and oil extraction in Nigeria. From a risk standpoint, programmes such as REDD+ trigger interest not only from environmental but also interaction with political, market and business risks (UNEP 2011: 591). With respect to governance, the authors argue that a shift from crap governance towards good governance is important in managing resources. The authors also argue that the analysis of resource-related conflicts often ignores social solidarities and conclude that the use of a systems perspective is likely to incorporate all voices.

A deeper understanding of risk and governance issues within contexts that involve multiple stakeholders including government, industry and social partners around the management of natural resources is a common thread within all three chapters. Systems dynamics, systems thinking and systems perspective, as promoted through Applied Systems Analysis (ASA), are useful for understanding such complex arrangements. From an evidence-informed policy making point of view, Chapters 2, 3 and 4 provide valuable information that is relevant for the local and regional implementation of global policy commitments relating to natural resources.

## References


Beck, B., Das, D. K., Thompson, M., Chirisa, I., Eromobor, E., Kubanza, S., Rawal, T., & Burger, E. (2018). Cities as forces for good in the environment: A systems approach. In P. Mensah, S. Hachigonta, D. Katerere, A. Roodt (Eds.), Systems analysis approach for complex global challenges. Heidelberg: Springer.

Bennett, M., & Smyth, S. (2016). How capital markets can help developing countries manage climate risk. *Boston College Environmental Affairs, 43,* 251.

Brits, A., Burke, M., & Li, T. (2014). Improved modelling for urban sustainability assessment and strategic planning: Local government planner and modeller perspectives on the key challenges. *Australian Planner, 51*(1), 76–86.

Hsu, T. W., Shih, D. S., Li, C. Y., Lan, Y. J., & Lin, Y. C. (2017). A study on coastal flooding and risk assessment under climate change in the Mid-Western Coast of Taiwan. *Water, 9,* 390.

Janowiak, M. K., Iverson, L. R., Fosgitt, J., Handler, S. D., Dallman, M., Thomasma, S., et al. (2017). Assessing stand-level climate change risk using forest inventory data and Species distribution models. *Journal of Forestry, 115*(3), 222–229.

Montmasson-Clair, G., & Ryan, G. (2014). The impact of electricity price increases on the competitiveness of selected mining sector and smelting value chains in South Africa. Pretoria.

Mudombi, S. (2018). Risk, Resilience and adaptation to global change. In P. Mensah, S. Hachigonta, D. Katerere, A. Roodt (Eds.), Systems analysis approach for complex global challenges. Heidelberg: Springer.

Musango, J. K., Brent, A. C., & Bassi, A. M. (2014). Modelling the transition towards a green economy in South Africa. *Technological Forecasting and Social Change, 87,* 257–273.

Nasiritousi, N., Hjerpe, M., & Linne´r, B. (2016). The roles of non-state actors in climate change governance: Understanding agency through governance profiles. *International Environmental Agreements, 16,* 109–126.

Perrot, R. (2015). The Trojan horses of global environmental and social politics. In L. Mytelka, V. Msimang, R. Perrot (Eds.), On behalf of Mapungubwe Institute for Strategic Reflection. Johannesburg: Real Africa Publishers.

Umejesi, I., Thompson, M., Marcello, M., Vellemu, E. (2018). Extract of Africa: towards the equitable and ecologically sound governance of mining and drilling. In P. Mensah, S. Hachigonta, D. Katerere, A. Roodt (Eds.), Systems analysis approach for complex global challenges. Heidelberg: Springer.

United Nations Economic Commission for Africa. (2015). Inclusive green economy policies and structural transformation in Ethiopia, United Nations, Addis Ababa, Ethiopia.

United Nations Economic Commission for Africa. (2016). Greening Africa's industrialization: Economic report on Africa, United Nations, Addis Ababa, Ethiopia.

United Nations Environment Programme. (2011). Towards a green economy: Pathways to sustainable development and poverty eradication.

Wamsler, C. (2015). Mainstreaming ecosystem-based adaptation: Transformation toward sustainability in urban governance and planning. *Ecology and Society, 20*(2), 30.

World Economic Forum (2016). The Global Risks Report 2016 (11th edn.). World Economic Forum. [Online]. Available at http://www3.weforum.org/docs/GRR/WEF_GRR16.pdf. Accessed Sep 2017.

# Chapter 2
# Cities as Forces for Good in the Environment: A Systems Approach

**M. Bruce Beck, Dillip K. Das, Michael Thompson, Innocent Chirisa, Stephen Eromobor, Serge Kubanza, Tejas Rewal and Everardt Burger**

**Abstract** *Background*: The various elements of infrastructure in cities and their systems of governance—for transport, buildings, solid waste management, sewerage and wastewater treatment, and so on—may be re-worked such that cities may become forces for good (CFG, for short) in the environment. The chapter is a study in the lessons learned from implementing and pursuing research into how a systems approach can be employed to meet the challenges of achieving CFGs. *Methodology*: Four case studies in CFG are presented within the framework of the methods and computational models of Systems Dynamics (SD): transport infrastructure for the Kanyakumari city-region in India, resource recovery from wastewater infrastructure in the city of Harare, Zimbabwe, environmental injustice in the handling of solid

The original version of this chapter was revised: Corresponding and chapter author names have been corrected and a deleted term has been included. The erratum to this chapter is available at https://doi.org/10.1007/978-3-319-71486-8_18

M. B. Beck (✉)
Department of Civil & Environmental Engineering,
Imperial College London, London, UK
e-mail: mbrucebeck@gmail.com

D. K. Das · S. Eromobor · E. Burger
Department of Civil Engineering, Central University of Technology,
Bloemfontein, South Africa

M. Thompson
Risk and Resilience Program, International Institute for Applied Systems Analysis,
Laxenburg, Austria

I. Chirisa
Department of Urban and Regional Planning,
University of Zimbabwe, Harare, Zimbabwe

S. Kubanza
School of Geography, Archaeology, and Environmental Studies,
University of the Witwatersrand, Johannesburg, South Africa

T. Rewal
Department of Architecture and Planning,
Indian Institute of Technology, Roorkee, India

 
P. Mensah et al. (eds.), *Systems Analysis Approach for Complex Global Challenges*,
https://doi.org/10.1007/978-3-319-71486-8_2

municipal wastes in Kinshasa, Democratic Republic of Congo, and improving the use of energy in university campus buildings in Bloemfontein, South Africa. *Application/Relevance to systems analysis*: The chapter presents the successes and the difficulties of undertaking Applied Systems Analysis (ASA) in demanding urban contexts. *Policy and practice implications*: Policy for CFG derived from ASA often appears to be a matter of determining better technological innovations and engineering interventions in the infrastructure of cities, while practice often demands that infrastructure improvements follow from social and institutional improvements. *Conclusion*: The first of three conclusions is that combining the rigorous, logical, non-quantitative, more discursive and more incisive style of thinking derived from the humanities, particularly, social anthropology, with better computational modelling will yield better outcomes for ASA. Secondly, in a global context, cities—as opposed to nation-states—are increasingly becoming the locations and scale at which today's environmental, economic, and social "problems" might best be "solved". Third, and last, we conclude that South Africa, while it may not have a long tradition of problem-solving according to ASA, has for us emphasised (through our experience of the South African YSSPs) the limitations of an historical over-reliance on hard, quantitative methods of systems analysis.

## 2.1 Introduction

In the 2007 Winter edition of *Options*—a quarterly publication of the International Institute for Applied Systems Analysis (IIASA)—a short article was published (Crutzen et al. 2007). It bore the title "Grand Challenges for Engineering. Turning Cities into Forces for Good in the Environment". It had originally been commissioned as a background essay for the 2006/2007 US National Academy of Engineering's Blue-ribbon Panel, which was deliberating upon what should be the "Grand Challenges for Engineering in the 21st Century" (NAE 2016). The essay's authors, all Institute Scholars at the time, were Paul Crutzen (Nobel Laureate in Chemistry), M. Bruce Beck (Engineer), and Michael Thompson (Anthropologist). It asked (Crutzen et al. 2007):

> How can the built infrastructure of the city be re-engineered to restore the natural capital and ecosystem services of the nature that inhabited the land before the city arrived there, in "geological time"?
>
> How can this infrastructure be re-engineered to enable the city to act as a force for good, to compensate deliberately and positively for the ills of the rest of Man's interventions in Nature?
>
> How can cities of the Global South avoid adopting the same technological trajectory as those of the Global North? Can they, as it were, "leap-frog" the Global North by forgoing the entire human-waste-into-the-water-cycle phase, thereby ending up one step ahead?
>
> More profoundly, how can the engineering of city infrastructure be deployed expressly so that those at the bottom of the pyramid of dignified human development may be brought to a level where they care to engage in a debate over such a grand challenge for this century — of cities as forces for good — beyond their desperate needs of survival for just today and tomorrow?

In the decade since publication of the *Options* article, a substantial body of work has evolved and been developed under the rubric of Cities as Forces for Good in the Environment (CFG, for short). Much of it can be found archived at www.cfgnet.org.

While all four of the challenges are germane to the discussion of this chapter, those of the third and fourth are especially relevant. The authorship of the chapter reflects work conducted collectively during the 2013–2014 and 2014–2015 South Africa Young Scientists Summer Programmes (SA-YSSP), hosted in Bloemfontein, South Africa, under the auspices of IIASA.

This chapter is a further contribution to the ongoing programme of CFG research. Above all, however, it illustrates how Applied Systems Analysis (ASA) can be put to effective and distinctive use in problem-solving and question-setting in this broad area of making cities less unsustainable. Accordingly, the chapter begins with a succinct working definition of what we understand the essential nature of ASA to be. Four case studies are then set out in order to convey an impression of ASA at work. We could have arranged the sequence of their presentation in a number of ways, but have chosen to order them along the axis of decreasing scale: from the city nested in a regional context, through the city-scale itself, to the smaller scale of educational campuses (themselves nested in cities). We begin therefore with a study of transport systems, followed by one of resource recovery from urban wastewater infrastructure, then another of solid waste management (but more especially, the issue of failing governance in the provision of this service), and the chapter closes with a case study of sustainable energy use in building complexes on educational campuses. In this last, we can usefully think of a campus as a force for good in the environment (cfg), at work and nested within the city. Put cryptically, this would be a case of CFG{cfg}.

It will be apparent how these studies entail features ranging from the more quantitative aspects of engineering systems analysis across to the (in part) discursive nature of anthropological enquiry and analysis. Our four case studies refer to four specific locations: the district of Kanyakumari, Tamil Nadu, India; city of Harare, Zimbabwe; city of Kinshasa, Democratic Republic of the Congo (DRC); and the city of Bloemfontein, South Africa.

Expectations of this chapter must be placed in context. For a period of just three months, young scientists came to Bloemfontein to acquire an appreciation of what ASA is, and then begin immediately applying what they were learning (as they went through the experience of the SA-YSSP) to some of the most recalcitrant problems in some of the most challenging circumstances around the world. Inevitably, the case studies ended up "works in progress". What this means is that the chapter is about how ASA can (and is) being put to work in the utmost demanding of real-life circumstances. It is not about summarizing final research outcomes from the four case studies.

## 2.2 An Understanding of Applied Systems Analysis

For many the term Applied Systems Analysis (ASA) may suggest analysis of a more quantitative nature, with a reliance on the use of computational models, in particular. Indeed, so it is. The use of models is pervasive, as much in our own field of environmental systems analysis as in any other. And there is no shortage of contemporary "grand challenges" for future research in respect of computational modelling, as expressed in a 2009 White Paper prepared for the US National Science Foundation (Beck et al. 2009).

Whatever it is in the universe that grips our interest (the city here), let us "go for it". Let us cut out the city from everything else with which it interacts, thus dividing the world (for the purpose of technical analysis) into the *system* and its *environment*. Some aspects of the city's socioeconomic life constitute the system (its metabolism, for example; Wolman 1965; Barles 2007a, 2007b; Beck et al. 2013). Some aspects of the remainder of the world make up the system's environment, for instance, regional or national infrastructure, flows of goods and resources into the city, the atmosphere, surrounding water bodies (both upstream and downstream of the city), and so on. From within the environment emerge disturbances impinging on the system; and likewise, from the system arise things that disturb the environment. But if the city is everything to us, it matters not (primarily) how what has been defined as the environment absorbs those disturbances from the system. Likewise, it does not behove us to work out fully or exactly how the disturbances that affect the system are generated by its environment, but we shall surely want and need to make some astute assumptions about the nature and characteristics of those disturbances, especially as they evolve into the future.

What *is* of concern is an understanding of the mechanics of how one thing relates to another within the system. A more or less complex computational model of those mechanics is constructed and run in various ways, forecasts of future behaviour of the city are made for a variety of policy-technology interventions, from which a specific decision is arrived at and implemented in practice. Thus runs ASA—in short and in part—and sufficient evidence of this will follow later in this chapter (as well as can be found in other chapters of this book).

The skill in all of this, of course, is to avoid the "paralysis of systems analysis": that everything is related to everything else (which, in the end, it is); so how on earth—might the paralysed analyst protest—can we simply, effectively, and incisively extract the system of interest to us?

There are a couple of other essential features of ASA, however. They have to do with Systems Thinking.

***Systems Thinking—Multi-disciplinary***

First, there is that perhaps overworked phrase, of being "holistic" in one's analysis. All things about the problem to be addressed are taken into account, even the proverbial kitchen sink. In this multi-disciplinary context, the problem is inspected from a great variety of perspectives, in order eventually to generate a seamlessly composed solution (all the better than otherwise for being so comprehensively well-rounded). An instance of multi-disciplinarity can be found above in the backgrounds of the authors of the *Options* article. An example of the outcome of this kind of holistic thinking and analysis is the *Sustainability Concepts Paper* (Beck 2011), although we should note (and warn) that it took nine years to compose. Our Harare case study of Sect. 2.3.2 below is also an illustration and beneficiary of such holism—though self-evidently one that attracted but a few months' thought!

We should also hasten to acknowledge a difficulty with the words "problems" and "solutions". There are messy issues referred to as "wicked problems", which do not submit to being "solved" once and for all. Unlike with the familiar "tame problems", where there is general agreement on what the problem is (the hole in the ozone layer, for instance), wicked problems are characterised by conflicting definitions of what the problem is (climate change, for example) and these do not converge as the policy process progresses (Ney 2009; Verweij 2011). So our use of the words problem and solution is primarily for the great convenience of shorthand.

***Systems Thinking—Cross-disciplinary***
Holism is not *all* of Systems Thinking. Its second feature may be summed up as the "cross-disciplinary communications busbar". At its very simplest, given a question (say Q) with a ready-made answer (say A) in one discipline, is it possible to map this Q across to an identically crafted question in a second discipline, where (in that other discipline) that question has no answer? For if such is possible, the previously unanswered question in the second discipline may, in fact, be answered by the suitably cross-disciplinary transcription of A from the first discipline. Doing this is to be engaged in sparking both novel A's *and* novel Q's, back and forth across disciplines. Our introduction of the Kinshasa case study illustrates the process, albeit very briefly (in Sect. 2.3.3).

Much more fully, in *Man and Nature, A Complex But Single System* (Thompson 2002) ideas about the nature of social dynamics (and implicitly resilience) are mapped one-to-one from Anthropology to Ecology—to the betterment of both. When it comes to the subject of this chapter (cities, urban infrastructure, and CFG), further elaboration and implementation of the cross-disciplinary busbar are in progress. Specifically, the goal is to transcribe ideas of governance and resilience in the human environment (Man) to practical engineering realisations of resilience in designing and operating the built environment (Built). In this, the cross-disciplinary busbar is working to transfer answers (A's) from Ecology, through Anthropology, and into Engineering, thus to answer some of Engineering's unanswered questions. Things, just as much, should flow the other way. There is surely another article in the offing: *Man and Nature* and *Built, A Complex But Single System.*[1]

[1]There is also, or alternatively, a book in the offing (it is being prepared by one of us; MBB). It addresses this cross-disciplinary facet of ASA, especially in respect of developing the pan-systems concepts of resilience, learning, and adaptation. And our grasp of those concepts—hence the significance of this footnote—owes a very great deal to an avenue of thinking, analysis, and research opened up some four decades ago, in the early, formative years of IIASA. This was on the occasion of publication of the book *Adaptive Environmental Assessment and Management* (Holling 1978), with its first cross-disciplinary steps, from ecology (Nature) to institutional

All together—holism, the cross-disciplinary communications busbar, computational models, *and* ever deeper disciplinary analysis—should add up, as they say, to more than the sum of these elementary parts.

### 2.2.1 Overall Procedure of ASA for the Cities Case Studies

Given this brief account of the wider purview and reach of Systems Thinking, we now set out those particular aspects of ASA most prominent in our case studies. Specifically, each case study is reported roughly according to the following template, with its six parts.

***(A) Initial Crafting of the "Problem"***

In a sense, nothing is new under the sun. Someone, somewhere, will have published their analysis of how to make cities similar to Kinshasa, Harare, and Bloemfontein less unsustainable. What matters here, however, is not the recitation of all the painful and distressing challenges of which we are already abundantly aware. The review of the literature must combine penetrating insights from the experience of others into the specific issues to be addressed in the particular, if not unique, circumstances of each city at the focus of our four case studies. Armed with the template here being defined, preliminary constructive criticism is to be developed of the presence/absence or success/failure of ASA in the other literature studies under review. A good review is not merely a list of who else has done what. It has an intellectual spine, to which are attached the literature citations, strictly according to the forward-moving logic of the spine. Development of that coherent spine will often be the first challenge of ASA.

***(B) Identification of the System, its Environment, and the System's Sub-systems —Prelude to Formal Analysis***

To reiterate, the choice and definition of the system are driven by the issue (problem) we are seeking to resolve (solve). All systems have sub-systems, and the sub-systems have sub-sub-systems, and so on, just as in scaling down from CFG to cfg. With each step down into greater detail, defining the sub-system tacitly includes some relatively more detailed features in the analysis and excludes others. At each stage, some things are made important, others unimportant. Thinking in the opposite direction, the system is contained within a supra-system, the supra-system in a supra-supra-system, and so forth. Given our system-centric perspective, however, the supra-system and supra-supra-system are intuitively facets of the environment surrounding the defined, cut-out system.

How the analytical dividing line is drawn around the system and its various sub-systems is, at bottom, a matter of the *subjective* judgement and choice of the

behaviour (Man). Further substantial steps—great strides, one might better say—were taken subsequently in Holling (1986).

Systems Analyst. Such choices are born of a combination of the preliminary sculpting of the problem to be addressed (Step (A)) and some anticipation of the nature of the formal analysis to follow (Step (D), in particular). Crucially, someone has to decide what to include and what not. Few prior criteria may exist to discriminate a good from a bad subjective choice, which distinction we should fully expect to emerge only after the event.

In our four case studies, in the light of the experience of our second author (DKD), the prior supposition was that formal analysis would in all probability be conducted within the software framework of Systems Dynamics (SD) modelling (Forrester 1969; Sterman 1982; Coyle 1996; Shen et al. 2009; Das and Sonar 2013). After all, if everything has to happen within a three-month period, one had better have the insurance policy of a clear eventual form and methodology of systematic analysis to be applied. Apart from this inherent prejudice, to which inclinations we are *all* subject, SD has the advantage of being a device by which to make a first formal, eventually (quasi-)quantitative engagement with the problem at hand. It is, moreover, something that can be done even in "data-poor" situations. Other preludes to subsequent applications of other formal procedures of ASA are possible. In some instances, the analyst might instead have her/his eye on the eventual construction and use of belief networks (Varis 2002) or Bayesian nets (Borsuk et al. 2004) or the classical methods of Decision Making Under Uncertainty (DMUU; Reichert et al. 2007, 2015). The point is that such anticipations of the next step can influence the way in which the analyst chooses to define the system in the first place. We are well aware of the trap: given the hammer as the solution, all problems will be crafted as that of a nail needing to be driven into wood.

***(C) Preliminary Analysis of Logic of Visual-analog Representation***

Specifically, the prelude to more detailed, computational SD analyses is the construction of a causal loop diagram for the system (as we shall illustrate in the case study of Kanyakumari in Sect. 2.3.1). This is a visual, largely qualitative representation of how the analyst believes the system behaves in respect of the interactions among those entities that matter to resolving the problem crafted in Step (A). In this, to be pragmatic and without pretence (and rather as in the construction alternatively of a decision tree), the very process of assembling the loop diagram can be more instructive and enlightening than the end-product itself. The merit of the loop diagram (as likewise for belief networks and Bayes networks, for instance) is the manner in which it can impose some logic on how elements from the natural, human, and built environments[2] interact with one another (Nature, Man, Built). It is biased (and distinguished from the other visual-analog representations) in the sense that the loop diagram is tied to the presumption of the system's behaviour being analysed in due course through the numerical solution of differential equations (which is integral to the SD software). Intuitively, however, we do

[2]Using the word "environment" in its more familiar sense, i.e., not in the technical sense of "everything surrounding the defined system".

not generally consider the affairs of human-centred socioeconomic systems or sub-systems (the affairs of Man) to lend themselves to analysis through the quintessential calculus of the natural and (more so) built environment.

Formal analysis of the causal loop diagram entails the following. Having assembled and meshed together the constituent cause-effect (disturbance-response) couples of what affects what, the sense of positive or negative influence of cause over effect may be assigned. The positive (negative) designation reflects the judgement that an increase in the magnitude of the disturbance induces an increase (decrease) in the magnitude of the response. Perhaps most significant is the identification of feedback loops that may not have been apparent a priori. It matters greatly whether the loop has a stabilizing effect (negative feedback) or de-stabilizing effect: the positive feedback of, say, a small increase in a disturbance evoking a chain of responses propagating around the loop in a manner that will return to increase the disturbance yet further. In particular, analysis of the causal loop diagram can identify those feedback loops that may dominate the behaviour of the system as a whole, hence condition the likely shape of the policy-technology interventions to be tested subsequently (in Step (D)).

***(D) Computational Analysis with a Model***

In many instances, analysts are catapulted into this step (as it were) almost from the beginning of the ASA. They may be dealing with problems largely identical in form and structure to those addressed many times previously. Or it may be that only a handful of pre-formed computational models are available for implementing the formal quantitative analysis. For example, very few Integrated Assessment Models (IAMs) are available for computing the economic consequences of climate change (e.g., Nordhaus 2014; Anthoff and Tol 2014), with the result that the nature of the problem may have to be tailored to one or the other IAM (witness Mercer 2015; CISL 2015).

We have judged this to be anything but the case in our studies of cities reported in this chapter. That being so, the open "free" form of an SD analysis is quite an advantage. In outline, one may proceed from the network of entities and their interactions in the causal loop diagram; through a stock-flow diagram, i.e., a set of boxes representing the stock of some entity and links between the boxes representing logical flows of entities; and so to an icon-based interface to a set of ordinary differential equations formed with the software in the computer and ready for computational numerical solution. Standard Runge-Kutta and Euler methods are used for numerical solution of these equations.

Answers are thus generated to a set of "What If?" questions: what happens if nutrient recovery technology is installed in Harare's sewerage sub-system; what would happen if the system of governance for solid waste management in Kinshasa could be "clumsified" (Thompson and Beck 2014). Ease and simplicity are good descriptions of the process of applying an SD analysis.

***(E) Screening and Analysis of Computational Results***

The purpose of the computational exercises with the SD model is, of course, to assess what might happen in the future, given this, that, or the other form of policy-technology intervention in the system. The computer becomes the laboratory world of policy formation, not least because we cannot reason through these things in our heads, neither consistently nor many (repetitive) times over. Having generated all the answers to all the "What If?" questions, the pairs of interventions and future outcomes are assessed against a set of decision criteria, in order to arrive at one specific, preferred intervention. To over-simplify, we might say we are answering some "How Best?" questions at this stage in the ASA.

***(F) Implementing the Decision and Handling Uncertainty***

Too many times has this been said:

> If the problems of unsustainable cities were ones of technology and engineering alone, then surely they would have been fixed long ago.

The inference, for resolving these tough, urgent challenges (those of water, in particular), was that it was good governance that was needed, above all else, if not indeed *exclusively* so (Beck 2011; Beck and Villarroel Walker 2011; Beck 2016).

If therefore the contemplated intervention is a matter primarily of engineering and technology, the ASA should entail an integral assessment of any institutional (governance) issues that may facilitate or stifle actually implementing the intervention to good effect in practice, in real life (see also Beck et al. 2011). As will doubtless become quickly apparent, the sole purpose of our Kinshasa case study (in Sect. 2.3.3) is indeed to search for interventions intended to change institutional arrangements, not interventions intended to change the "mechanics" of systems for mobility (as in Kanyakumari), or for prudent, sustainable manipulation of the resources of water, energy, and nutrients flowing into, through, and out of the campus, the city, or the region (in Harare and Bloemfontein). But the two, policy and technology, should not be separated. Not for nothing, therefore, have we referred above to "policy-technology" interventions.

Steps (A) through (F) of the ASA have been presented linearly. This is unavoidable in a written document. In practice, messy substance will tend to predominate over the neatness of form of a linear procedure, with several iterations back and forth among the stages actually being required, including perhaps continual adaptation of what exactly is the problem to be solved. This, we shall discover, is especially so for our second case study of Harare (Sect. 2.3.2).

In retrospect, given the highly untidy and awkward nature of the problems in our four case studies, it is perhaps surprising the word uncertainty has been mentioned just once (other than its announcement in the title of this sub-section). Its previous appearance was as the distinguishing and defining last letter in the acronym DMUU. Much attention has been focussed on the subject of uncertainty in the areas of interest to this paper (for example, Beck 2014). Suffice it to say, had we been blinded by all the uncertainty, there may well have been no progress in any of our case studies. Yet it is important to point out that some of the deepest and most intractable sources of uncertainty in applying ASA to the study of policy interventions for moving towards less unsustainable cities are those arising from the starkly contrasting and mutually irreconcilable beliefs analysts and experts (and those famous other "stakeholders") hold about the way the system works (Patt 2007). Plural, mutually opposed rationalities thrive, as very well illustrated in a 2009 study of how to restore the quality of waters in the Bagmati River as it passes through Kathmandu, Nepal (NWCF 2009).

The theory of plural rationalities, or Cultural Theory (CT), suggests that not just one causal loop diagram or just one SD model of the city may exist, but four (Thompson et al. 1990). So strongly may the truth of each rationality be asserted by its proponent, we have no DMUU, but the much more uncertain and problematic case of Decision Making Under Contradictory Certainties, or DMUCC (Thompson and Warburton 1985; see also Beck 2014). An SD-supported analysis of such a state of affairs strikes us as a most worthy research question and one yet to have been addressed anywhere (to the best of our knowledge).

## 2.3 Case Studies

Our first case study benefits from relatively extensive, prior assessment of the problem to be addressed. Consequently, during the 2014–2015 South Africa Young Scientists Summer Programme (SA-YSSP), work on it was able to progress as far as implementing Steps (D) and (E) in the procedure of ASA set out in Sect. 2.2.

### *2.3.1 City-Region Transport: Kanyakumari, Tamil Nadu, India*

Across India in general, lack of appropriate road infrastructure is a primary hindrance to economic development and growth. In 2009, only about 16% of India's roads had 4 lanes or more (NHDP 2014). In a 2007 survey, average speeds for buses and trucks were found to be just 30–40 km/h (19–25 mph), with the obvious environmental and resource consequences of inefficiencies in fuel-consumption rates and increases in gaseous, polluting emissions of $NO_X$, $SO_X$, methane, CO and $CO_2$ (WHO 2009; NHDP 2011). The country has the same per capita rate of deaths from road accidents as the USA, such that, with India's larger population, this amounts to 105,000 road-related deaths each year (WHO 2009). Transport, we may conclude, is an issue in need of some problem-solving capacity at the national scale.

At smaller scales, the design of networks for mobility in and around the city matters for sustainability and resilience, as well as health and well-being (ICSU 2011; Salingaros 2005; Mehaffy and Salingaros 2014; Arup 2014). In the region of

Kanyakumari, Tamil Nadu, India, however, the more urgent priority is for effective transport links to be established between the sites of production of goods and the sites of their consumption—*without* the anticipated economic benefits therefrom condemning the inevitable environmental impacts thereof to the customary status of unaccounted-for economic externalities (in this, the reader may find of some interest the report *Natural Capital at Risk. The Top 100 Externalities to Business*; Trucost 2013). In Kanyakumari such environmental impacts are wide-ranging and diverse: on climate change, air quality, water quality, soil quality, biodiversity, land-use/habitat change, and noise.

***(A) Initial Crafting of the "Problem"***

In this most mature of our four case studies, the problem is therefore this: environmental sustainability must be manifest in the development of city-region transport, from vehicle emissions control to green supply chain practices (even supply *circle* practices; Beck 2011).

***(B) Identification of the System***

The city-region of Kanyakumari comprises, first, the cities/towns of Nagecoil, Kuzhithurai, Colacel, Padmanabhapuram, and Kanyakumari itself and, second, a network of roads, i.e., National Highways (NH) 47, 7, 47B, State Highways (SH) 45, 46, 179, 189, 194, and a set of major district roads such as M187, 188, 216, 412, 930, 932, 943, 979, which link the cities and their "hinterlands". The region occupies an area of roughly 1684 km$^2$ (approximately 62 km by 27 km). Running through essentially the middle of the region, National Highway NH 47 connects Nagercoil and Kaliyakkavilai. It is the backbone of the system, in terms of both its physical and socioeconomic sub-systems. Its influence touches upon almost every aspect of the behaviour of the region. The majority of passenger flows and flows of goods pass along NH 47. It provides access to the majority of the region's economic activities (industry, financial institutions, academic institutions), with thus an impact on the system's economic sub-system, through incomes and the generation of employment opportunities.

More specifically, flows crossing the system boundaries are accordingly those of traffic along the trunk roads and the fuel imported with the trucks, buses, and automobiles, as inputs. Spent fuel gives rise to flows of $NO_X$, $SO_X$, methane, CO, and $CO_2$ as outputs to the atmosphere. The system's sub-systems are those of road transport, the socioeconomic metabolism of the city-region, and the various segments of natural environment within the region. The challenges attaching to the spent fuel outputs and the other impacts on the natural environment of the road transport sub-system render many such city-regions in India unsustainable (Pucher et al. 2005).

***(C) Preliminary Analysis of Logic of Visual-analog Representation***

Preliminary analyses of the causal loop diagrams for the Kanyakumari case study (Fig. 2.1) indicate that a reduced level of service (LOS) of the major arterial roads, i.e., their congestion, has deleterious consequences for both the economic

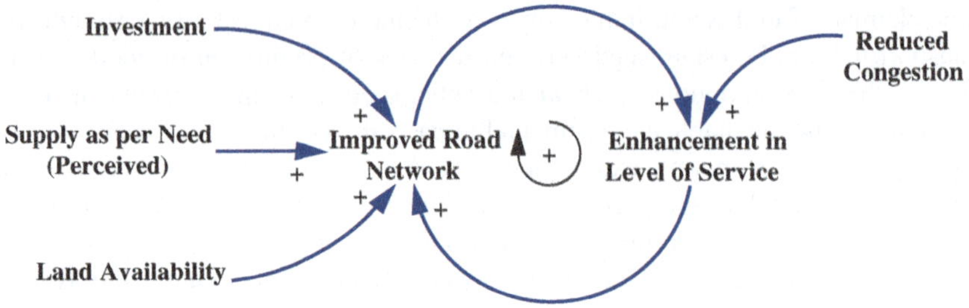


**Fig. 2.1** Causal loop diagrams for part of Kanyakumari city-region transport system

and environmental sub-systems of the region. To rectify this, through an improved road network, two things are required: investment and available land. In detail, increases (+) in the variables of "Investment", "Supply", and "Land Availability" in Fig. 2.1 will all promote an "Improved Road Network". Proceeding then around the loop in Fig. 2.1, the logic runs as follows. As the road network is improved (+), in concert with "Reduced Congestion" (also a +), an "Enhancement in Level of Service" is achieved. In turn, this enhanced LOS feeds back into further improvements in the road network, as signalled by the + of the loop feeding back from "Enhancement in Level of Service".

Given investment and available land, together with upgrades in some of the lower-order roads to higher-order status, hence reduced congestion, thus a higher LOS (all from Fig. 2.1), not only should travel times decrease, but also carbon emissions should decrease, along with reduced costs of transportation—where these latter are the subjects of analysis in other causal loop diagrams (not shown here). A further ramification should be an increase in GDP (from a higher regional productive capacity), with this increase in GDP feeding back to yet more efficient transportation and yet greater economically productive capacity—a virtuous and positive feedback loop in the system, in every sense. That, then, is the aspiration.

In all of this, the need for local government institutions to release land for use in transport has been highlighted. In particular, from the perspective of the natural environment and habitat, land severance and wastage are important. Similarly, traffic volume has a bearing on fuel consumption. Certain changes in the volume and type of traffic—such as smart systems, greater share of public transportation, modal split, increased non-motorized transportation, and encouragement of pedestrian mobility—can help to reduce fuel consumption, *ergo* carbon emissions.

***(D) Computational Analysis with a Model***

Results from computational work with an SD model suggest that policy interventions should combine increased road infrastructure investment with a change in the hierarchy of links in the road network (i.e., upgrades of roads) and a reduction in vehicle flow volumes. If this can be achieved, the gap between the available road network and that required to meet demand may be reduced by 40%; otherwise, the gap might instead increase by 136% by 2041 (Rewal 2016). The same policy intervention should also improve the LOS index significantly: from roads currently being used at capacity (actual traffic volume equals design capacity volume) during normal hours, to markedly uncongested; and from over-saturated during peak hours to uncongested.

***(E) Screening and Analysis of Computational Results***

The expectation is that the foregoing policy intervention should lead to something like a 15% increase in GDP of the Kanyakumari city-region over the base-case "business-as-usual" (BAU) scenario, with no such intervention. Carbon emissions should be reduced by about 24% relative to the BAU, fuel consumption by 7%, and wastage of land by over 60% (again, relative to the BAU).

What, then, might be the next steps towards crafting Kanyakumari as a CFG? One attractive option could be to prise open some opportunities for prudent re-arrangements of the transport sub-system that are designed expressly to contribute towards the city-region becoming a net generator of ecosystem services. Another equally ambitious possibility might be to formulate plural causal loop diagrams (and accompanying SD models) in order to examine policy options for transport systems in Kanyakumari from a richer and healthier diversity of perspectives. Figure 2.1 above, for example, has probably been constructed according to just one view of the problem—it might be recognised as the perspective of the hierarchist rationality in the Cultural Theory of Thompson et al. (1990)—when a further three are also possible and legitimate, with each being very different in their structure and function to those of Fig. 2.1. To undertake such analyses might be a suitable point of departure into both engaging with local communities in what has elsewhere been called Adaptive Community Learning (Beck et al. 2002; Beck 2011) and formulating policy in a clumsified manner (Thompson and Beck 2014).

### 2.3.2 Nutrient Recovery from Urban Wastewater: Harare, Zimbabwe

The *Options* article on Cities as Forces for Good asked this (Crutzen et al. 2007):

> How can cities of the Global South avoid adopting the same technological trajectory as those of the Global North? Can they, as it were, "leap-frog" the Global North by forgoing the entire human-waste-into-the-water-cycle phase, thereby ending up one step ahead?

From a systems perspective, and in the big picture, we shall see in this case study of Harare a further development of the idea of the supply circle (introduced above in the Kanyakumari case study, albeit largely in passing there).

Thus, the water that is the foundation of the water-based paradigm of sanitation and sewerage in cities of the Global North serves the purpose of conveyance: to remove, in the interests of personal and public health, the residuals of our human metabolism from the confined spaces of households, as quickly and efficiently as possible. In doing so, that water (as the transport medium) entrains many nitrogen- (N), phosphorus- (P), and carbon-based (C) materials—the potential feedstocks of fuels and fertilizers—and conveys them to a wastewater treatment plant. There, on the downside of the city, they have historically been considered pollutants, to be got rid of. Indeed, in the case of the N-based materials, much energy was (and still is) consumed in wastewater treatment plants in order (ultimately) to shunt the N into the atmosphere, as biochemically "unreactive" nitrogen gas. In this wider supra-system context, it is apparent that a very great deal of energy is consumed on the upside of the city in taking the nitrogen gas out of the atmosphere in order to produce N-based fertilizers, in turn to produce the foodstuffs that households in the city subsequently consume.

Perceived thus from the wider, more global, perspective, this does not make much good sense (as discussed extensively in Beck 2011). Indeed, historically, the introduction of the British sanitary invention of the water closet (WC) into the city of Paris in the late 1870s led eventually to the collapse (by the 1910s) of a mature commercial system of urban nutrient recycling (Barles 2007a, 2007b). For decades previously, that system had sustained an effective symbiosis between the city and the surrounding agricultural hinterland of Paris—albeit, perhaps, with less than complete security of public health in the city.

Today, there is a growing interest in turning back this history, but only up to a point: to the point of *not* compromising the security of public health in cities, so very hard-won over the 20th century (in cities of the Global North). This change genuinely qualifies as a change of paradigm (Larsen et al. 2013). It includes the change from sub-systems of wet (water-based) sanitation in the system of the city to sub-systems of dry sanitation. Some of the most interesting, pioneering experiments in implementing the latter can be found in Africa, specifically in Ouogadougou, the capital of Burkina Faso (Dagerskog et al. 2010; see also Drechsel and Erni 2010). From a systems perspective, once more, such a sanitation-food nexus within cities (for this is how it has been described) clearly short-circuits the energy- and water-intensive extra-city, outer cycling of food-associated N, P, and C materials around some regional (if not global) supra-system.

Such a change of paradigm, which we might summarise as that of achieving a "smarter urban metabolism" (Beck et al. 2013), holds out the promise of some significant advantages. It is one possible response to the challenge set out above in the *Options* article.

***(A) Initial Crafting of the "Problem"***

We approach our second case study of Harare, Zimbabwe, with just such a prejudice: that we should seek to understand the city's current metabolism in terms of multiple resource flows (energy, water, C, N, and P) into, through, around, and among multiple sectors of the city's economy (water, energy, food, waste, and forestry). Our approach is called a Multi-sectoral Systems Analysis (MSA) and has been applied to Atlanta, USA, London, UK, and Suzhou, China (Villarroel Walker and Beck 2012, 2014; Villarroel Walker et al. 2012, 2014, 2017; Beck et al. 2013). In covering so many sectors of urban infrastructure, and in examining the interactions among multiple resource flows, MSA has a self-evident holism about its approach. This is illustrative of not only the multi-disciplinary style of Systems Thinking outlined above, but also (and quite directly so) the current focus in various policy forums of the discussion and analysis of water-food-energy nexus issues (Beck and Villarroel Walker 2013a, 2013b; Villarroel Walker et al. 2014, 2017).

In reality, nothing is neat and tidy, not even when expressed as crisply as leap-frogging the "human-waste-into-the-water-cycle phase" of urban development (in the challenge of the *Options* article). The trouble is that cities, and the way we customarily conceive of them and their infrastructure, are not naturally presented in terms of the networks of resource flows into, around, and out of the city. Instead, we think of infrastructure as segmented conventionally as follows: into transport, just as was done in the preceding case study of Kanyakumari (in Sect. 2.3.1); into power supply and distribution; food supply and distribution; water treatment and supply; sewerage and wastewater treatment; refuse collection and solid waste management; and so on. Thus, in the city of Harare, Zimbabwe, we find stresses and unsustainability manifest in a tangled mesh of sub-systems and sub-sub-systems: a wet sanitation sub-system; an informal dry sanitation sub-system; a heavily polluted shallow lake (Chivero); a potable water supply sub-system; an agricultural irrigation infrastructure; urban agricultural enterprises; peri-urban livestock management; industrial production units, with and without effluent treatment sub-sub-systems; and nearby mines for the extraction of ores for fertilizer production. The list could easily continue. Everything can indeed seem to be related to everything else. Worse still, the familiar way in which we are conceiving of the system does not reveal the nature of those key resource flows, the re-engineering and "re-wiring" of which might be pivotal in enabling Harare to move incrementally towards greater sustainability. What is more, all this will have to be done while improving and maintaining security of public health in the city.

The city has the commonly perceived "super-structure" of various customary and familiar segments of infrastructure. It has a much less immediately obvious "sub-structure": of unfamiliar, but vital and important, resource flows through those conventional infrastructure segments.

***(B) Identification of the System—And Re-crafting the "Problem" (Step (A))***

Let us begin then by focussing in on Lake Chivero, on the downside of the city. Imagine upstream of it the city of Harare, whose population has grown substantially in recent decades. Yet the city is still served by a wet-based sanitation and sewerage system (typical of cities in the Global North) installed in former times to cater for a much smaller population. C, N, and P materials flow into the city (in food) and significant portions of these pass through the sewer network, to be arrested in part by what are called Biological Nutrient Removal (BNR) technologies, at the downstream wastewater treatment plants.[3] Not all of the C, N, and P materials are

[3]But let not that convention slip by unnoticed: the "R" for *Removal* in BNR. There are limits to its desirability; and these limits should become much more prominent in our thinking and problem-solving.

removed, however. Much of them passes out of the city's infrastructure into the surrounding rivers and into Lake Chivero, where sizeable amounts may accumulate, as what are conventionally judged to be pollutants, hence condemned to having no beneficial use whatsoever.

Meanwhile, to one side, almost beyond the periphery of our (holistic) Systems Thinking, there are industries in the vicinity of the city extracting ores for the commercial production of fertilizers. Those industries fuel resource flows that loop around and into the city, and on to Lake Chivero, coming almost full circle—but so conspicuously and utterly *not* full circle! This is no supply circle, around which may flow pre-consumption and post-consumption resources (nutrients N, P, and C, water, energy, and so forth). It is no circular economy (in the sense of Mo et al. 2009).

In short, and from a systems perspective, the problem of Step (A) has evolved across this Step (B), to be re-expressed as: nearby mineral extraction becomes nearby pollution, mediated by the city's metabolism. Moreover, what was once called a sewage farm (with a hint of beneficially productive capacity) is now referred to as a wastewater treatment works—something intended as a barrier to protect the natural environment from the immense polluting potential of the city. On this account, Harare seems an unlikely candidate for a city on its way to becoming a force for good in its environment.

Having identified the system as a system of resource flows manipulated for better and for worse by sub-systems of the city—its population, their moral positions (Thompson 2003), the city's infrastructure, and its surrounds—the "problem" to be solved can be re-crafted yet again. It is this:

> Overturn conceptually (in the mind's eye) not just the historical commitment to water-based urban sanitation infrastructure, by "drying it out", but also the historical commitment to nutrient *removal* from the resulting water-diluted sewage.[4] Then, in succession, find policy and technology interventions to *recover* nutrients commercially from their historic accumulations in the shallow depths of Lake Chivero and, in due course, from the ever richer fluxes of nutrients (not waste*water*) arriving at the city's nutrient *recovery* facilities (not wastewater treatment plants), made indeed all the richer by making Harare's sanitation sub-systems, as we say, progressively ever "drier" in nature.

Needless to say, analysis of causal loop diagrams and SD models should eventually follow from this problem re-statement.

### *(C) and (D) Quantification of Resource Recovery and Recycling Potential*

The nutrients that enter Lake Chivero from Harare are delivered there by the city's water-based sub-system of sewerage and wastewater infrastructure. Their quantity in the lake is highly significant. Our desk-top analyses and first trial

[4]The International Water Association (IWA) once had a Specialist Technical Group (SG) titled "Nutrient *Removal*". It is now called "Nutrient Removal *and Recovery*" and comes under a new umbrella initiative on "Resource Recovery" together with a second SG on "Resources Oriented Sanitation". A state-of-the-art compendium on resource recovery from water has now been published as a result of this new initiative (Holmgren et al. 2016).

computations with an SD model indicate that some 8000 tonnes of N-based materials (estimated as elemental N) have already accumulated in the lake's sediments, of which over half could be extracted and recovered over a 5-year period. To gauge the significance of this, we estimate that 100 tonnes of these materials (as N) might assist in the production of over 30,000 tonnes of food in the city-region. Such impressive quantities, when checked and further verified, emphasise rather dramatically the significance of turning a recalcitrant problem of pollution into the benefit of resource recovery—provided the economic and institutional aspects thereof are conducive to such strategic change (see, for example, Beck (2016) and the Afterword in Thompson (2017)).

In sharp contrast to the study of transport for Kanyakumari, one can see how this second case study of Harare has been much less about problem-solving than about problem formulation, problem re-formulation, and re-formulation yet again. Its next small step towards the distant goal of CFG could self-evidently be to engage in some further "back-of-the-envelope" problem-solving calculations to gauge the extents to which nutrients in the material flows around the Harare-Chivero system might be sustainably recoverable. In retrospect, and again in contrast to the Kanyakumari case study, any subsequent formal computational analysis should probably *not* be undertaken within the Systems Dynamics (SD) framework, but instead within that of the material flow analyses of (most obviously) the MSA already applied to Atlanta, London, and Suzhou.

### *2.3.3 Governance, Environmental Injustice, and Managing Solid Wastes: Kinshasa, Democratic Republic of Congo*

Like Harare, Kinshasa has some conventional infrastructure as a legacy from former times, specifically, that which provides the service of managing solid wastes. The problem, however, is that the service extends to just a privileged minority in the city. There is Environmental Injustice (EI), as some observers have called it (Dodds and Hopwood 2006; Walker 2009; Leonard and Pelling 2010). In this third case study, the problem is essentially bound up with the fourth challenge articulated in the *Options* article on Cities as Forces for Good (Crutzen et al. 2007):

> [H]ow can the engineering of city infrastructure be deployed expressly so that those at the bottom of the pyramid of dignified human development may be brought to a level where they care to engage in a debate over such a grand challenge for this century—of cities as forces for good—beyond their desperate needs of survival for just today and tomorrow?

It is not that the "engineering mechanics" and technologies of the infrastructure might need to be re-configured (as they were in Harare), but that the current structure of governance for Solid Waste Management (SWM) is not fit for purpose.

In 1960, the year of independence for the Democratic Republic of Congo (DRC), waste management in Kinshasa was regulated by the country's health code.

The code was supported by an inter-departmental decree, which set the standards of protection for urban sanitation and SWM in the city. These policies were intended to contain and reduce the spread of endemic diseases and other communicable diseases in Kinshasa and, indeed, there is evidence of their success (Mbumba 1982; De Maximy 1984; Pain 1984; Kubanza 2004, 2006, 2010; Tshishimbi 2006). In 1975, however, when the service of urban SWM was transferred to the Ministry of Environmental Affairs, a decline in environmental standards was observed (Pain 1984; Tshishimbi 2006; Kubanza 2006). In the decades since, civil wars, armed conflicts, socially highly disruptive looting, illegal settlements within the city, weak institutions of local government, financial constraints, and loss of political will have variously led to SWM coming within the remit of non-governmental organisations (NGOs) and community-based organisations (CBOs) (Tukahirwa et al. 2010). In other words, the conventional public- and private-sector actors have disappeared from the scene, to be substituted by actors from civil society.

***(A) Initial Crafting of the "Problem"***

There is an element of cross-disciplinary Systems Thinking (as opposed to holistic thinking) in our crafting of the Kinshasa SWM problem. It runs as follows:

> The essential goal is to address EI and, through ASA, to develop pragmatic policy interventions to redress it. To do so, we come to the view that theories of EI (with their origins in the discipline of Sociology, in particular, the sociology of Law) should first be transcribed into the framework of the theory of plural rationalities from Anthropology, i.e., Cultural Theory (CT; Thompson et al. 1990). Our argument is as follows: there are questions (Q) in the sociology of Law that may lack some answers (A), which answers *might* be found by re-casting the Q's of EI within the framing of CT. Specifically, the concerns raised by those who focus on EI can be transposed into CT's plural "ideas of fairness" (Verweij 2011; and as elaborated below). CT can then be deployed to diagnose the expected low deliberative quality of governance (Ney 2009) for SWM in Kinshasa and accordingly point to options for elevating deliberative quality, through specific changes to the structure of governance. In other words, we seek a prognosis of how to rid Kinshasa of EI in respect of MSW.

Indeed, looking well beyond the scope of this chapter, we wonder whether CT might generate different, novel insights regarding EI when transcribed back into its native disciplinary domain (the sociology of Law).

CT is introduced in the first few pages of Chap. 4 (in this volume) "Extract of Africa: Towards the Equitable and Ecologically Sound Governance of Mining and Drilling", on which one of us (MT) is the lead author. Its four forms of social solidarity (or "ways of organizing") are there set out by way of the four kinds of goods that have long been familiar to economists and political scientists: public, private, common-pool, and club. The idea—and it also holds for solid wastes—is that, while the inherent physical properties of the objects are not irrelevant, category membership is under-determined by those physical properties. Solid wastes, of course, are usually seen as "bads", not "goods", except in those circumstances where some actors—they are disparagingly referred to as "scavengers" or worse—regard them as mis-categorised "valuable resources". In Cairo, for instance, solid waste/valuable resource management is almost completely in the hands of the Zabbadin, Coptic Christians, for the most part, who make a good living by collecting and skillfully sorting out the household and business refuse, with the edible portion then being fed to their pigs: unclean animals in the eyes of those citizens (the majority) who are of the Muslim faith.

This "malleability" creates some difficulties for those who frame things in terms of environmental injustice (EI), in that the Zabbadin are simultaneously treated as a despised minority—irrefutably unjust—while being quite well rewarded, economically, for their socially valuable contribution, which is admirably just from the individualist perspective. EI, CT suggests, will always be a contested concept (as is sustainability; Thompson 2011; Beck 2011), because each of the four forms of solidarity has its own idea of fairness, with none of them being reducible to any of the others (see Davy 1997; Thompson 1998; also Verweij 2011).

These ideas of fairness are as follows:

- The *individualist* voice is pro-market (and pro-private goods). A fair outcome, its proponents are convinced, is one in which those who put most in get most out, and they see a "level playing field" as a crucial pre-condition for that. Equality of opportunity is therefore the individualist idea of fair process, with outcome fairness being the matching of reward to contribution.
- The *hierarchist* voice is pro-control (and pro-public goods) and is much concerned with order and status: who has the right to do what and to whom. Distribution should thus be by rank and station, and this idea of fairness then requires that, if that orderly distribution is not happening, the hierarchy should step in so as to ensure that each gets the desserts appropriate to his or her position within the layered totality.
- Both the individualist and the hierarchical ideas of fairness are anathema to those who speak with the *egalitarian* voice. Egalitarian actors are levellers (and therefore in favour of common-pool goods). Absolute parity—before, during, and after—is their idea of what is fair. People, they insist, should start off equal and end up equal.
- Those who cannot come up with the entry costs for the market, find themselves categorised as deviants/undeserving poor by the hierarchy, and lack the cohesion and commitment that would enable them to rise up against all sources of inequality, constitute the *fatalized* and largely voiceless margin. As the discards from all and sundry rain down on them they cope as best they can. "Not in this world" is the fatalist verdict on fairness.

The proponents of EI, we can now see, are very much of the egalitarian persuasion: highlighting, in outraged tones, the gross unfairnesses that are being visited upon the fatalized margins in order to castigate the unresponsive and exploitative hierarchies and markets. This, of course, is a perfectly valid and legitimate voice—a voice, moreover, that all too often finds itself ignored and sometimes persecuted—but, as well as needing to be heard, cultural theorists argue, it also needs to be constructively engaged with the other voices. Only then will we be in an institutional position to move towards what are called "clumsy solutions". Kinshasa, at present, is rather a long way from that position. The country's elite—politicians, musicians, businessmen and so on—have turned the capital city into a club good. This elite, in encompassing both status and economic "clout"—the bureaucracy and the bourgeoisie—has excluded the egalitarian voice, while also failing to look after the "lowerarchs" or to level the playing field. In consequence, none of the ideas of fairness are being delivered, and much of the waste that is generated in the wealthier districts ends up being dumped in the poor suburbs, hence their environmental degradation.

Just as in the foregoing Harare case study, therefore, the ambition should be to up-end the labelling of certain urban material flows as "wastes", whether they be solid, aqueous, or gaseous, and to have them viewed anew as resources. Some of our colleagues in the CFG Network—faculty from the Centre for Industry and the Circular Economy, part of the Environmental Systems Analysis Group in the School of Environment, Tsinghua University, Beijing—have already undertaken work on assessing the scope for "urban mining" of the valuable resources previously viewed as solid "wastes" (Wen et al. 2015). They have been collaborating with the city of Suzhou and private-sector actors to implement a system for recycling food oil "wastes" from restaurants (http://cfgnet.org/archives/1667). It exploits the Internet of Things (IoT). A real-time market for household "waste" trading and recycling is likewise under development (http://cfgnet.org/archives/1667; see also Zhang and Wen 2014). It combines classical and modern technologies: the bicycle and the mobile phone.

While these companion developments err towards emphasising the engineering and technological elements of infrastructure adaptation, the present case study of SWM in Kinshasa is, as we have said, devoted instead to re-working the structure of its governance (for infrastructure operation, that is).

***(C) Preliminary Analysis of Logic of Visual-analog Representation***

There exist clear causal feedback relations among the various factors that contribute to the current plight of solid waste management in Kinshasa. The city's large population, the de facto segregation of the homes of the rich and poor into geographically separated areas, and the city's growing economy generate a high volume of solid waste—and lead to the EI already described.

Further analysis suggests that, if three of the four social solidarities, i.e., the municipal authorities (hierarchists), private companies (individualists), along with the NGOs, CBOs, and other community and social organisations (egalitarians), could only but engage with each other, the "management" (or rather "*non*-management") in solid waste management could be made a more equitable and productive matter of participatory governance. This is *not* entirely a matter of wishful thinking. We may point to the experience of Nepal (Gyawali et al. 2017), where it seems that some sort of "insurgency" is needed (preferably non-violent insurgency, though Nepal has experienced both), if the excluded voice is to force its way into the policy debate. For instance, the journey to the now celebrated community forests in Nepal began with impoverished villagers chasing away the foreign aid providers who had been authorised by the Nepal government to convert what they wrongly saw as ownerless waste land into a eucalyptus plantation. The impasse was eventually resolved by the two sides sitting down and discussing their different ideas of what a healthy forest should consist of, and with them quite quickly agreeing that, whatever it was, it did not include any eucalyptus trees! In Kinshasa, the fear that this sort of confrontation might be imminent could well spur the authorities into redressing the current EI, which, as we have now noted above, is unfair by all the different ideas of fairness (in CT). However, in regimes that are not democratic, the gaining of access by the excluded voices is seldom straightforward.

If the ideal form of urban SWM could be realized in Kinshasa, we should call it participatory resource recovery governance (PRRG). An environmental policy tailored to very local circumstances—together with some financial support from the government public sector and private companies, and the deployment of social awareness campaigns designed to reduce the generation of "waste" at source (and to emphasise the economic resource value of the misnomer of "wastes")—could succeed in shifting things towards our PRRG. In it, all stakeholders would share equitably the responsibility of resource recovery and environmental protection, if not restoration. The PRRG should reduce the volume of "waste" to be "disposed of" in the poor suburbs and stimulate business opportunities. Indeed, if there were reductions in waste generation in the poor suburbs, space could be liberated afresh for a re-greening and opening out of their local environments. In short, PRRG promises less EI in Kinshasa.

Looking once more to the distant future, when the notion of CFG might conceivably sustain some meaningful association with the city of Kinshasa, we take heart from the following. Positive developments, such as those imagined above, *have* come to pass in other places, notably in the informal (as opposed to the formal) sector of Nepal—a country in which, again notably, plural rationalities on innovation, technology, and development thrive and enjoy boisterous, healthy contestation (Gyawali et al. 2017). It is just possible that the voice of one of us among the authorship of this chapter (SK) might emerge to assist in sparking the beginnings of like incremental changes in Kinshasa.

### 2.3.4 Energy Metabolism of Campus Buildings: Central University of Technology, Bloemfontein, Republic of South Africa

The new 11-storey headquarters building of the San Francisco Public Utilities Commission (SFPUC) has a wetland ecosystem in its atrium, as a live, full-scale demonstration of how the "wastewater" of the building's occupants is being recycled (Harrington 2012). The equally new 38-storey "Walkie-Talkie" office block in the City of London (20 Fenchurch Street) features a green roof garden.[5] That the garden should be open to the public (to be visited by appointment) was stipulated as a condition for approval of the building's planning application. Single buildings, and clusters of buildings, can be created as beacons of sustainability in the urban environment and there are many, many examples of this occurring. The programme of LEED buildings (http://www.usgbc.org/leed) makes a virtue out of our instinct for competition, to achieve the best exemplars of sustainable design and living, in practice.

Universities too can emulate this model of communicating and educating by example—in practice, in the midst of urban life. In Newcastle, in the north-east of the UK, the University is constructing an Urban Sciences Building (USB), as part of its new Science Central campus, next to its equally new Business School building, on the site of the former Newcastle Brewery, just a stone's throw away from St James's Park (the home of Newcastle United FC)—and all in the heart of the city (http://www.ncl.ac.uk/sciencecentral/urban/; last accessed 11 November, 2016). Principal themes of the work of the USB are major showcases of sustainable urban drainage systems and sustainable electrical power storage systems (http://www.ncl.ac.uk/sustainability/news/archivednews/energystoragetestbedisofficiallyswitched-on.html; last accessed 11 November, 2016). The USB is viewed as a Living Laboratory; it participates in a Europe-wide Network of Living Labs (ENOLL; http://openlivinglabs.eu/, last accessed 11 November, 2016); and some of our colleagues in the CFG Network are integrally involved in these initiatives.

Energy is likewise the focus of our fourth case study, of green building design on the campus of the Central University of Technology, Bloemfontein.

### *(A) and (B) Initial Crafting of the "Problem" and Identification of the System*

The analysis of two buildings constitutes our case study. Here, they will be treated as separate systems, in isolation, as just two free-standing elements—and as therefore but the very simplest of beginnings of a "cfg" (within a CFG). Each building's sub-systems comprise, first, its physical dimensions and attributes and, second, those of its features that influence the quality of its internal environment (illumination, temperature, freshness of air, and so on). What is to be analyzed is a system already in place and in operation. As our narrative reveals, therefore, this fourth case study harks back to the origins of ASA in the 1940s war-time operations research (OR), such as that of developing the system of the combined novel technologies (sub-systems) of radar and the Hurricane fighter aircraft—and as they were rather more dramatically failing and succeeding in "real time". Here too, in our Bloemfontein case study, ASA caters for analysis, in particular, of the interactions among the building's physical and environmental sub-systems.

The first building is three storeys high and has a large atrium at its centre. Its lecture rooms are clustered along the sides of a corridor and have large windows to

[5]See "First view of Europe's highest roof garden — an oasis on top of the Walkie Talkie tower"; *Evening Standard*, posted 6 June, 2013, www.standard.co.uk.

the outside, which remain shut for most of the year. The lecture rooms have access therefore to daylight (supplemented by artificial lighting), but only limited ventilation: openings are confined to those in the internal walls of the lecture rooms. The corridor configuration—lecture rooms on one side, large windows to the atrium on the other—enables sufficient illumination of the corridor, but not the classrooms. The walls between class rooms and the corridor have no glazing. A portion of the building has a two-sided corridor with staff offices on either side. These offices do not have sufficient access to natural lighting, hence are served by electrical lighting as well as artificial ventilation.

The second building is an administrative building. It too has a large central atrium with a translucent roof, which illuminates the entrance lobby. The east and west wings of the building comprise corridors with offices on either side. Both corridors have access to natural light from the nearby atrium, but the offices suffer from a lack of natural cross-ventilation and are therefore equipped with mechanical ventilation systems.

The self-evident problem for each building is this: what would an ASA contribute to minimising the consumption of energy for non-natural illumination and ventilation of the two buildings?

***(C) Preliminary Analysis of Logic of Visual-analog Representation***

Evaluation of the two buildings against the South Africa Green Star rating scheme reveals that neither scores highly. The first building attains a score of 13 (out of 30), the second just 4.

From the initial crafting of what the problem is, energy consumption is clearly the central issue here. Analysis of the causal loop diagram for each building, in anticipation of an eventual analysis using an SD computation, suggests that altering building form could have a profound effect on the efficiency of energy consumption in each building. Many factors are intimately interwoven together. Energy consumption is influenced by indoor quality and the use to which the buildings are put. Building orientation, hence access to natural light, also affects energy consumption, again as a function of the use to which the building is put. Energy consumption, furthermore, is influenced by building geometry, yet building shape and size are determined by the building's uses and its architectural value. In turn, openings in a building are related to its shape, size, and orientation. They influence the requirement for artificial lighting and artificial ventilation (air-conditioning) and these two features influence indoor quality, which is also affected by the building materials used in its roofs, floors, walls and openings. Altogether, we are nearing the impression that "everything is indeed affected by everything else"!

The simple rectangular shape of the second building (with its central atrium), although rated lowly, performs notably better in its access to daylight. Less complex forms, with an appropriate orientation, adequate and astutely located openings, can improve energy efficiency through the better use of natural illumination and ventilation (hence reduced energy consumption). Building form has consequences for the uses to which it is put. In addition, a relatively narrower rectangular shape oriented along a north-south axis improves indoor environmental quality markedly, as well as reducing energy consumption. Buildings with a relatively higher volume, but a very high window-to-face ratio (or window-width to wall-length ratio) may achieve a relatively low consumption of energy.

In sum, our ASA indicates that energy consumption can be reduced by 15–22% relative to the base case of current campus building design.

Looking back over our four case studies, this, on campus buildings in Bloemfontein, has been the best defined and the most tightly focussed, in this instance, on the technical aspects of energy consumption. Looking forward, to CFG {cfg} (as we have put it cryptically), the grand challenge now might be to consider the energy metabolism of buildings as inextricable from their water and nutrient metabolisms, in the archetypal manner, therefore, of the "systems approach" to problem-solving. The idea of a campus as a force for good in its surrounding city environment had its origins in the settings of the campuses of the University of Georgia (in Athens, Georgia) and the Georgia Institute of Technology (in Atlanta, Georgia). The imagination there was that the arrangements of the buildings on campus, and their interactions one to another, might be changed progressively, one small step at a time, in order to achieve this kind of performance: water of poor quality is taken from the surrounding campus environment and passed and used around the campus so as, in the end, to be returned to the campus environment as water of a superior quality. This, if it could be achieved, would be called water "upcycling". The term, and the idea, are due to McDonough and Braungart (2002), in their case in respect of designing a textile manufacturing facility. If one wanted to begin addressing the water-food-energy nexus on a university campus, here, in Bloemfontein, the book by Larsen et al. (2013) would be a very good starting point.

## 2.4 Conclusions: Lessons Learned

All authors of this chapter learned a great deal from their participation in the South Africa Young Scientists Summer Programmes (SA-YSSP). For their part, the supervisors (DKD, MBB, and MT) learned how difficult it can be for them to pass on one very particular, but essential, skill to their younger co-authors: that of the nature and practice of the more qualitative—yet penetratingly systematic—logic that must accompany and complement deployment of the much more obvious and familiar quantitative, computational analyses generally associated with Applied Systems Analysis (ASA). In 1972, when IIASA was founded, these quantitative methods were on their way to reaching the subsequent zenith in their early breakthrough applications. The methods had worked so successfully for the military aerospace systems of the 1960s. It seemed self-evident back then that they could do the same for the natural environment and for the civil production and infrastructure systems of food and agriculture, healthcare provision, energy supply, and so on. It was easy in those early days to look down on the prospect of ASA being applied for better governance. For where might be the hard, quantitative differential equations and constrained optimisation in that?

To be more specific, three important lessons have been learned.

### 2.4.1 Anthropology for Engineers

The primary lesson learned is this. It may be relatively easy, and it is highly attractive, of course, to pick up and use computational methods such as those of Systems Dynamics. It is a palpable sign of activity and progress, of something being done. There are well established procedures for learning about these methods, for instruction and training in them, and for supervising their assimilation in those undergoing the learning. And indeed, they can work well, provided (at bottom) the problems to which they are applied are relatively well defined, witness our case studies of Kanyakumari (physical infrastructure of transport) and Bloemfontein (physical infrastructure of buildings). It also tends to be the case that roads and buildings are always among the first segments of city infrastructure to receive investment and be constructed. They are the obvious, very visible symbols of progress. Sanitation infrastructure, sewerage, and infrastructure for pollution control and waste disposal are not—never mind the opportunities for valuable resource recovery we have promoted herein (witness the Harare and Kinshasa case studies). In these latter instances, it is much more difficult to cut a clear path to a practically implementable policy solution, especially through the tangled mess of the conflicting social forces in play and the governance that must be exercised over them (to the benefit of all).

But what is needed here is *not* just a matter of "better governance" (more systematic qualitative logic) instead of "better engineering and technology" (more computational methods), or vice versa. The two must be melded together in a mutually supportive and stimulating manner. In 1979 one of us (MT) published *Rubbish Theory* (Thompson 1979). Nearly forty years on it is to be re-published, with barely any changes at all, except for the inclusion of an Afterword co-authored by MT and MBB. "Anthropology [from MT] for Engineers [MBB]" is its self-explanatory sub-title (Thompson 2017). How to pass on the skill of "doing" the anthropology for engineering is likely to remain a challenge, however, just as it was in our experience of the SA-YSSP. One of us (the engineer) should know. He remains still a student of anthropology, nearly twenty years on from when he first began picking up the subject!

How to compress the time taken for cross-disciplinary learning while engaging in cross-disciplinary problem-solving (as introduced in Sect. 2.2 above) is surely a grand challenge, but perhaps one to which ASA itself has something highly significant to contribute.

### 2.4.2 Cities, not Nations

A second lesson to be learned is this. Cities and resource extraction (as opposed to resource recovery) seem to be the key African/Global South concerns. By focusing on those with our younger co-authors, we have been able to remedy some of what we see to be the main disabilities of ASA in the Global North—or at least the way the purpose and process of ASA may be misconstrued. For readily misconstrued they can be, precisely because of the fact that we would be unable to comprehend the substance and nature of global problems, such as climate change, *without* the "flagship" computational models the public now increasingly associates with ASA. And when addressing such global problems, the world is portrayed as segmented into its nations: coloured in various shades of green for sustainable countries; and in various shades of red for unsustainable countries.

One of the popular misconceptions about ASA is an excessive focus on the global, supra-supra-supra-system level and the resulting inability of such globally framed analyses to say anything useful about the lower levels (the system of the city, especially), which is where the solutions (if they exist) will be found. In respect of climate change, and very much in the spirit of Cities as Forces for Good, Fink (2012) writes of cities as "geo-engineering building blocks" and as "forces for climate repair". Katz and Bradley's book *The Metropolitan Revolution* makes the yet broader case for the vital problem-solving role of cities, and (we judge) very convincingly so (Katz and Bradley 2013). Writing of the current crisis in the use of statistics, and all of the quantification that now comes with such, Williams (2017) reminds us of the etymology of the word "statistic", going back to the 1700s (and beyond). In short, the collection of quantitative data and their quantitative analysis were tailor-made for the centralized *nation-state*. Things became aggregated at the national level. But what today counts for him is this (Williams 2017):

> The key geographic units involved are no longer nation states. Rather, it is cities, regions or individual urban neighbourhoods that are rising and falling.

Countries are not uniform (as the coloured maps of the globe suggest). There are city/countryside "gradients" in the heterogeneity of a nation, not least in respect of population distribution. And these gradients are generating all sorts of flows—foodstuffs, heat, energy, water, greenhouse gases, migrants, and so on—that would simply not exist if these gradients were not there (nor, if people were spread out evenly, would there be much in the way of infrastructure). Cities are the focus for developing and applying policies for changing our *demands* for water, energy, and food, as opposed to their supply, on which (presently) so much ASA is focussed instead. At the yet smaller scale, households in cities are where personal, lifestyle changes can be implemented (Dake and Thompson 1999) in order to drive collective, aggregate change in the city's metabolism. The household should not discard "waste", but be designed and operated so as to generate post-consumption resources; these would be tailored for passing on to the next downstream beneficiary; and thereafter duly returned eventually to the household as pre-consumption resources, thanks in part to Nature (and thus to complete the circle).

### *2.4.3 Hard Thinking About Soft Systems*

Our third lesson, like the first, requires a historical context. When IIASA was founded, the idea was that scientists and mathematicians from East and West would sit down together and work across their various disciplinary and political boundaries so as to come up with systematic understandings that, by teasing out what was and was not physically possible, would enable "policymakers" to feed the world, supply it with energy, keep it watered, and so on. Hence the words "Applied Systems Analysis" (and their political neutrality) in the Institute's title. There was mischievous talk then of an IIASA version of the *orthogonal* axes of Fleming's left-hand rule: we were to undertake research that was (a) "good science", (b) "policy relevant", and (c) would span the then East-West divide (but alas, perhaps not all in one project at the same time!).

We believe there remains today a strong and enduring bias towards what is called "hard systems analysis", and away from the sort of "soft systems analysis" that starts from assuming that whatever it is that is "out there" is always mediated through "perspectives", "lenses", "*Weltanschauungen*", "social constructions", and the like (see, for instance, Checkland 1985). Indeed, the primary argument of Williams (2017) is not just that cities may be more important units of analysis than nation-states, but also that qualitative evidence can be more believable than quantitative evidence in guiding the way we make decisions. Statistics is facing something of a crisis, Williams argues. He is, of course, responding in the wake of the political upsets of 2016 (in the UK and USA), in the processes of which experts were denigrated, not least statisticians and economists. And ASA, just as obviously, generates vast volumes of the expert-interpreted quantitative evidence. If Williams is correct (for we should always be wary of any up-to-the-minute commentary), the public at large may be putting we the experts—we the community of ASA practitioners—very much on notice. Neglect the non-quantitative aspects of systematic reasoning and assessment and we risk the credibility and value of all our quantitative analyses being entirely undermined.

South Africa, though it may be something of a novice ASA-wise, has therefore provided a welcome and long-overdue corrective to the over-reliance on hard, quantitative systems analysis. In particular, if we are to have effective, democratic and clumsy governance, then, as our case studies show, we will need the softer elements of Systems Thinking. In effect, as in our first lesson, the qualitative and the quantitative need to be both re-balanced and better combined. Although labelled "soft" or "qualitative", the skill of such argumentation and analysis are anything but easy to acquire: "hard thinking" indeed—including very hard thinking about the hard systems of urban engineering infrastructure, as in the Afterword of Thompson (2017).

## References


Anthoff, D., & Tol, R. S. J. (2014). FUND—climate framework for uncertainty, negotiation and distribution: Technical description (Version 3.9) (available online at www.fund-model.org/versions).

Arup, O. (2014). City Resilience Framework. Report, The Rockefeller Foundation and Ove Arup & Partners International, (April), 24 pp.

Barles, S. (2007a). Feeding the city: Food consumption and flow of nitrogen, Paris, 1801–1914. *Science of the Total Environment, 375*, 48–58.

Barles, S. (2007b). Urban metabolism and river systems: An historical perspective—Paris and the Seine, 1790–1970. *Hydrology and Earth System Sciences, 11*, 1757–1769.

Beck, M. B. (2011). *Cities as forces for good in the environment: Sustainability in the water sector* (xx + 165 pp). Warnell School of Forestry and Natural Resources, University of Georgia, Athens, Georgia, (ISBN: 978-1-61584-248-4) (online as http://cfgnet.org/archives/587).

Beck, M. B. (2014). Handling uncertainty in environmental models at the science-policy-society interfaces. In M. Boumans, G. Hon, & A. C. Petersen (Eds.), *Error and uncertainty in scientific practice* (pp. 97–135). London: Pickering & Chatto.

Beck, M. B. (2016). Understanding the science of ecosystem services: Engineering infrastructure for urban water services. *Applying systems thinking: An outreach paper* (viii + 37 pp). International Water Association, The Hague, The Netherlands (online as http://www.iwa-network.org/ecosystem-services-not-so-much-the-water-as-whats-in-it/).

Beck, M. B., Fath, B. D., Parker, A. K., Osidele, O. O., Cowie, G. M., Rasmussen, T. C., et al. (2002). Developing a concept of adaptive community learning: Case study of a rapidly urbanizing watershed. *Integrated Assessment, 3*(4), 299–307.

Beck, M. B., Gupta, H., Rastetter, E., Shoemaker, C., Tarboton, D., Butler, R., et al. (2009). *Grand challenges of the future for environmental modeling* (White Paper). National Science Foundation, Arlington, Virginia (ISBN: 978-1-61584-248-3; online as http://cfgnet.org/archives/249).

Beck, M. B., Thompson, M., Ney, S., Gyawali, D., & Jeffrey, P. (2011). On governance for re-engineering city infrastructure. *Proceedings of the Institution of Civil Engineers, Engineering Sustainability, 164*(ES2), 129–142.

Beck, M. B., & Villarroel Walker, R. (2011). Global water crisis: A joined-up view from the city. *Surveys and Perspectives Integrating Environment and Society*, 4.1, [Online] Online since 27 December 2011 (available online at http://sapiens.revues.org/1187).

Beck, M. B., & Villarroel Walker, R. (2013a). On water security, sustainability, and the water-food-energy-climate nexus. *Frontiers of Environmental Science and Engineering, 7*(5), 626–639.

Beck, M. B., & Villarroel Walker, R. (2013b). Nexus security: Governance, innovation, and the resilient city. *Frontiers of Environmental Science and Engineering, 7*(5), 640–657.

Beck, M. B., Villarroel Walker, R., & Thompson, M. (2013). Smarter urban metabolism: Earth systems re-engineering. *Proceedings of the Institution of Civil Engineers Engineering Sustainability, 166*(5), 229–241.

Borsuk, M. E., Stow, C. A., & Reckhow, K. H. (2004). A Bayesian network of eutrophication models for synthesis, prediction, and uncertainty analysis. *Ecological Modelling, 173*(2–3), 219–239.

Checkland, P. (1985). The approach to plural rationality through soft systems methodology. In M. Grauer, M. Thompson & A. Wierzbicki (Eds.), *Plural rationality and interactive decision processes* (pp. 8–21). Berlin: Springer.

CISL. (2015). *Unhedgeable risk. How climate change sentiment impacts investment* (57 pp). Cambridge: Cambridge Institute for Sustainability Leadership (available online at www.cisl.cam.ac.uk/publications/sustainable-finance).

Coyle, R. G. (1996). *System dynamics modelling: A practical approach*. London: Chapman and Hall.

Crutzen, P. J., Beck, M. B., & Thompson, M. (2007). *"Cities", Options, International Institute for Applied Systems Analysis*. Laxenburg, Austria, 8.

Dagerskog, L., Coulibaly, C., & Ouandaoga, I. (2010). The emerging market of treated human excreta in Ouagadougou. *Urban Agriculture Magazine*, *23*, 45–48 (April) (posted at www.ruaf.org).

Dake, K., & Thompson, M. (1999). Making ends meet, in the household and on the planet. *GeoJournal, 47,* 417–421.

Das, D., & Sonar, S. G. (2013). Perspective impacts of information technology industry in development of Pune City in India. *Journal of New Generation Sciences, 1*(3), 1–17.

Davy, B. (1997). *Essential injustice: When legal institutions cannot resolve environmental and land disputes*. Vienna: Springer.

De Maximy, R. (Ed.). (1984). *Kinshasa, Ville en Suspens, Dynamique de Croissance et Problème d'Urbanisme*. Paris: De l'Orstom.

Dodds, L., & Hopwood, B. (2006). BAN Waste, environmental justice and citizen participation in a policy setting. *Local Environment, 11*(3), 269–286.

Drechsel, P., & Erni, M. (2010). Analysing the nexus of sanitation and agriculture at municipal scale. *Urban Agriculture Magazine*, *23*, 11–12 (April) (posted at www.ruaf.org).

Fink, J. H. (2012). Cities as geoengineering building blocks. In R. J. Dawson, C. L. Walsh & C. G. Kilsby (Eds.), *Earth systems engineering 2012: A technical symposium on systems engineering for sustainable adaptation to global change*. Centre for Earth Systems Engineering Research, Newcastle University, UK, pp. 59–64.

Forrester, J. W. (1969). *Urban dynamics*. Massachusetts: MIT Press.

Gyawali, D., Thompson, M., & Verweij, M. (2017). *Aid, technology and development: The lessons from Nepal*. London: Earthscan-Routledge.

Harrington, E. (2012). Building a coastal wetland in the heart of a city. *Nature, 486,* 189.

Holling, C. S. (Ed.). (1978). *Adaptive environmental assessment and management*. Chichester: Wiley.

Holling, C. S. (1986). The resilience of terrestrial ecosystems: Local surprise and global change. In W. C. Clark & R. E. Munn (Eds.), *Sustainable development of the biosphere* (pp. 292–317). Cambridge: Cambridge University Press.

Holmgren, K. E., Li, H., Verstraete, W., & Cornel, P. (2016). *State of the Art Compendium Report on resource recovery from water*. The Hague: International Water Association.

ICSU. (2011). Health and Wellbeing in the Changing urban environment: A systems analysis approach. In *Interdisciplinary science plan* (42 pp). Paris: International Council for Science (ICSU). (http://www.icsu.org/publications/reports-and-reviews/health-and-wellbeing/).

Katz, B., & Bradley, J. (2013). *The Metropolitan revolution. How cities and metros are fixing our broken politics and fragile economy*. Washington, DC: Brookings Institution.

Kubanza, N. S. (2004). The consequences of the failure of urbanisation of the City of Kinshasa (Honours Dissertation, Department of Sociology and Anthropology, University of Kinshasa, DRC).

Kubanza, N. S. (2006). *The resurgence of endemic diseases in the City of Kinshasa, Democratic Republic of Congo*. MSc Research Report, Department of Sociology, University of Kinshasa, DRC.

Kubanza, N. S. (2010). *Perception and issues of solid waste management in South Africa, A case study of Johannesburg*. Masters Research Report, Department of Development Planning, University of the Witwatersrand, Johannesburg, South Africa.

Larsen, T. A., Udert, K. M., & Lienert, J. (Eds.). (2013). *Source separation and decentralization for wastewater management*. London: IWA Publishing.

Leonard, L., & Pelling, M. (2010). Mobilisation and protest: Environmental justice in Durban, South Africa. *Local Environment, 15*(2), 137–151.

Mbumba, N. (Ed.). (1982). *Kinshasa 1881–1981, 100 ans après Stanley, Problèmes et Avenir d'une Ville*. Kinshasa, DRC: Centre de Recherches Pédagogiques.

McDonough, W., & Braungart, M. (2002). *Cradle to cradle: Remaking the way we make things*. New York: North Point Press.

Mehaffy, M. W., & Salingaros, N. A. (2014). The biological basis of resilient cities. *Ecologist* [Online] (www.theecologist.org/green_green-living; posted 25 January, 2014).

Mercer. (2015). *Investing in a time of climate change* (103 pp). London: Mercer LLC (available online at www.mercer.com/ri/climate-change-study).

Mo, H.-P., Wen, Z.-G., & Chen, J. (2009). China's recyclable resources recycling system and policy: A case study in Suzhou. *Resources, Conservation and Recycling, 53,* 409–419.

NAE. (2016). *Grand challenges for engineering: Imperatives, prospects, and priorities: Summary of a forum* (42 p). Washington, DC: US National Academy of Engineering, National Academies Press.

Ney, S. (2009). *Resolving messy policy problems: Handling conflict in environmental, transport, health and ageing policy*. London: Earthscan.

NHDP. (2011). National highways development project maps, NHDP Project Phases—I, II & III. Ministry of Road Transport and Highways, Government of India, September 2011.

NHDP. (2014). National highways development project: An overview. Government of India, pp. 1–2 (Retrieved 7 June 2014).

Nordhaus, W. (2014). Estimates of the social cost of carbon: Concepts and results from the DICE-213R model and alternative approaches. *Journal Association of Environmental and Resource Economists, 1,* 273–312. https://doi.org/10.1086/676035.

NWCF. (2009). *The Bagmati: Issues, challenges and prospects*. Technical Report, prepared by Nepal Water Conservation Foundation (NWCF) for King Mahendra Trust for Nature Conservation, Kathmandu, Nepal.

Pain, M. (Ed.). (1984). *Kinshasa, la Ville et la Cité*. Paris: De l'Orstom.

Patt, A. (2007). Assessing model-based and conflict-based uncertainty. *Global Environmental Change, 17,* 37–46.

Pucher, J., Korattyswaropam, N., Mittal, N., & Neenu, I. (2005). Urban transport crisis in India. *Transport Policy, 12,* 185–198.

Reichert, P., Borsuk, M. E., Hostmann, M., Schweizer, S., Spörri, C., Tockner, K., et al. (2007). Concepts of decision support for river rehabilitation. *Environmental Modelling and Software, 22*(2), 188–201.

Reichert, P., Langhans, S. D., Lienert, J., & Schuwirth, N. (2015). The conceptual foundation of environmental decision support. *Environmental Management, 154,* 316–332.

Rewal, T. (2016). Planning for optimum transportation system for sustainable development in Kanyakumari District, Tamil Nadu (PhD Dissertation, Indian Institute of Technology, Roorkee).

Salingaros, N. A. (2005). *Principles of urban structure*. Amsterdam: Techne Press.

Shen, Q., Chen, Q., Tang, B., Yeung, S., Hu, Y., & Cheung, G. (2009). A system dynamics model for the sustainable land use planning and development. *Habitat International, 15,* 25–33.

Sterman, J. (1982). *Business dynamics: Systems thinking and modelling for a complex world*. Boston: McGraw Hill.

Thompson, M. (1979). *Rubbish theory: The creation and destruction of value*. Oxford: Oxford University Press.

Thompson, M. (1998). Waste and fairness. *Social Research, 65*(1, Spring), 55–73.

Thompson, M. (2002). Man and nature as a single but complex system. In P. Timmerman (Ed.), *Encyclopedia of global environmental change* (Vol. 5, pp. 384–393). Chichester: Wiley.

Thompson, M. (2003). Stoffströme und moralische Standpunkte. In M. Fansa & S. Wolfram (Eds.), *Müll: Facetten von der Steinzeit bis zum Gelben Sack*, Mainz am Rhein, Philipp von

Zabern (English translation as "Material Flows and Moral Positions". *Insight*, Cities as Forces for Good (CFG) Network; online as http://cfgnet.org/archives/531).

Thompson, M. (2011). Sustainability is an essentially contested concept. *Surveys and Perspectives Integrating Environment and Society*, 4.1, [Online] Online since 23 November 2011 (available online at http://sapiens.revues.org/1177).

Thompson, M. (2017). *Rubbish theory: The creation and destruction of value* (Second and extended edition). London: Pluto.

Thompson, M., & Beck, M. B. (2014). *Coping with change: Urban resilience, sustainability, adaptability and path dependence*. Working Paper 13, Foresight Future of Cities, UK Government Office for Science (GOS), December, p. 45 (Available from https://www.gov.uk/government/publications/future-of-cities-coping-with-change).

Thompson, M., Ellis, R., & Wildavsky, A. (1990). *Cultural theory*. Boulder, Colorado: West View.

Thompson, M., & Warburton, M. (1985). Decision making under contradictory certainties: How to save the Himalayas when you can't find out what's wrong with them. *Applied Systems Analysis, 12,* 3–34.

Trucost, (2013). *Natural capital at risk: The top 100 externalities to business*. London: Trucost.

Tshishimbi, E. (2006). *Travail des Enfants et des Jeunes dans la Ville de Kinshasa, Galiléo*. Kinshasa, DRC: Galiléo.

Tukahirwa, J., Mol, A., & Oosterveer, P. (2010). Civil society participation in urban sanitation and solid waste management in Uganda. *Local Environment, 15*(1), 1–14.

Varis, O. (2002). Belief networks: Generating the feared dislocations. In M. B. Beck (Ed.), *Environmental foresight and models: A manifesto* (pp. 169–205). Amsterdam: Elsevier.

Verweij, M. (2011). *Clumsy solutions for a wicked world*. Basingstoke: Palgrave.

Villarroel Walker, R., & Beck, M. B. (2012). Understanding the metabolism of urban-rural ecosystems: A multi-sectoral systems analysis. *Urban Ecosystems, 15,* 809–848. https://doi.org/10.1007/s11252-012-0241-8.

Villarroel Walker, R., & Beck, M. B. (2014). Nutrient recovery. Nexus innovation impact analysis. *Insight*, 1. BeCleantech Initiative, Sustainability Specialist Group, International Water Association, www.becleantech.org, August, (2014), p. 30 (see also www.cfgnet.org/archives/1528).

Villarroel Walker, R., Beck, M. B., & Hall, J. W. (2012). Water—and nutrient and energy—systems in urbanizing watersheds. *Frontiers of Environmental Science and Engineering, 6*(5), 596–611. https://doi.org/10.1007/s11783-012-0445-4.

Villarroel Walker, R., Beck, M. B., Hall, J. W., Dawson, R. J., & Heidrich, O. (2014). The energy-water-food nexus: Strategic analysis of technologies for transforming the urban metabolism. *Journal of Environmental Management, 141,* 104–115.

Villarroel Walker, R., Beck, M. B., Hall, J. W., Dawson, R. J., & Heidrich, O. (2017). Identifying key technology and policy strategies for sustainable cities: A case study of London. *Environmental Development*, http://dx.doi.org/10.1016/j.envdev.2016.11.006 (published on-line 23 November, 2016).

Walker, G. (2009). Beyond distribution and proximity: Exploring the multiple spatialities of environmental justice. *Antipode, 41*(4), 614–636.

Wen, Z.-G., Zhang, C., Ji, X., & Xue, Y. (2015). Urban mining's potential to relieve China's coming resource crisis. *Journal of Industrial Ecology, 19*(6), 1091–1102.

WHO. (2009). *Time for Action*. Global Status Report on Road Safety, World Health Organisation, Geneva (www.who.int/violence_injury_prevention/publications/global_reports/en/).

Williams, D. (2017). How statistics lost their power—and why we should fear what comes next. The Long Read, *The Guardian*, (19 January) (posted at https://www.theguardian.com/politics/2017/jan/19/crisis-of-statistics-big-data-democracy).

Wolman, A. (1965). The metabolism of cities. *Scientific American, 213*(3), 179–190.

Zhang, H., & Wen, Z.-G. (2014). The consumption and recycling collection system of PET bottles: A case study of Beijing, China. *Waste Management, 34*(6), 987–998.

# Chapter 3
# Risk, Resilience and Adaptation to Global Change

**Shakespear Mudombi**

**Abstract** *Background*: The significance and threat of global change is increasingly being acknowledged. Understanding and responding to it is of critical importance. Early action is more beneficial than delay. Responding to global change entails both adaptation and mitigation. This chapter focuses on the former. It sets out to contribute to the understanding of what global change is, and its implications for Africa in general and South Africa in particular. Understanding the risks that are present is vital for the formulation and implementation of appropriate responses to such risks. For South Africa, responding to global change is a priority and it is one of the grand challenges that have been identified in its policy documents. The chapter is based on extensive literature review. *Methodology:* An extensive literature review including policy documents and published scientific literature was conducted. *Application/Relevance to systems analysis:* Understanding and responding to global change requires the need to acknowledge that processes, risks and the impacts occur in multiple stressor and multiple scale contexts. The complexities associated with global change as well as the potential for maladaptation and unintended consequences motivate the need to apply systems thinking. *Policy implications:* South Africa as part of the global system, will also be impacted by global change and its associated risks. Hence, the need for the country to be proactive. Some of the factors that can promote resilience and adaptation to global change include: taking a "glocal" approach, promoting information generation and dissemination, enabling relevant and responsive institutions, promoting flexibility and learning, building the asset base of households and communities, promoting stakeholder buy-in and stewardship in programmes, enhancing ecological infrastructure, and forging partnerships and collaborations. The state and other stakeholders should strive to enable the creation of a favourable environment that can foster appropriate resilience and adaptation to global change. *Conclusion:* Strides are being made in terms of understanding and responding to global change. However, due to complexities involved, more effort is still needed to establish and further understand global change processes. There is need for more multi-disciplinary stakeholder partnerships in order to realise synergies. Continued effort should be directed at creating awareness and building positive perception of the

S. Mudombi (✉)
Trade and Industrial Policy Strategies (TIPS), Pretoria, South Africa
e-mail: mudombi.shakespear@gmail.com

 
P. Mensah et al. (eds.), *Systems Analysis Approach for Complex Global Challenges*,
https://doi.org/10.1007/978-3-319-71486-8_3

need to adapt amongst various stakeholders. Proper assessment methodologies should be employed to evaluate various adaptation options before their implementation in order to avoid maladaptation. Global change should be embraced at the local level in the context of multiple stressors that tend to increase vulnerability.

## 3.1 Introduction

The Earth System is comprised of land, oceans, atmosphere and poles, plus the interacting physical, chemical, and biological processes, and the associated natural cycles (the carbon, water, nitrogen, phosphorus, sulphur and other cycles) (IGBP 2016). Significant changes have been observed in the Earth System, which has brought to the fore the term global change. Global change refers to planetary-scale changes in the Earth System. These changes include: atmospheric circulation, ocean circulation, climate, the carbon cycle, the nitrogen cycle, the water cycle and other cycles, sea-ice changes, sea-level changes, food webs, biological diversity, pollution, health, fish stocks, human society, amongst others (IGBP 2016; Muccione and Schaepman 2014).

Though the term global change is at times used to refer to global climate change, it is important to take note that it is much broader. Climate change is not the same as global change, but it is a component of global change; the Earth System has many other components and processes. Thus, global change includes changes in many aspects of the Earth System, including the climate (IGBP 2016; NAS 2000). Erisman et al. (2015) observes that current global-change risk assessments generally target single stressors, such as the climate, while paying less attention to wider impacts on land degradation, food and energy production, water supply and environmental hazards. In this regard, global change response efforts should not focus on the climate alone, rather should integrate many other components of the Earth System (NAS 2000). Much as this chapter takes cognisance of the fact that global change is not global climate change, it is important to highlight that many of the examples and the literature that is cited are mostly climate change related. This is so, mainly because a lot of research has been undertaken in the climate change field.

Although there are natural drivers of global change, it is noteworthy to point out that humans are increasingly contributing to global change. The Amsterdam Declaration on Global Change that was issued in 2001 highlights that the Earth System has already gone past the general natural variability (Moore et al. 2001). Humans are having profound impacts on the global environment, with detrimental effects on the climate, species, ecosystems, and human health (Camill 2010). This occurs as a result of population growth, pollution, energy and resource use, land use, agriculture, urbanization, transport and economic activities (Muccione and Schaepman 2014; Steffen et al. 2004). Two distinct aspects of human-induced global change are: first, humans are causing accelerated changes, and secondly, they are bringing new kinds of changes; which then interact to further compound the negative effects on the Earth System (NAS 2000).

## 3.2 The Interlinkages Between Risks, Vulnerability and Stressors

### 3.2.1 *Risks*

An important component of the discussion around global change is the notion of risk. Risk is defined as the probability of a negative event and its negative impacts (OECD 2014). Risk can be viewed as the likelihood of experiencing harm or loss (Mitchell and Harris 2012). Associated with risk are shocks and stresses. A shock is a sudden event that often has negative impact on the vulnerability of a system and its parts, while a stress is a long term trend, that worsens the vulnerability of the actors (OECD 2014).

Broadly, global change is associated with global risks. These risks do not have geographic boundaries, as their cascading effects stretch far and wide, with impacts that can affect several countries or industries (WEF 2016b). In other words, a global risk is not a threat to a particular region alone but covers many regions. The Global Risks Report 2016 ranks the failure of climate-change mitigation and adaptation as the most impactful global risk; it is also ranked as the third most likely to occur, while, water crises is ranked as the third most impactful and ninth most likely to occur (WEF 2016a, b).

In this context, individuals, households, and communities are battling with climate- and water-related challenges. The current and future projections paint a gloomy picture. The Global Climate Risk Index 2016 reports that between 1995 and 2014, greater than 525,000 people died worldwide and losses of more than US$ 2.97 trillion were experienced as a direct result of over 15,000 extreme weather events (WEF 2016b). In South Africa, it is estimated that about 5.8 million people will be affected by extreme rainfall events (Van Huyssteen et al. 2013). The year 2016 was recorded as the hottest year globally, with 43 °C being recorded in Pretoria (WWF-SA 2017). South Africa is a water scarce country, and projections point to further scarcity. Based on current usage trends, the country will likely face 17% water deficit by 2030, and the shortages will be worsened by climate change (WWF-SA 2017). Thus, delayed and inappropriate action on global change and its associated events will have significant detrimental effects on the socioeconomic development of the country.

### 3.2.2 *Vulnerability*

An important aspect in this discussion is the issue of vulnerability to global change and its associated risks. Vulnerability can be understood as the propensity or tendency to be negatively affected (Mitchell and Harris 2012). It is the manifestation of susceptibility to harm, and exposure to hazard (OECD 2014) or the propensity to suffer harm from exposure to external stresses and shocks (SRC 2015).

Vulnerability must be understood as a dynamic characteristic that is influenced by larger scale economic and environmental changes (Leichenko and O'brien 2002).

Some of the factors that influence vulnerability are location, access to information and resources, the quality of infrastructure, housing type, density of the built environment, economic wellbeing of a community, and socioeconomic and political status (van Donk and Gaidien 2014). Most of the factors that create potential for harm are inherent in social systems (Cutter et al. 2008), hence the strong and complex inter-linkages between local drivers of vulnerability and exposure (World Bank 2013).

The political and economic system affects the allocation and distribution of resources in a society, and can be a key source of vulnerability (Van Huyssteen et al. 2013). For South Africa, the historical socioeconomic and political marginalisation of many people in the country reinforces their vulnerability (GGLN 2014). Patterns of colonial development and apartheid legacies impact on urban resilience; Cape Town illustrates the continued high levels of social, spatial and structural inequalities (Rodina and Harris 2016). Van Huyssteen et al. (2013) also note that past and current urban planning as well as high levels of inequality in the country have seen people staying in unsafe and vulnerable locations such as floodplains, hillsides, and coastlines.

The negative effects of hazards tend to be both regressive and heterogeneous, thereby contributing to higher inequality (World Bank 2013). At the same time, the unequal distribution of vulnerability is worsened by pre-existing inequalities (Adger 2006). The negative effects of global change worsen the conditions of people who are already suffering. Hence, poverty underlies most of the vulnerability of communities and households. For instance, vulnerable low-income households and the unemployed are likely to face more severe climate related impacts (DEA 2011). Accordingly, tackling poverty will help to reduce vulnerability, and on the other hand reducing vulnerability helps to reduce poverty. Vulnerability does not only come from direct global change related impacts. In some cases, there are response measures that can have indirect negative impacts on the livelihoods of people. For example, South Africa may be economically vulnerable to commitments adopted at the international and national level to lower greenhouse gas (GHG) emissions as it is highly dependent on electricity generated from coal which is considered a dirty fuel (DEA 2011).

Population size, human settlements, and the availability of support resources are critical issues in the discussion on resilience and adaptation to global change. If the population in a particular area is not proportional to available resources, this reduces the ability of such a community to cope with and adapt to change. This seems to be the case with informal settlements or slums, which are increasingly being part of the urban landscape across the world. Areas such as these tend to be more vulnerable as they have inadequate access to the most important basic infrastructure and services. In this regard, South Africa is experiencing growing urbanisation and forecasts indicate further growth (Fig. 3.1). This urbanisation is also associated with expansion of informal settlements, which is partly driven by the high levels of inequality and marginalisation in the country.

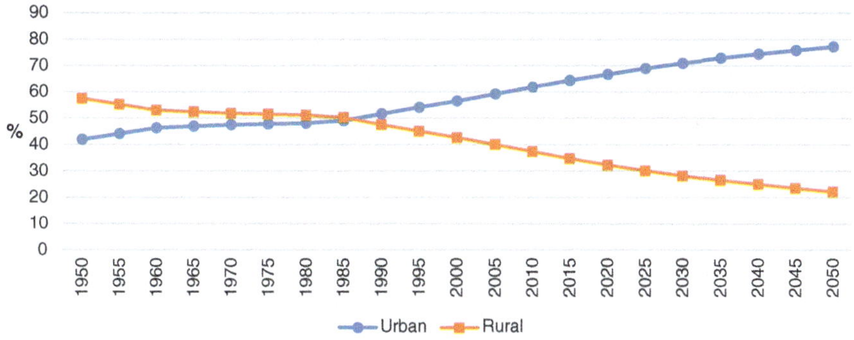


**Fig. 3.1** Percentage of population residing in urban and rural areas for South Africa, 1950–2050. *Source* Author based on United Nations, Department of Economic and Social Affairs, Population Division (2014)

Informal settlements experience multiple pressures that combine with climate change impacts to worsen their pre-existing vulnerabilities and inequalities (Taylor and Peter 2014). In recent years, the vulnerability of many South African communities is becoming more evident. For example, Joubert and Martindale (2013) observed the vulnerability to flooding of informal settlements in Cape Town; the people living there generally had limited capacity to cope. This is also the case in Johannesburg. The Mail and Guardian (2014) reported that,

> *About 70 shacks in the Vusimuzi informal settlement near Tembisa, in Ekurhuleni, were flooded, and about 25 displaced families were moved to the community hall ... About 90 shacks in Kliptown, Soweto, had also been flooded due to heavy rains in the area.*

### 3.2.3 *Multiple Stressors*

Most environmental challenges are likely to show combined action of several driving forces, acting at varying spatial and temporal scales (NAS 2000). This implies that many risks and shocks are closely related and interact with each other. Erisman et al. (2015) refer to them as networked risks—a sudden change in one can have a domino effect on others. In other words, a change in one has an effect on others, presenting a compounding effect. The interaction of such risks, shocks, and trends is sometimes referred to as multiple stressors. They act and impact differently but tend to have a reinforcing effect upon each other, thereby further worsening the situation. Casale et al. (2010) describes the situation as "entangled crises", whereby development efforts to disentangle one thread or another off the knot is difficult and generally fails.

The interconnectedness between stressors tends to blur the distinction between them, which in some cases lead to wrong diagnosis and action. Unfortunately, most of the past studies did not take vulnerability as a pre-existing state generated by multiple factors and processes that influence the ability to respond to stress (Eriksen and Kelly 2007). Tackling the multiple stressors does not require a single lens that is specific to a particular stress, but a combination of complementary lenses that are able to diagnose multiple stressors. This calls for an integrative approach rather than focusing on a single type of hazard (Van Huyssteen et al. 2013). This is to say—all the key risks and stressors need to be considered collectively (World Bank 2013), as there are feedbacks between various processes.

In the same context, it is crucial to understand the effects of socioeconomic and biophysical processes on global change, and how global change impacts on the socioeconomic and biophysical processes (NAS 2000). The biophysical stress may further worsen existing socioeconomic and political stresses and vice versa (Vogel 2011). For example, Nel et al. (2014) observed that farmers in Eden District in Western Cape, South Africa were impacted by the continued occurrence of droughts, floods and wildfires; this had severe knock-on effects on their farming and the whole local economy, which further increased their vulnerability.

O'Brien et al. (2004) suggested that through the lens of vulnerability, in areas that face multiple stressors, climate change may be the stressor that pushes people or ecosystems "over the edge". This indicates that climate change compounds existing stressors and also brings with it new ones. For instance, climate change and water risks are closely related to food insecurity risks (WEF 2016b). Also, weather-related hazards, intensified by climate change, converge with local drivers of exposure and inherent vulnerability to amplify disaster risk (World Bank 2013). With a changing climate, it is important to understand both the potential 'big, extreme' events and also the regularly occurring 'smaller' events (Vogel 2011).

Fourie et al. (2015) identified a relationship between wave action, coastal erosion and shoreline retreat at Monwabisi Beach near the City of Cape Town. It was observed that the beach is experiencing extreme rates of coastline erosion which is damaging local infrastructure. Fourie et al. (2015) concluded that the vulnerability of the beach to erosion is due to many factors such as the number and height of big wave events, waves coming from a more southerly direction, the underlying geological substrate, and the impacts of local infrastructure on the geological substrate.

The additional risks for water security as a result of climate change, have knock-on effects on highly water dependent sectors such as agriculture, electricity production, mining and manufacturing (DEA 2011). Southern Africa has high dependence on the natural environment for livelihoods, which makes it more vulnerable to impacts (Davis 2011). Similarly, climate change can also have direct and indirect negative health impacts, for example, the reduced water availability associated with droughts can cause health hazards associated with poor water sanitation and the food insecurity can result in nutritional deficiencies (WHO 2014). The DEA (2011a) notes that South Africa has a notable proportion of people particularly the poor, who already face complex health challenges which are likely to be worsened by climate change-related health risks, for example the spread of vector-borne diseases such as malaria, rift valley fever and schistosomiasis.

The majority of natural biomes in South Africa are diverse and can be sensitive to changes in the climate (Midgley 2011). Moreover, the widespread presence of invasive alien plants (IAPs) can also undermine the resilience of ecosystems and communities to withstand risks and hazards. IAPs have been noted to draw more water compared to native plants which negatively impacts on the sustainability of such environments. This is a common challenge in South Africa. Estimates show that about 9000 plant species have been introduced in the country, of which about 161 species are deemed invasive, they tend to spread at a fast rate and consume more water (DEA 2012). IAPs also tend to increase the risk of fires, which becomes worse in drought conditions. In a study conducted in Eden District in Western Cape, South Africa, Nel et al. (2014) found that allowing the spread of IAPs into untransformed vegetation could halve monthly river flows experienced during drought as well as double fire-line intensities.

## 3.3 Resilience

The resilience concept has been gaining traction in both research and development practice. This increased attention to resilience is partly due to the current thinking about sustainable futures in the face of growing risk and uncertainty (Mitchell and Harris 2012). In this context, resilience-building is envisioned to expedite holistic, positive and lasting solutions in communities and nations who are most at risk of harm (Mitchell 2013). The Global Risks Report calls for a 'resilience imperative', which requires an urgent need to explore new avenues and more opportunities to mitigate, adapt to and build resilience against global risks and threats (WEF 2016a, b). The definition of resilience varies and in some cases there are contestations (Klein et al. 2004; Mitchell and Harris 2012; Adger 2000).

Resilience is the ability of a social system (household, community, nation, or region) to respond and recover from shocks, which includes those inherent characteristics that enable the system to absorb impacts and cope with an event, and post-event adaptive processes that facilitate the ability of the social system to re-organize, change, and learn in response to a threat (Cutter et al. 2008). In other words, resilience is the capacity of a system and its component parts to anticipate, absorb, accommodate, or recover timely and efficiently from the effects of a shock or stress (Mitchell and Harris 2012). That is, being able to deal with change and continuing to develop (SRC 2015).

Klein et al. (2004) state that while resilience has been defined in many different ways, it is important that it is used to define specific system attributes which are: the amount of disturbance a system can allow and remain within the same state; and the extent to which the system is capable of self-organisation. Resilience has two attributes, namely inherent and adaptive (Cutter et al. 2008). The inherent attribute means the system functions well during normal periods, i.e. non-disaster periods, while adaptive relates to the flexibility in response during disaster period, which allows the system to function well. Bahadur et al. (2010) identified ten characteristics of resilient systems namely: high level of diversity; effective governance/institutions/control mechanisms; acceptance of uncertainty and change; community involvement and inclusion of local knowledge; preparedness, planning and readiness; high degree of equity; social values and structures; non-equilibrium system dynamics; learning; and adoption of a cross-scalar perspective of events and occurrences.

Resilience is a system-level concept; it avails a framework that integrates how multiple systems interact across temporal and spatial scales (Anderies et al. 2013). The use of 'system' in the context of resilience stems mainly from ecological theory (Bahadur et al. 2010). It considers multiple risks, shocks and stresses and their impacts on natural systems as well as people's livelihoods; also taking cognisance of the slow drivers of change that have impact and non-linearity (Mitchell and Harris 2012). It is important to point out that resilience is a dynamic process (Cutter et al. 2008; Mitchell and Harris 2012), meaning that it is not static but always evolving.

Adger (2000) distinguishes the resilience concept in ecology from social resilience, but also takes note that they are closely related. Ecological resilience relates to the application of the concept to ecological systems; this is where the concept was first used. Social resilience entails applying the concept to social systems, which considers how individuals and social groups respond and it has economic, spatial and social dimensions (ibid). The latter is the focus of this chapter.

Although resilience is generally perceived as having good purpose, some authors have suggested that there are cases in which it is undesirable. Resilience is generally associated with stability, however, this attribute might not always be desirable from an evolutionary perspective (Adger 2000). Mitchell and Harris (2012) assert that the 'dark side of resilience', occurs when it results in the persistence of a negative attribute, the system becomes fixed and less responsive to future threats. This means that in some cases, resilience might result in the system losing its ability to be flexible, or to adjust and be modified in response to harm or a disturbance.

## 3.4 Adaptation

Adaptation is an important response to global change. This is particularly so because some of the impacts associated with global change are already being experienced. Adaptation is a process, action or outcome in a system that helps the system to better cope with, manage or adjust to some changing condition, stress, hazard, risk or opportunity (Smit and Wandel 2006). It involves taking the right measures to reduce the negative effects or exploiting the positive ones, by making the appropriate adjustments and changes (UNFCCC 2007). Adaptation is generally meant to cushion against the negative effects of global change. However, it can be an opportunity to meet other developmental objectives. Davis (2011) states that proactive responses can harness opportunities for human development. For example, as noted by South Africa's Department of Environmental Affairs (DEA) that well planned adaptation responses can be properly linked to sustainable development policies, whereby issues such as unemployment and poverty are addressed simultaneously (DEA 2011).

Adaptation involves cascading decisions across a landscape made up of various agents (Adger et al. 2005). The agents include individuals, households, communities, sectors, regions, and countries. Successful adaptation is dependent on three elements i.e. timely recognition of the need to adapt, an incentive to adapt, and the ability to adapt (Ikeme 2003; Fankhauser et al. 1999). There is a need to understand what types and forms of adaptation are feasible, the stakeholders involved, and what is required to facilitate or encourage their development or adoption (Smit and Skinner 2002). Adaptation involves a variety of measures. For instance, adaptation projects implemented across the world under the Global Environment Facility (GEF) were categorised into 10 categories namely: capacity building, management and planning, practice and behaviour, policy, information, physical infrastructure, warning or observing system, green infrastructure, financing, and technology (Biagini et al. 2014).

Resilience and adaptation are closely related and complementary concepts. However, there is lack of conceptual clarity on their relationship—"whether resilience pertains to an idealised form of adaptation or whether the terms can be used interchangeably" (Bahadur et al. 2010, 19). Similarly, adaptive capacity is an important concept as well. Engle (2011) highlights that adaptive capacity is a mutual thread between vulnerability and resilience frameworks. Adaptive capacity is defined as the ability to plan, prepare for, facilitate and implement adaptation options (Klein et al. 2004). The OECD (2014) defined adaptive capacity as "the ability of a system to adjust, modify or change its characteristics and actions to moderate potential future damage and to take advantage of opportunities, so that it can continue to function without major qualitative changes in function or structural identity". Increasing adaptive capacity helps a system to respond to varying ranges and magnitudes of impacts (Engle 2011). But, having adaptive capacity does not assure that it is used appropriately (Klein et al. 2004).

Despite the fact that many adaptation options have beneficial outcomes, some options may result in unintended consequences. The adoption and implementation of adaptation measures will have local (that is, specific to the project area) effects, as well as non-local or non-target group effects. This can be viewed as the external costs of adaptation measures, since such measures may impact negatively outside of the target area and/or group. This can happen by increasing the vulnerability of the target area/group or other areas/groups. This failed adaptation is termed maladaptation and has been defined by Barnett and O'Neill (2010) as 'action taken ostensibly to avoid or reduce vulnerability to climate change that impacts adversely on, or increases the vulnerability of other systems, sectors or social groups'. Due to complexity, it is generally not easy to predetermine whether a particular adaptation will be sustainable or maladaptive (Engle 2011). This is one of the motivations for applying systems thinking to global change, its management and policy planning.

It is not enough to discuss adaptation to global change without dwelling on potential barriers. Identifying such barriers can help to understand the process and assist in decision-making (Moser and Ekstrom 2010). There are many barriers to adaptation to global change which include the inability of natural systems to adapt; systemic constraints (technological, financial, cognitive and behavioural, and social and cultural); knowledge gaps for adaptation, and impediments to flows of information and knowledge that are critical for appropriate decision making (IPCC 2007).

Perceptions and socio-cognitive factors influence adaptation as they influence the willingness and ability of an individual to take action. How people perceive global change and the associated risks determine how they respond (Steffen et al. 2004). An individual may be aware or unaware of the risk/s posed by global change. In the same vein, an individual might be aware or unaware of the appropriate adaptation action/s required for that risk or those risks. Risk awareness informs risk perception, which if positive can act as a motivation to take action. Béné et al. (2016) found that how people perceived their own ability to deal with risky events influenced the type of response(s) they adopted. Grothman and Patt (2005) stressed[1] that focusing on socio-cognitive indicators (for example, perceived adaptive capacity) helps to make better predictions about future adaptation and vulnerability, and the overall adaptive capacity can be increased by improving the communication of risk and information on possible, efficient and cost-effective adaptation options.

Adaptation can be limited by the values, perceptions, processes and power structures within society (Adger et al. 2009). In that context, gender norms, roles and relations can either enable or constrain adaptive capacities (WHO 2014). Poor and marginalised groups (disabled, elderly, orphans, widows) are generally less resilient and have difficulties in absorbing and recovering from disaster impacts (World Bank 2013). In addition, compared to men, women might have lower mobility and cultural limitations that hinder them from moving away from risk-prone areas or to utilise shelters during extreme events, which is likely to increase their exposure and vulnerability to hazards (ibid).

An important aspect relating to power dynamics is the issue of powerful actors whose interest take precedence over all other important development objectives. The World Bank noted that vested interests prioritise short-term responses over long-term prevention (World Bank 2013). For example, political leaders might be interested in implementing projects that give them huge support from the electorate in the short-term, e.g. distributing food parcels, rather than implementing long-term projects that empower such people to be self-sufficient. In the same context, Mitchell (2013) observed that the politicisation of initiatives can have a negative impact on the execution of projects, especially long-term large infrastructure initiatives that require sustained effort and resources across multiple election cycles. Such initiatives may be discontinued when a new government comes into power.

[1]Models such as the Protection Motivation Theory and Model of Proactive Private Adaptation to Climate change are critical in describing and predicting the process of adaptation.

## 3.5 Enabling Responses to Global Change

The DEA (2011) asserts that South Africa will adopt the resilience approach to climate change-related extreme events, because resilience enables a holistic approach to disaster management. This view is supported by Erisman et al. (2015) who state that rather than managing many individual risks, resilience should be promoted in responding to adverse events, because it focuses on the whole system and targets long-term security. A resilience systems analysis provides actors with a shared view of the risk landscape; it enables people to have an understanding of the broader system, the key components, attributes, impacts; power dynamics; and it enables the creation of a shared vision of the need to build resilience (OECD 2014). Such a holistic approach is important as it considers various risks and their interaction. It is acknowledged that in certain cases, having an in-depth understanding of a particular risk is essential; nonetheless as greater attention is paid to that particular risk, its relationship with other risks/factors should not be forgotten.

The World Bank (2013) identified three major ways to deal with risks associated with disasters, *viz.* 'retreating' to reduce exposure to the hazard that is, relocating to safer locations, 'protecting' (people and assets) by reducing the hazard risk (for example, through resilient infrastructure), and 'accommodating' that is, active decision to live with the hazard but reducing the vulnerability to it. The option to retreat to safer locations seems to be applicable to hazards that are specific to small areas, however, in the context of global change this might not be appropriate as the associated risks and hazards impact on larger areas making it impossible for the relocation of a large number of people or assets. More relevant options for global risks are protecting the people and assets from the hazard risk, as well as accommodating the risks and hazards and working on reducing vulnerability through adaptation, while reducing future risk through mitigation. In this context, a number of important issues were identified which can help to reduce the risks and impacts from global change, and also enhance adaptation responses to it.

### *3.5.1 Taking a "Glocal" Approach*

Global change implies that the change occurs at the global level, with the effects or impacts manifesting at various levels. While looking at the global, it is vital to also view such impacts at the local level. In other words, zooming in and focusing on the local level will help to reveal the detailed picture that might be obscured by looking at global change as a global phenomenon in the strictest sense. Of importance is to consider local level impacts on communities and marginalised populations. Resilience building entails proactively understanding the risk landscape in each context and for different layers of society (Mitchell 2013), including the perspective of specific stakeholders (WEF 2016b). For instance, using a gender-disaggregated approach might help to understand the impacts of global change on different groups. This can help in the designing and implementation of appropriate adaptation measures that are inclusive.

Understanding local level impacts of global change requires the use of the 'glocal' approach.[2] This approach helps to understand how the impacts of global change are experienced at the local level, and how these interact with other factors on the ground (Eriksen 2004). There is need to understand the physical projections, as well as, assessing the levels of vulnerability generated by social, economic, and political processes interacting across geographic scales (Eriksen and Kelly 2007). The glocal approach is important as it presents the picture of global change at the global level and within it, inserting a zoomed picture of local level impacts. Erisman et al. (2015) seem to also support the glocal approach by suggesting that in delivering the global risk-network model there should be two shifts. On one hand, the risk narrative has to be reframed by putting the individual at the centre. On the other hand, risk modelling must take a wide focus, which includes both environmental and socioeconomic risks on the whole Earth system. Thus, incorporating risks at the local to the global level and understanding their linkages can help people adopt effective actions that enhance resilience (Erisman et al. 2015).

### 3.5.2 Information Generation and Dissemination

Information is an important basis for risk management, resilience building and adaptation to global change. Mudombi (2014) observed two serious constraints. First, there can be lack of access to the necessary and complementary information and knowledge on what is happening, what to do, and how to do it. Second, in cases when such information and knowledge are available, action is limited by lack of resources to do what is supposed to be done. The World Bank (2013) suggested that the first step should be to improve the understanding of the risks, and the second step is to develop adaptation options based on that information. Having relevant information and disseminating it in an appropriate way can help to build a positive perception of the risks which is likely to improve people's motivation to adapt.

Effective adaptation planning requires improved observations; improved regional, national and global data, as well as denser networks; the recovery of historical data; building of support among the user communities; and promoting greater collaboration between the providers and users of the information (UNFCCC 2007). Early warning systems are also important sources of the much needed information to respond to various types of global change related risks and hazards. The early warning information should be accurate and timely, and based on relevant data and robust analysis. It is necessary to complement that information by availing

[2]Eriksen (2004) noted that the term 'glocal' has been used particularly in relation to cities (for example, by Brenner 1998); it represents the idea that globalisation takes place through local manifestations and forces on the ground.

supporting tools, technologies and resources needed to undertake the recommended actions.

When stakeholders have a better understanding of the challenges and are equipped with appropriate technical skills, their collaboration can result in optimal sustainable win-win solutions (Mudombi et al. 2017). Taking a multidisciplinary approach that involves natural, social and human disciplines is crucial (Muccione and Schaepman 2014). Therefore, there is need to intensify efforts to transfer the growing knowledge base to various stakeholders, while at the same time empowering the next generation of scientists with the essential skills to undertake Earth System science (Steffen et al. 2004).

A good research base will generate the much needed information to improve people's understanding. The information should be packaged and disseminated in an appropriate format and manner. Research can enhance adaptation by providing more reliable information about the risks and its impacts, as well as developing and testing improved adaptation options and technologies (Fankhauser et al. 1999). Thus, innovation is an important enabler that can facilitate global change adaptation because of the need to formulate as well as adopt new and appropriate technologies and strategies (Mudombi 2014).

Moore et al. (2001) suggested the need for a new system of global environmental science in order to understand global change. This new system should enable greater integration across disciplines, and collaboration within and across national boundaries. There are some examples that have been or are being undertaken in the country. For example, a lot of work is being undertaken by the South African Environmental Observation Network (SAEON), whose responsibilities focus on three mandates namely observation, information and education (SAEON 2009). In addition, the DST has supported a number of programmes, such as the development of the South African Risk and Vulnerability Atlas (SARVA), to bridge the gap between science and policy by improving access to information on impacts and risks associated with global environmental change (Davis 2011).

### *3.5.3 Relevant and Responsive Institutions*

Institutions, institutional arrangements, and institutional capacity are critical in facilitating adaptation. Institutions are the norms and rules that shape human interactions; they can be formal or informal (SRC 2015). Eriksen (2004) highlighted that institutional factors can limit or enhance local capacity to carry out appropriate adaptation measures. The resilience of a social system depends on the institutional rules which govern that system (Adger 2000). Effective institutions and institutional structures can strengthen resilience in a system as well as enhance community cohesion. Ideally, the institutions should be decentralised, flexible, locally appropriate, and facilitate system-wide learning (Bahadur et al. 2010).

Responding to global change requires a new set of institutions or institutional arrangements that are capable of embracing risks in a holistic manner. The impacts of global change will affect multiple sectors in multiple ways and the current institutional and governance systems are largely sector-based thus limiting how they are able to respond (World Bank 2013). Proper coordination between actors at various levels in the global change space is critical. This should be present at all levels (international, national, regional, community, and household level). Thereby helping to reduce conflicts and duplication of roles, while at the same time building synergies and complementarities as well as ensuring efficient allocation and usage of limited resources. Institutions should be flexible enough to proactively respond to global change, as well as being firm enough to take concrete steps towards building resilience.

Related to the broad institutional framework, is the social capital within a particular community

Social capital relates to social relations among individuals and the norms and social trust they generate which enhances coordination and cooperation for their mutual benefit (SRC 2015). Adger et al. (2007) state that human and social capital are key determinants of adaptive capacity at all levels. Social capital related aspects that can contribute to resilience building include: social cohesion, mechanisms of reciprocity, 'positive' social norms, strong social fabric, local 'good' governance, and the local capacity for collective action (Béné et al. 2016). However, it is not always the case that social capital is beneficial, in some cases it can be a constraint. Hence, the need to understand the various forms of social capital and the conditions under which they enhance people's resilience at different levels (Béné et al. 2016).

### 3.5.4 Flexibility and Learning

Resilience seeks to enable systems to be capable of learning, self-organising, and adapting to change (Anderies et al. 2013; Folke 2006). Ability to adjust (flexibility) to changing circumstances and timeframes is necessary for adaptation (DEA 2011). Flexibility ensures that the system can adjust appropriately in the face of risk or disturbance. In this regard, the uncertainty, change, non-linearity, randomness of events in a system should be embraced, with the policy shifting from seeking to control change and creating stability, to enabling the capacity of systems to respond to change (Bahadur et al. 2010). The important ingredients for resilience entail "learning to live with change and uncertainty, nurturing diversity, combining different knowledge systems for learning and renewal, and creating opportunities for self-organisation and cross-scale linkages" (GGLN 2014). Learning and experimentation through adaptive and collaborative management allows different types and sources of knowledge to be valued and included in developing solutions (SRC, n.d.).

Learning is critical in terms of ensuring that a system can adopt the good aspects from past experience, while avoiding the bad ones. This also entails learning from the experience of other areas or systems, what can have good results or bad results. While formal learning is necessary, this should be complemented by other forms of learning. This includes promoting conditions that nurture social learning, which is an essential ingredient in enhancing the adaptive capacity of communities (Mudombi et al. 2017). Social learning can facilitate a shared understanding of the challenge and the recognition of the need and motivation to work together in tackling the challenge (Mudombi et al. 2017).

### *3.5.5 Building the Asset Base of Households and Communities*

Assets are an important factor in determining how people respond. Assets include different forms of capital namely human, social, physical, financial, and natural capital. However, it is important to acknowledge that assets by themselves are not sufficient in ensuring resilience. Béné et al. (2016) found that the importance of assets should be understood by making a distinction between response and recovery. In their study, assets appeared to be more important in the recovery process of households affected by shocks and stressors, rather than in the response process. Assets can enable people to have a wider set of livelihood options, which is an important basis of livelihood diversification. Regrettably, in many developing countries, the local population has limited livelihood options which makes it difficult for them to transition to a sustainable future when they have to first meet pressing survival needs (Schlesinger 2006).

### *3.5.6 Ownership, Participation, and Stewardship*

Wider participation of various stakeholders is required in order to have their buy-in and support. Broad and well-functioning participation creates trust among stakeholders (SRC, n.d.). Strengthening resilience in vulnerable communities should be bottom-up, taking into cognisance the important role that the affected people can play in strengthening their own resilience and adapting to change. Ziervogel et al. (2017) warn that building resilience in African urban settings should not be based on externally defined pathways and approaches, rather the primary focus should consider the physical and social complexities, as well as the development of critical infrastructure and governance systems necessary and appropriate to the local settings. This brings to the fore the concept of 'negotiated resilience', which is a process of building resilience through considering the interests and needs of diverse groups, including the marginalised (Ziervogel et al. 2017).

Participation, inclusion, and stewardship go hand in hand. Adaptation and sustainability should be prioritised in different aspects of life, in order to ensure good management of the Earth's environment, as well as meeting socioeconomic development objectives. The state can play a crucial role in creating an enabling environment that allows community resilience to flourish (van Donk and Gaidien

2014). At the global level, Moore et al. (2001) highlighted the need for an ethical framework for stewardship and strategies for the management of the Earth system. In order to meet this objective, the stewardship thinking should be entrenched amongst the public, and various stakeholders including both private and public sector leaders.

### 3.5.7 Enhancing Ecological Infrastructure

Investing in infrastructure, both physical 'hard' infrastructure and other softer forms of infrastructure is necessary for building resilience and supporting adaptation. In this context, ecological infrastructure is very important and in recent times increasing attention is being paid to building and restoring it. Ecological infrastructure can be understood as an interconnected network of natural areas and open spaces that holds valuable natural and biodiversity assets that are necessary for sustainable livelihoods (DEA 2012). This form of infrastructure is important to buffer and minimise the impacts of hazards associated with global change. The ecological infrastructure helps in the provision of various ecosystem services. An example of a programme to strengthen ecological infrastructure is the uMngeni Ecological Infrastructure Partnership (UEIP) in KwaZulu-Natal. The UEIP involves collaboration and partnership between various organisations with the aim of enhancing greater water security through improving and maintaining ecological infrastructure (Colvin et al. 2015).

The World Bank (2013) asserts that ecosystem-based solutions tend to be cost effective and enable flexibility in adapting to changing hazard patterns over time. The benefits go beyond better livelihoods but also entail the maintenance of flora and fauna. In reference to a community in Eden District in Western Cape, South Africa, Nel et al. (2014) noted that the multiple co-benefits of ecosystem management and restoration are significant, for instance, clearing invasive alien trees could reduce the effects of drought, wildfire and flood hazards, while at the same time creating employment.

### 3.5.8 Forging Partnerships and Collaborations

There is a need to further leverage the participation of a wide range of stakeholders to mutually address global risks, as these are beyond the domain and capacity of just one actor (WEF 2016b). Furthermore, there must be significant reform and renewal in the state in providing leadership, as well as promoting a culture of learning, deliberative and collaborative engagement, and development partnerships (GGLN 2014). Fostering public–private partnerships can assist in harnessing required resources, improving the uptake of a multiscale approach, stimulating innovation from stakeholders and ensuring that the needs of users are at the centre

(Erisman et al. 2015). Partnerships are important not only because they facilitate learning and co-generation of outputs, but also ensure complementarity by bringing actors with similar or different sets of capabilities and combining them to achieve more. For instance, the private sector usually brings more professionalism, better capabilities research and management, and resources, while the local government and civil society have experience, demonstrated impact, better understanding and operational capacity at the local level (Mitchell 2013).

There are efforts to promote partnerships to enhance resilience in some communities in South Africa. For example, Santam (a private insurance company), the Department of Cooperative Governance (CoGTA), and the South African Local Government Association (SALGA) forged a partnership through the Business Adopt-a-Municipality (BAAM) programme. Santam helped in disaster management, improving sustainability, and service delivery in vulnerable municipalities by providing support for fire-fighting, flood and storm water management (Santam 2016). In the initial phase, Santam supported 5 municipalities (four local and one district municipality) across various provinces in the country. The programme was later expanded to 10 district municipalities which comprise of 54 local municipalities (Santam 2016). There are two-way benefits associated with this approach; building the capacity of municipalities helps the municipalities to deal with risk and disasters, which is also beneficial to the insurers as there can be a reduction in insurance claims.

Moore et al. (2001) observed that while new partnerships among academic, industrial and government research institutions are being forged, there is a need to formalise, consolidate and strengthen these initiatives. Collaboration should also be between countries. South Africa in the National Climate Change Response White Paper explicitly stated that cooperation and collaboration is critical to dealing with climate change risks.

> *All states in the Southern African sub-region ... often face similar risks due to climate change and may also have similar adaptation needs. South Africa will therefore strive to develop climate change adaptation strategies ... in collaboration with its neighbours where appropriate, and seek to share resources, technology and learning to coordinate a regional response. A regional approach that achieves climate resilience will have significant socio-economic benefits for South Africa* (DEA 2011, 16).

## 3.6 Conclusion

Global change and its associated risks are getting increased attention from various stakeholders across the world. In South Africa, global change is one of the grand challenges that have been enunciated in policy documents. Moreover, significant amounts of resources have been allocated to enhance its understanding, in particular, how it relates to the country and region. Based on the literature review, the chapter explored issues of risk, resilience, and adaptation in the context of global change. It is evident that terms like 'resilience' and 'adaptation' mean differently to various stakeholders, however at the core of their meaning they point to desired outcomes in relation to responding to global change. In this discussion, it was also revealed that the scale at which these issues are assessed is important. For instance, the factors that determine resilience, vary at different spatial and temporal scales. Spatial scale is associated with the relationship between local and global contexts, whereas the temporal scale relates to short-term versus long-term dynamics.

Some of the factors that were highlighted as important in promoting resilience and adaptation to global change include: taking a "glocal" approach, promoting information generation and dissemination, enabling relevant and responsive institutions, promoting flexibility and learning, building the asset base of households and communities, promoting stakeholder buy-in and stewardship in programmes, enhancing ecological infrastructure, and forging partnerships and collaborations.

Strides are being made in terms of understanding and responding to global change. However, due to complexity, more effort is still needed to establish and understand several relations between global change processes. Global change science has a lot of work to undertake in a limited time space (Schlesinger 2006), which calls for more multi-disciplinary stakeholder partnerships in order to realise synergies. Adaptation to global change is of paramount importance, hence continued effort should be dedicated to creating awareness and building positive perception of the need to adapt amongst various stakeholders. Proper assessment methodologies should be employed to evaluate various adaptation options before their implementation in order to avoid maladaptation. Global change should be embraced at the local level in the context of multiple stressors that tend to exacerbate vulnerability. Resilience building and adaptation are likely to cushion households, communities and nations from the effects of global change, and it is everyone's duty to take appropriate action in that regard.

## References

Adger, W. N. (2000). Social and ecological resilience: Are they related? *Progress in Human Geography, 24*(3), 347–364.

Adger, W. N. (2006). Vulnerability. *Global Environmental Change, 16,* 268–281.

Adger, W. N., Arnell, N. W., & Tompkins, E. L. (2005). Successful adaptation to climate change across scales. *Global Environmental Change, 15,* 77–86.

Adger, W. N., Dessai, S., Goulden, M., Hulme, M., Lorenzoni, I., Nelson, D. R., et al. (2009). Are there social limits to adaptation to climate change? *Climatic Change, 93,* 335–354.

Adger, W. N., Agrawala, S., Mirza, M. M. Q., Conde, C., O'Brien, K., Pulhin, J., Pulwarty, R., Smit, B., & Takahashi, K. (2007). Assessment of adaptation practices, options, constraints and capacity. Climate Change 2007: Impacts, Adaptation and Vulnerability. In: M. L. Parry, O. F. Canziani, J. P. Palutikof, P. J. van der Linden and C. E. Hanson, (Eds.), *Contribution of working group II to the fourth assessment report of the intergovernmental panel on climate change* (pp. 717–743). Cambridge, UK: Cambridge University Press.

Anderies, J. M., Folke, C., Walker, B., & Ostrom, E. (2013). Aligning key concepts for global change policy: Robustness, resilience, and sustainability. Ecology and Society, *18* (2). doi:http://dx.doi.org/10.5751/ES-05178-180208.

Bahadur, A. V., Ibrahim, M., & Tanner, T. (2010). *The resilience renaissance? Unpacking of resilience for tackling climate change and disasters*. Institute of Development Studies: Strengthening Climate Resilience. Brighton.

Barnett, J., & O'Neill, S. (2010). Maladaptation. *Global Environmental Change, 20,* 211–213.

Béné, C., Al-Hassan, R. M., Amarasinghe, O., Fong, P., Ocran, J., Onumah, E., et al. (2016). Is resilience socially constructed? Empirical evidence from Fiji, Ghana, Sri Lanka, and Vietnam. *Global Environmental Change, 38,* 153–170.

Biagini, B., Bierbaum, R., Stults, M., Dobardzic, S., & McNeeley, S. M. (2014). A typology of adaptation actions: A global look at climate adaptation actions financed through the global environment facility. *Global Environmental Change, 25,* 97–108.

Camill, P. (2010). Global change. *Nature Education Knowledge* 3(10).

Casale, M., Drimie, S., Quinlan, T., & Ziervogel, G. (2010). Understanding vulnerability in Southern Africa: comparative findings using a multiple-stressor approach in South Africa and Malawi. *Regional Environmental Change, 10,* 157–168.

Colvin, C., Cartwright, A., McKenzie, M., Dent, M., Maherry, A., & Mhlongo, T. (2015). Enhancing ecological infrastructure in the uMngeni catchment through private sector action and engagement. Green Fund Research Report.

Cutter, S. L., Barnes, L., Berry, M., Burton, C., Evans, E., Tate, E., et al. (2008). A place based model for understanding community resilience to natural disasters. *Global Environmental Change, 18,* 598–606.

Davis, C. L. (2011). *Climate risk and vulnerability: A handbook for Southern Africa*. Pretoria: Council for Scientific and Industrial Research.

DEA. (2011). *National climate change response white paper*. Pretoria: Department of Environmental Affairs.

DEA. (2012). 2nd South Africa environment outlook. A report on the state of the environment (Executive Summary). Pretoria: Department of Environmental Affairs.

DST. n.d. Ten-Year innovation plan. Department of Science and Technology, Republic of South Africa.

Engle, N. L. (2011). Adaptive capacity and its assessment. *Global Environmental Change, 21,* 647–656.

Eriksen, S. (2004). Building adaptive capacity in a 'glocal' world: Examples from Europe and Africa. *The ESS Bulletin 2*(2).

Eriksen, S., & Kelly, P. M. (2007). Developing credible vulnerability indicators for climate adaptation policy assessment. *Mitigation and Adaptation Strategies for Global Change, 12,* 495–524.

Erisman, J. W., Brasseur, G., Ciais, P., van Eekeren, N., & Theis, T. L. (2015). Put people at the centre of global risk management. *Nature, 519,* 151–153.

Fankhauser, S., Smith, B., & Tol, R. S. J. (1999). Weathering climate change: Some simple rules to guide adaptation decisions. *Ecological Economics, 30,* 67–78.

Folke, C. (2006). Resilience: The emergence of a perspective for social–ecological systems analyses. *Global Environmental Change, 16,* 253–267.

Fourie, J. -P., Ansorge, I., Backeberg, B., Cawthra, H. C., MacHutchon, M. R., & van Zyl, F. W. (2015). The Influence of wave action on coastal erosion along Monwabisi Beach, Cape Town. *South African Journal of Geomatics*, *4*(2), 96–109.

GGLN. (2014). *Community resilience and vulnerability in South Africa*. Good Governance Learning Network: State of Local Governance Publication. Cape Town.

Grothmann, T., & Patt, A. (2005). Adaptive capacity and human cognition: The process of individual adaptation to climate change. *Global Environmental Change, 15,* 199–213.

IGBP. (2016). Earth system definitions. *International Geosphere-Biosphere Programme (IGBP)*. http://www.igbp.net/globalchange/earthsystemdefinitions.4.d8b4c3c12bf3be638a80001040.html.

Ikeme, J. (2003). Climate change adaptational deficiencies in developing countries: The Case of Sub-Saharan Africa. *Mitigation and Adaptation Strategies for Global Change, 8,* 29–52.
IPCC. (2007). Summary for policymakers. In *Climate Change 2007: Impacts, Adaptation and Vulnerability. Contribution of Working Group II to the Fourth Assessment Report of the Intergovernmental Panel on Climate Change*. Cambridge: Cambridge University Press.
Joubert, L., & Martindale, L. (2013). *Rising waters: Working together on Cape Town's flooding*. The Flooding in Cape Town under Climate Risk (FliCCR) Project: African Centre for Cities, University of Cape Town.
Klein, R. J. T., Nicholls, R. J., & Thomalla, F. (2004). Resilience to natural hazards: How useful is this concept? EVA Working Paper 9. Potsdam: Potsdam Institute for Climate Impact Research.
Leichenko, R. M., & O'brien, K. L. (2002). The dynamics of rural vulnerability to global change: The case of Southern Africa. Mitigation and Adaptation Strategies for Global Change 7, 1–18.
Mail & Guardian. (2014). Rain causes flooding Havoc in Jo'burg. *Mail & Guardian Online*. https://mg.co.za/article/2014-03-07-rain-causes-flooding-havoc-in-joburg.
Midgley, G. F. (2011). Climate change, species and ecosystems. In *Observation on environmental change in South Africa*. Section 2. SUN MeDIA Stellenbosch.
Mitchell, D. (2013). Risk and resilience: From good idea to good practice (A scoping study for the experts group on risk and resilience). WP 13/2013. OECD.
Mitchell, T., & Harris, K. (2012). *Resilience: A risk management approach*. Overseas Development Institute: Background Note. London.
Moore, B., Underdal, A., Lemke, P., & Loreau, M. (2001). The Amsterdam declaration on global change. In *Challenges of a Changing Earth: Global Change Open Science Conference*. Amsterdam.
Moser, S. C., & Ekstrom, J. A. (2010). A framework to diagnose barriers to climate change adaptation. *PNAS, 107*(51), 22026–22031.
Muccione, V, & Schaepman, M. (2014). Global change. Terminology brief series. University Research Priority Program Global Change and Biodiversity. University of Zurich.
Mudombi, S. (2014). Analysing the contribution of ICTs in addressing climate change amongst communal farmers from two districts of Zimbabwe. Pretoria: University of South Africa. http://uir.unisa.ac.za/bitstream/handle/10500/14668/thesis_mudombi_s.pdf?sequence=1&isAllowed=y.
Mudombi, S., Fabricius, C., Patt, A., & Bulitta, V. Z. (2017). The use of and obstacles to social learning in climate change adaptation initiatives in South Africa. *Jàmbá: Journal of Disaster Risk Studies 9*(1). doi:http://doi.org/10.4102/jamba.v9i1.292.
NAS. (2000). *Global change ecosystems research*. Washington, D.C: National Academy of Sciences.
Nel, J. L., Le Maitre, D. C., Nel, D. C., Reyers, B., Archibald, S., van Wilgen, B. W., et al. (2014). Natural hazards in a changing world: A case for ecosystem-based management. *PLoS ONE, 9*(5), e95942. https://doi.org/10.1371/journal.pone.0095942.
O'brien, K., Eriksen, A., Schjolden, A., & Nygard, L. (2004). What's in a word? conflicting interpretations of vulnerability in climate change research. CICERO Working Paper 4. Oslo: Center for International Climate and Environmental Research (CICERO).
OECD. (2014). *Guidelines for resilience systems analysis*. OECD Publishing.
Rodina, L., & Harris, L. M. (2016). *Resilience in South Africa's urban water landscape*. The Conversation Africa: Opinion Piece.
SAEON. (2009). About SAEON. *South African Environmental Observation Network*. http://www.saeon.ac.za/about-saeon.
Santam. (2016). Building resilience through partnerships. https://www.santam.co.za/blog/intermediary-advice/building-resilience-through-partnerships/.
Schlesinger, W. H. (2006). Global change ecology. *Trends in Ecology & Evolution, 21*(6), 348–351.
Smit, B., & Skinner, M. W. (2002). Adaptation options in agriculture to climate change: A typology. *Mitigation and Adaptation Strategies for Global Change, 7,* 85–114.

Smit, B., & Wandel, J. (2006). Adaptation, adaptive capacity and vulnerability. *Global Environmental Change, 16,* 282–292.

SRC. n.d. Applying resilience thinking: Seven principles for building resilience in social-ecological systems. Stockholm Resilience Centre.

SRC. n.d. (2015). Resilience dictionary. *Stockholm Resilience Centre*. Accessed May 15. http://www.stockholmresilience.org/research/resilience-dictionary.html.

Steffen, W., Sanderson, A., Tyson, P. D., Jäger, J., Matson, P. A., Moore, F. Oldfield, et al. (2004). *Global change and the earth system: A planet under pressure (Executive summary).* Berlin: Springer-Verlag.

Taylor, A., & Peter, C. (2014). Strengthening climate resilience in African cities: A framework for working with informality. Working Paper. African Centre for Cities, Climate and Development Knowledge Network.

UNFCCC. (2007). *Climate change: Impacts, vulnerabilities and adaptation in developing countries*. Bonn: United Nations Framework Convention on Climate Change.

United Nations, Department of Economic and Social Affairs, Population Division. (2014). World urbanization prospects: The 2014 revision. CD-ROM Edition. https://esa.un.org/unpd/wup/CD-ROM/.

Van Donk, M., & Gaidien, G. (2014). In search of community resilience. In *Community Resilience and Vulnerability in South Africa*. State of Local Governance Publication. Cape Town: Good Governance Learning Network.

Van Huyssteen, E., Le Roux, A., & Van Niekerk, W. (2013). Analysing risk and vulnerability of South African settlements: Attempts, explorations and reflections. *Jàmbá: Journal of Disaster Risk Studies 5*(2). doi:http://dx.doi.org/10.4102/jamba.v5i2.80.

Vogel, C. (2011). Climate change risk, adaptation and sustainability. In *Observation on environmental change in South Africa*. Section 1. SUN MeDIA Stellenbosch.

WEF. (2016a). *Resilience Insights*. Geneva: Global Agenda on Risk & Resilience, World Economic Forum.

WEF. (2016b). The global risks report 2016 (11th Edition). Geneva: World Economic Forum.

WHO. (2014). *Gender, climate change and health*. Geneva: World Health Organization.

World Bank. (2013). Building resilience: Integrating climate and disaster risk into development. Lessons from world bank group experience. Washington DC: The World Bank.

WWF-SA. (2017). Scenarios for the future of water in South Africa. Cape Town: World Wide Fund for Nature—South Africa (WWF-SA).

Ziervogel, G., Pelling, M., Cartwright, A., Chu, E., Deshpande, T., Harris, L., et al. (2017). Inserting rights and justice into urban resilience: A focus on everyday risk. *Environment & Urbanization, 29*(1), 123–138. https://doi.org/10.1177/0956247816686905.

# Chapter 4
# Extract of Africa: Towards the Equitable and Ecologically Sound Governance of Mining and Drilling

**Ikechukwu Umejesi, Michael Thompson, Maria Marcello and Emmanuel Vellemu**

**Abstract** *Background*: Economists and political scientists have long recognised four kinds of goods: private, public, common-pool and club, the assumption being that category membership is determined by the physical properties of the goods themselves. But in the theory of plural rationality—the approach taken in this chapter—where specific goods end up is, to the extent that that is not determined by their physical properties, the outcome of a never-ending struggle between four kinds of social solidarity: individualism (which works towards privatisation), hierarchy (which favours the creation of public goods), egalitarianism (which is supported by common-pool goods) and fatalism (whose upholders enable club goods by the ease with which they can be excluded). *Methodology*: The study uses historical surveys and case-studies of different contexts where natural resource governance has upset harmonious relationship between different stakeholders. *Application/Relevance to systems analysis*: Our argument is that policy and governance, particularly in Africa, have allowed (indeed encouraged) individualism and hierarchy to dominate, thereby drowning out the other two institutional "voices". The result, as we show by way of a continent-wide historical survey and three case-studies—REDD+ in the Democratic Republic of Congo, acid mine drainage in South Africa and oil extraction in Nigeria—has been "crap" governance (in contrast to good governance, which requires that all four voices are both heard and responded to by the others). Put another way, the "resource curse" is not the inevitable consequence of a country being heavily reliant on extractive

I. Umejesi (✉)
Department of Sociology, University of Fort Hare,
East London Campus, East London 5200, South Africa
e-mail: IUmejesi@ufh.ac.za

M. Thompson
Risk and Resilience Program, International Institute for Applied Systems Analysis (IIASA), Laxenburg, Austria

M. Marcello
Department of Environmental Engineering, Louisiana State University, Baton Rouge, USA

E. Vellemu
Institute for Water Research, Rhodes University, Grahamstown, South Africa

 
P. Mensah et al. (eds.), *Systems Analysis Approach for Complex Global Challenges*,
https://doi.org/10.1007/978-3-319-71486-8_4

industries; it stems from insufficient "clumsiness": exemplified, as Kofi Annan has recently pointed out, by just two solidarities—multinational companies (individualism) and political leaders (hierarchy)—colluding to swindle the citizens out of their just rewards from their natural resources. *Policy implications*: Of policy implication, is the bringing-in of the two currently excluded voices, and we conclude by showing how, in relation to our case-studies, this can be achieved. *Conclusion*: Analyses of resource-related conflicts, especially in Africa, have often ignored the voices of the excluded social solidarities. Analysing this problem through a systems perspective will allow the incorporation of all voices as a way of constructing a more harmonious system in natural resource governance.

## 4.1 Introduction

Economists and political scientists have long recognized four kinds of goods: *public*, *private*, *common-pool* and *club* (Samuelson 1948; Snidel 1991). This typology is derived from two sets of distinctions: excludability versus non-excludability and high versus low jointness of consumption. A club, for instance, can exclude non-members, and a lighthouse is as bright for the nth user as it is for the first. The assumption, which is seldom made explicit, seems to be that category membership is determined by the inherent physical properties of the goods themselves: lighthouses, for instance, which are often seen as epitomising public goods, shine for everybody (non-excludability) and their brightness is non-rivalrous (high jointness of consumption). But, in the theory of plural rationality—the theory we will be relying on in this chapter—category membership is under-determined by those physical properties (Verweij 1999; Thompson 2000). Even lighthouses can be privatized (indeed, that is how they started off) and private goods will soon disappear in a social setting where everyone has become convinced that, as Prince Kropotkin famously put it, "All property is theft" (Kropotkin and Marshall 1995: 239).

Where specific goods end up (for a time, at least) is, to the extent that this is not being determined by their inherent physical properties, the outcome of a never-ending struggle between four forms of social solidarity: *individualism* (which works towards privatization), *hierarchy* (which favours the creation of public goods), *egalitarianism* (which is supported by common-pool goods) and *fatalism* (whose upholders enable club goods by the ease with which they can be excluded from them). The result, once we introduce all the institutional dynamics and politics that have been excluded by the "inherent physical properties" assumption, is a *contested terrain*. On a contested terrain, a proposal—to build a lighthouse, say—will not go unchallenged; there will always be a plurality of voices.

- *Hierarchy*. The proposed lighthouse is a collective good. Everyone will benefit and no one will be excluded. Since individuals will not contribute, the government has to step in. Otherwise the development of the entire region will suffer.
- *Individualism*. The market can do all this. Ships of countries or firms that do not contribute can be refused right of passage through territorial waters.

Also, technological developments—radar and eventually satellite-based positioning systems—will render the proposed lighthouse obsolete (probably before it has even been built). In the meantime, marine pilots (for whom there is a ready market) will fill the gap.

- *Egalitarianism*. We need to look at things holistically. The lighthouse will spoil an area of outstanding natural beauty and will consume lots of energy. And oil spills, and other problems related to increased shipping and reduced on-board vigilance, may destroy the fragile coastal ecosystem. A lifeboat, maintained and crewed by the local fishermen and drawing on their "hands-on" knowledge, is what is needed, not the unsustainable "top-down" lighthouse or the high-tech, growth-crazed innovations.
- *Fatalism*. Why should we contribute to the financing of this lighthouse? None of its economic benefits will come our way. These, as always, are sewn-up: by the construction companies, the bent civil servants, the shipping magnates and the bankers.

Our argument in this chapter will be that time and again, and particularly in Africa, policy and governance have allowed (indeed, often encouraged) individualism and hierarchy to dominate, thereby excluding the other two solidarities: egalitarianism and fatalism. The result has been an undemocratic and unconstructive state of affairs that is remarkably similar to how things were in Britain, more than two centuries ago, at the time of "the enclosures".[1]

> The law locks up the hapless felon
> Who steals the goose from off the common,
> But lets the greater felon loose
> Who steals the common from the goose.
> [Anon. First recorded in Tickler, London, 1 February 1821, but probably in oral circulation for a decade or so before that.]

"The law", of course, is hierarchy, and we see it acting here in cahoots with "the greater felon" (individualism), thereby propelling the unfortunate commoners (egalitarianism) towards the marginalised condition of "the hapless felon" (fatalism). The result is *crap governance*,[2] and if that is to be avoided the hierarchy will need to be firmly upholding the rights of the commoners, as it does, to give a couple of examples, with the sheep pastures in the English Lake District (the largest common in Europe which, moreover, has been in existence for 1000 years now) and with the

---

[1]Enclosure (sometimes spelt inclosure) was the process by which areas of customarily open and commonly-owned land were fenced in and "privatized", usually by large land-owners and often in the face of stiff resistance by the commoners. Though it went on for centuries, it reached its peak in the later years of the eighteenth century and the early years of the nineteenth. The "battle of Otmoor"—a 4000 acre common in North Oxfordshire—was perhaps the most celebrated instance, with commoners being arrested by the yeomanry (part-time military) and transported in carts to Oxford Jail. Even today, some 4% of England and Wales is common land (Stamp and Hoskins 1963).

[2]In contrast, of course, to good governance, where we have an institutional set-up in which none of the solidarities is excluded and where each is then responsive to (rather than dismissive of) the others (see Ney 2009; Thompson 2008).

19,000 community forests in Nepal (where the Forest Act of 1993 enshrines the user rights of each as the inalienable property of its villagers).[3] Nor, *contra* the influential "tragedy of the commons" argument (Hardin 1968), are common-pool goods any less viable than any of the other kinds of goods. As Susan George (1998: xii) tartly observed, "Hardin may or may not be a competent biologist but he knows little about history and actual human behaviour". And she goes on to point out that "dozens of historical and contemporary examples", to which we can add our English sheep pastures and Nepali community forests, "demonstrate that common property is not over-exploited *so long as group members retain the power to define the group and to manage their own resources*" (emphasis in original).

Two of these four solidarities—hierarchy (which institutionalises inequality and sets all sorts of bounds on competition) and individualism (which institutionalises equality of opportunity, and promotes competition)—constitute the classic dualism that has long been at the heart of social science, where they are usually referred to as "hierarchies and markets" (e.g. Williamson 1975; Lindblom 1977; see Chap. 9 of Thompson 2008 for the 50 permutations that are logically possible if the plurality is fourfold). The theory of plural rationality's novelty therefore lies, first, in its recognition of the other two solidarities—egalitarianism (which institutes equality, of result, and restricts competition) and fatalism (which institutes inequality and promotes unfettered competition)—and, second, in its making explicit the different *social constructions of reality* that make each of these ways of organising viable within an environment that contains the others. The upholders of these solidarities, though they will be disposed to behave in very different ways, and to resort to very different justifications for those divergent behaviours, are all perfectly rational, given their social constructions: their distinctive sets of convictions as to how the world is and people are.[4] And, of course, the way in which each strives to shape up goods into one of the four kinds that are possible is one instance of that. Moreover, for as long as social science is content to go along with the classic hierarchies-and-markets dualism, it will be reinforcing the exclusion of the other two solidarities rather than acting to remedy that defect: the "treason of the clerks", as it is sometimes called (Scully 1976: 5). And the uncritical acceptance of Hardin's "tragedy of the commons", across so much of social science (particularly in economics and game theory), is a prime example of just that. In other words, there is some balance that needs to be redressed, and that is what we will try to do in this chapter.

---

[3]See Rebanks (2015) for the former and Gilmore and Fisher (1991) for the latter.

[4]Briefly, these are: *nature* can be counted on to bounce back from any insult (individualism), is stable within limits that can be identified by experts (hierarchy), everywhere fragile (egalitarianism) and operating without rhyme or reason (fatalism). And *man* is everywhere self-seeking (individualism), flawed but redeemable by firm, long-lasting and nurturing institutions (hierarchy), inherently caring-and-sharing (egalitarianism) and fickle and untrustworthy (fatalism). In consequence, hierarchical actors are intent on *managing* the globe, individualist actors on *commoditizing* it, egalitarian actors on *tabooing* it and fatalist actors on *coping* with it. And, since each of these intentions is incompatible with the others, the contested terrain is here to stay! For a more detailed treatment see Thompson et al. (1990).

## 4.2 Setting the African Scene

Mineral resource production is the major export revenue earner for many countries in Africa. Indeed, it is in the extractive sector that Africa holds the comparative advantage over the other continents.

> For many African countries, mineral exploration and production constitute significant parts of their economies and remain keys to future economic growth. The African continent is richly endowed with mineral reserves and ranks first or second in quantity of world reserves of bauxite, cobalt, industrial diamond, phosphate rock, platinum-group metals (PGM), vermiculite and zirconium, among others (ACMIH 2015: 9).

A further breakdown shows that, as at 2009, African countries produced 46% of the global total of those minerals; out of that African total, Botswana produced 35%, the Democratic Republic of Congo 34%, South Africa 17% and Angola 8%. The continent produced 21% of the global total of gold, of which South Africa produced 56%, Ghana 13%, Tanzania 10% and Mali 8%. Africa's uranium production amounted to 16% of the global total, 46% of that coming from Namibia, 44% from Niger and 10% from South Africa. With bauxite, Africa accounts for 9% of the global total with most of that coming from Guinea and the rest from Ghana, while the continent produces 62% of the world's platinum and palladium (almost all of which comes from South Africa and Zimbabwe). Earnings from mineral fuels—petroleum and coal—made up more than 90% of the export revenues of Algeria, Equatorial Guinea, Nigeria and Libya in 2013 (ACMIH 2015: 9–10).

While the mining and drilling industries have become the major revenue earner across the African continent, they have also become its Achilles heel, particularly in relation to environmental disasters and degradation, conflicts over land rights and between local communities and the state. And these woes are often exacerbated by corporate licentiousness, by corruption among public office-holders and by, at times and in places, a descent into the "failed state" category: unwelcome accompaniments that are often spoken of collectively as "the resource curse" (Obi 2008: 11–12; Ghazvinian 2005).

But does that curse arise from the reliance on these resources, or is it a consequence of flawed governance? The latter, we will argue; after all, Norway is the world's second largest exporter of oil, yet it is difficult to detect any signs that it is cursed. And the same, closer to home and as we will see, seems to hold for Botswana. So, taking this optimistic answer, we will direct our attention to the dynamic interactions of the four solidarities, since it is systemic variations in those interactions that, in any specific instance, determine the quality of governance. In our first example, we look at the climate change debate, much of which is currently conducted at the global (not the continental, national or village) level and focus specifically on one of the mitigation frameworks: REDD+ (Reducing Emissions from Deforestation and Degradation, with the plus sign indicating that other considerations—biodiversity, for instance, and the protection of indigenous communities—are also included). Africa, in containing some of the largest (and often tropical) forests in the world, has become a major target for these mitigation efforts, with the Virunga forest belt in the Democratic Republic of Congo featuring prominently in current REDD+ discussions, along with forest belts in Cameroon and in the south-east of Nigeria. Although undoubtedly well-intentioned, REDD+ is intensely hierarchical. It commoditizes nature, for instance, by setting up highly bureaucratized arrangements that look like markets in sequestered carbon, and it does not then factor local communities into these carbon-trading mechanisms. In other words, despite all its rhetoric about "stakeholders", "safeguards" and "capacity-building", it excludes the egalitarian voice while itself mouthing a "pidgin" version of the individualist's voice (Monbiot 2015).

In our second example, we explore the unwelcome legacy of coal-mining in the Upper Oliphants River catchment in South Africa's Mpumalanga Province, where acid mine drainage (AMD) from abandoned mines is having a markedly negative impact on the local communities. This environmental disaster, which is set to run continuously, unless there is some technological breakthrough that can eliminate the acid, stems from the long-entrenched hierarchy/individualist land-use practice which, while responsive to the interests of the state actors and the mining companies, has excluded the local communities and thereby loaded them with a host of environmental "negative externalities".

In our third, and final, example we show how this same source of crap governance—a hegemonic alliance of hierarchy and individualism that silences the other voices—underlies the perennial conflict in the resource-rich Niger Delta region of Nigeria. Again, we have a conflict—in this case highly destructive of both lives and livelihoods—that hinges on a state land-use policy that promotes resource exploitation and at the same time disenfranchises the local communities on whose ancestral lands and waterways oil is being exploited.

But first, let us take a look at the nature of states in Africa, and at the historical processes by which they have come to be the way they are.

## 4.3 The African State and Extractive Industry: How Did it All Begin?

Most of Africa's nation states are structured in such a way that power, especially power over natural resources, resides with their central governments. In other words, the land itself, the mineral resources that lie beneath it, and the revenues from those resources once they have been extracted, all belong to the central government, which then disburses those revenues to the provinces on the basis of some formula of its choosing. In Nigeria, for instance, the revenue sharing formula between the three tiers of government gives the federal government the lion's share of the revenue from mainly oil exports at the expense of the states and local government areas. The formula for allocation is: Federal Government 52.68%, 36 State Governments 26.72%, 774 Local Government Areas 20.60% (Lukpata 2013: 36). This framework, in most instances, leads to the exploitation and deprivation of the local communities within which those resources are located. At the same time, it creates a space for intense rivalry between politicians, who jostle for the positions that will give them control over the central government's huge revenues. In consequence, whoever controls the state controls the resources: a dysfunctional and inequality-heightening state of affairs that, as we will see, has its roots in the continent's colonial past.

During the colonial era, indigenous communities lost their sovereignty to the modern states that were evolving as a direct consequence of colonial rule. Local communities did not willingly give up their sovereignty; they had it sucked away from them as the states, under the protective might of Queen Victoria, Kaiser Wilhelm et al., became ever more powerful: powerful enough to impose those straight-sided and unnatural-looking boundaries that, to this day, are such a striking feature of the continent's map. Inevitably, this growth of the state resulted in a decline in the influence that the formerly largely autonomous communities wielded over their common property resources, especially their land. In other words, it was not the "tragedy of the commons"; it was, as Susan George has pointed out, the progressive erosion of those communities' ability "to define the group and to manage their own resources" (Wolverkamp 1999: 5). Nor is Susan George alone in making the diagnosis.

Indeed, Albert Jay Nock, writing during the maturing years of colonial rule, put it in almost law-like form: "Every assumption of State power, whether by gift or seizure, leaves society with so much less power; there is never, nor can there ever be, any strengthening of State power without a corresponding… equivalent depletion of social power" (Nock 1935: 1). And Bertrand de Jouvenel was of much the same mind. History, he declared, "is the picture of a concentration of forces growing to… the state, which disposes, as it goes, of ever ampler resources, claims over the community ever wider rights, and tolerates less and less any authority existing outside itself" (de Jouvenel 1952: 70). Of course, no trend goes on forever—Marx famously predicted the eventual "withering away of the state" (Ambedkar 2009: 35), and in recent years there has been much talk of globalization "hollowing out" the state (Ladi 2005: 18)—but the Nock-de-Jouvenel law seems to still be holding for Africa.

This colonising trend, unsurprisingly, has given rise to a negative attitude to the state among ordinary people, who see it as oppressive and exploitative (Norway referring back to our discussion of the resource curse, is very different; indeed it can be argued that the Norwegian state, perhaps uniquely within Europe, has never oppressed its citizens). Whereas the African state demands citizenship responsibilities from its local peoples, it is seen as giving back little or nothing to the communities, especially when it comes to infrastructural developments (Zolberg 1968). Hence the notion of the "distant state" (Zolberg 1968: 72–76). Moreover, the state's "distance" from its citizens has not lessened since the end of the colonial era: "they begin to see the nation-state as a curse", write Davidson and Munslow (1990: 10), "whereas before it was seen as an act of liberation, which initially of course it was". So the still-continuing contestation between African states and local communities—over land and mineral resources, in particular—is testimony that this power structure of African states has not shifted significantly since the end of colonialism.

David Lea sees this continuity across the colonial/post-colonial divide as having its roots in something more than just the struggle for mineral rights and access to land, important though that struggle is. There is, he argues, a fundamental clash between the "acquired rights" of the state/corporate bodies and the "aboriginal rights" of the local people; between a new sovereignty and an old one (Lea 1992). This is the oft-mouthed distinction among development theorists between "modern" and "traditional", with the grassroots who rise up against the state's superimposition of a new tenure system on the indigenous one then being stigmatized as standing in the way of development; a fine example of the hierarchical tendency to blame the victim (Umejesi 2016: 74). This institutional framework—acquired rights displacing aboriginal rights—is a general feature of the colonial era in Africa (but much less so in Asia—India and Malaya, for instance—perhaps because well-established systems of land-ownership were already in place). In both Nigeria and Uganda, for instance, the colonial state took possession of lands under the Crown Land regime, with Crown tenure then stipulating that the state now owned any acquired land for public good on behalf of the peoples of those colonies (Meek 1946). In this way, many a common-pool good was transformed into a public good[5]: to the detriment, of course, of many a commoner!

In this way, a foreign tenure system that had developed over centuries in Britain was imposed on the local people in Africa (and, to varying degrees, elsewhere in the imperial territories). In Britain, reasons Meek (1946: 87–88) "The king had complete freedom of disposal of crown lands, which were constantly being increased by confiscation, escheat or forfeiture". However, as Mabogunje (1979: 21) has observed, "This manner of acquisition and control of land by a centralised authority was new for indigenous communities in different countries in Africa where land and resources, from time immemorial, are more or less owned collectively". Only one voice, it seems, was raised in protest at this blanket imposition: an "establishment" voice, moreover. L.S. Amery—a former Dominion Secretary—argued that Western values "tend to judge distant problems in the light of its own experience and try to fit them into its own formulas, regardless of their relevance to local conditions" (Amery 1953: 181). Unfortunately, he was not heeded ("Poor chap's gone native") and the African colonies, and eventually their independent governments, became locked into the alien regime.

[5]Or more likely, given the cosy hierarchist/individualist alliance with its consequent exclusion of the egalitarian voice, a club good, with the hierarchical actors forgetting all about their duty to act as "trustees" and rooting after rent-seeking instead, and their individualist cronies doing well even when others (especially those commoners who find themselves squeezed out into fatalism) no longer benefit. Kleptocracy, in less mealy-mouthed words.

### 4.3.1 *Crown Lands but no Crown: The Post-colonial Paradox*

The massive development and growth of the extractive industry sector during, and particularly towards the end of, the colonial era witnessed the elaborate application of this alien regime in certain communities where mineral resources had been discovered. In Nigeria, for instance, the local communities whose lands were acquired for "the public good" seemed, initially, not to have understood the full implications; they thought that the state held their land in trust for their communities, and did not understood that they had been acquired permanently and were no longer common-pool goods. In this way, the seeds of resentment and distrust were sown (see Onoh 1997).

With the prospect of independence, the thinking among local communities was that their post-colonial governments would rectify the injustices of the colonial era, which typically included land thought to have been wrongfully taken, unemployment, socioeconomic inequalities and a lack of basic infrastructure in rural areas (Zolberg 1968). This was realistic enough, given that ordinary people had been promised fundamental changes in the way their states were governed. In Kenya and in what is now Zimbabwe, for instance, where grassroots disenchantment with discriminatory land policies that favoured European settlers had been entrenched for several decades, the expectation of radical reforms that would see people repossessing their land from the state and settlers was high (Betts 2006). And later, in South Africa, the dawn of democracy in 1994 brought hope of change, especially in land ownership. Unfortunately, these expectations of a better life after colonialism, or after apartheid in the case of South Africa, were dashed by the emerging states and their leaderships. Chinua Achebe—one of the leading intellectuals of post-independence Africa—has recounted how the hopes of the ordinary people went unfulfilled: "how passionate we felt and how good it was to be in a movement that could liberate us after centuries of denigration and deprivation" (Achebe 2009: 1).

This hope, in South Africa, was hinged on the belief that the intractable land ownership questions, stemming from the Native Land Act of 1913 (which appropriated 93% of arable land for the white minority population, leaving just 7% for black South Africans), would be significantly reviewed (Davenport and Saunders 2000). While black South Africans have incrementally gained more arable land since 1913, much is still owned by white farmers, by the government and by traditional rulers in the former homelands, who oversee lands that the government has returned to their communities. However, these traditional rulers, rather than acting as "trustees", have now become powerful and exploitative actors within democratic South Africa. Moreover, that power and exploitative propensity, thanks to the Mineral and Petroleum Resources Development Act of 2002 (MPRDA 2002), now reach beyond just land administration. Under the black economic empowerment (BEE) scheme, traditional rulers have been allowed to own shares in mining and drilling companies operating on their land. Communities such as the Bafokeng Nation, for instance, now have stakes in the North West Province's platinum belt, owning, *inter alia*, 13.4% of the shares in Implats: Imperial Platinum Limited. While hailed as novel and a model for Africa, this "community participation" scheme, studies indicate, has enabled traditional rulers to exploit this wealth (from both royalties and shares in the major mining companies) while leaving the local people poor and landless (Manson and Mbenga 2003; Mnwana 2014). In other word, "traditional" has become a somewhat Orwellian label, denoting, not the common-pool goods of old, but the club goods that are shaped up by an egalitarianism-excluding alliance of hierarchy and individualism.

While the citizenry felt bitterly disappointed by the direction taken by their newly-independent countries, the governments of those post-colonial states were much more concerned with strengthening their control over their countries' component parts than returning community land and mineral rights to the autonomous communities and kingdoms of the pre-colonial era (Davidson and Munslow 1990: 10). Indeed, to have taken this latter option would, it was reasoned, have lessened their chances of strengthening their authority over those component parts. Whether or not this reasoning was correct, that first option was chosen, with the result that post-colonial governments in Africa, almost without exception,[6] have failed to change the basic colonial institutions, of land and resource ownership, in particular. In this way, post-independence Africa has ended up by endorsing the colonial regime that it supplanted. In Kenya, for example, hopes of land redistribution were dashed: "What does independence mean?", demanded Wa Thiong'o (1972: 56). "Independence", he continued, "has not given them back their land". And much the same sense of disappointment has been noted in Nigeria, where the post-colonial leaders failed to reform the oil industry's ownership framework that they had inherited from the British. "It is therefore paradoxical", says Ebeku (2001: 3), "that the same persons who had so resented colonial statutes on mineral oils moved to retain the essence of those laws after independence."

Indeed, instead of reforming the colonial laws on ownership and control of oil resources in Nigeria, the post-colonial government actually strengthened its hold on land with more stringent and far-reaching laws, such as the Republican Constitution of 1963, the Petroleum Act of 1969 and the Land Use Decree of 1978. These new legislations progressively chipped away community access to land, along with the right to compensation when that land was acquired, while at the same time expanding the reach of the state. Nigeria, you could say, is now more colonial than it was when it was a colony!

[6]In South Africa, the post-Apartheid government through its Mineral and Petroleum Resources Development Act of 2002 (MPRDA) has allowed some form community participation through shareholding in the mining companies that operate in their communities. This is permitted where the land upon which mining takes place is a "tribal land" (Mnwana 2014).

### *4.3.2 Against the Commoditization of Land and Mineral Resources: The Excluded Voice of Egalitarianism*

The modern African state and the multinational mining and drilling corporations conceive of land in a purely economistic sense: as a commodity and a means of economic production. Both, for instance, routinely refer to indigenous communities as "mining fields" or "oil blocks". Local people, however, relate to land and its resources in more than an economic sense. Land, as Uchendu (1978: 64) has noted, is a medium for connection with "those invisible father-figures who bequeathed their land to a vast family which includes the dead, the living and the unborn" (see also Ogon 2006: 11). Hence in pre-colonial societies, such as the Igbo of south-eastern Nigeria, land was never sold; it was only mortgaged for a certain period of time and then returned to the family that owned it when the mortgager had redeemed it (Agukoronye 2001). The mortgaged land, as it was being redeemed, was personified in a proverb: "The land has returned from a journey to its family". A former British colonial official in Enugu, south-eastern Nigeria, noted that, among the local people, "ancestral land must never be bartered for money" Hair (1954: 56) and Ikejiofor (2009: 16) similarly holds that, among Igbo communities, "a market in land did not exist in the pre-colonial context".

Land, it can be argued, was a mystical resource: one that, while fulfilling economic, political and social needs, could not be reduced to them. But this conceptualisation was eroded by colonial rule, particularly through the process of urbanisation. "Urbanisation and the evolution of a market economy", argues Ikejiofor (2006: 139) "have led to the commodification of land". By "commodification of land" Ikejiofor means attaching financial value to it, either by developing a market in land or by paying "compensation" in the event of it being appropriated by the state. In 1915, the colonial authorities paid £250 to the local Igbo chiefs of Enugu-Ngwo for timber and farms on land where coalmines were being established (Hair 1954: 55; Isichei 1976: 203); prior to that there is no record of land being sold or bought.

Nor is this mystical perception of land, as something uncommoditized and therefore uncompensatable, limited to Nigeria, or to West Africa; it was widespread across the continent:

> Here [land is] historically debarred from sale, here redeemable if sold, here pledgeable but not rentable, mortgageable or saleable; land is nearly everywhere the subject of special protective strictures in local if not also imported custom. Despite what economic development planners may think and hope, land is seldom if ever just a commodity. In fact, some tenure reforms aimed at making land marketable have made it explosive. (Shipton 1994: 350).

And Ledum Mitee—president of MOSOP (Movement for Survival of Ogoni People) and successor to the executed Ken Saro-Wiwa—makes clear just why those tenure reforms can be so "explosive".

> To the Ogoni, land on which we live and the rivers that surround them are not just natural resources for exploitation, but have deep spiritual significance… This respect and reverence for land also means that forests are not merely a collection of trees in the forests and the abode of animals but also, and more intrinsically, a sacred possession. Trees in the forests cannot therefore be cut indiscriminately without regard for their sacrosanct status and their influence on the well-being of the entire community. Similarly, rivers and streams apart from their being the source for water for life are also intricately bound up with the life of the community and are not to be desecrated through pollution. (Mitee 2002: 2).

The point we are making here is that, although pre-colonial economies across Africa depended on land, the value of that land was not just economic; it was much more than just a factor of production. Hence the explosiveness—often leading, as in Nigeria, to civil strife and bloodshed—when tenure reforms strip away that land's non-economic value and force it into the marketplace. Not that people like the Igbo and Ogoni are anti-marketplace. Indeed, they are renowned marketers: in agricultural produce, fish, cloth, handicrafts and so on, but not land. Their transactions, you could say, are parcelled out among the different solidarities.

The egalitarian solidarity in placing all sorts of restrictions on competition tends to avert the misuse of the environment—you do not lightly chop down a tree that has someone's ancestor buried beneath it—and it promotes a keen sense of caution when (as has to happen, of course) its resources are being exploited. So, with the land belonging to the community, with it being non-fungible and with its users being accountable to the members of that community, both living and dead, we have a man-land relationship that is very different from that which is inherent in those tenure reforms. "Things fall apart", as they say, and they did, and are still doing, with the colonial and post-colonial redefinition of ownership—a redefinition that has shifted the control of land and resources from the communities to the state and the mining corporations. It is this institutional *disconfiguration* in land and resource ownership—a disconfiguration that remains indiscernible if we cleave to the social science orthodoxy and recognize only hierarchies and markets—that has resulted in the deprivation of ordinary people as they are forced out of egalitarianism and into fatalism: into an environmentally unsustainable and often disaster-prone mining regime, into massive official corruption and into the paradox of states making themselves ever more precarious through their self-strengthening efforts.

Now, having set the African scene, in terms of our theory's four forms of solidarity and their associated kinds of goods, together with the varying qualities of governance that their contentions can give rise to, let us turn to our three case studies.

#### 4.3.2.1 African States, Hierarchy and Climate Change Mitigation: Enter REDD+

Africa's total forest cover is estimated at 650 million hectares: 21% of its land mass, with this accounting for 16.8% of global forest cover (FAO 2015: 2). In 2015, the World Bank and some private sector partners—hierarchy, and individualism, but no egalitarianism—joined forces in the so-called "AFRIOO Initiative", which is aimed at restoring 100 million hectares of forest across the continent by 2030. It is estimated that this project will cost 1.6 billion US dollars. This AFRIOO Initiative, is also expected to build on the "Bonn Challenge", which aims to renew 150 million hectares by 2020, and on the "New York Declaration on Forests", which pushes the mark up to 350 million hectares by 2030 (WRI 2015: 1). Inherent in these initiatives, challenges and declarations is an assumption: that these global-level actors have some policy levers that they can pull on that, in then meshing with matching sets of levers in the ministries of all those forest-rich jurisdictions at the national level, will bring the entire continent into these ambitious targets, on time.

An increasing number of political scientists and anthropologists have their doubts about this sort of approach (e.g. Verweij 2006, 2011; Prins and Rayner 2007; Patt 2015), and we would add to those by pointing out that the sort of national-level compliance that is required pre-supposes a high quality of governance, which, as we have just seen, is precisely what is not there in Africa. Beyond that, as we will see, these global-level initiatives/challenges/declarations, as well as requiring the commoditization of the forests, are about as far-removed from the community level (which, as we have seen, is so effective in keeping the forests and other common-pool goods in place) as it is possible to be. Explosive, indeed! But hierarchical actors see that explosiveness as stemming from a lack of controls; those unruly logging companies, and those axe-wielding villagers, must be brought into line. Achieve that, they reason, and all those disruptive *improvised explosive devices* (IEDs) will be de-fused.

It is worth pausing here, we feel, so as to quickly set out a cautionary tale—it is from Nepal, another poor country that is currently on the receiving end of REDD+, which strongly suggests that the explosiveness simply cannot be got rid of in this way. Let us examine how an elegantly planned interventionist scheme fared in forest management in Nepal.

*The Bara Forest Management Plan: A Tale of Dharma Gone Wrong*

The Finnish-funded Bara Forest Management Plan set out to privatize a large tract of forest which, unbeknown to the aid-providers, was already a complex mosaic of property rights: private, public and common-pool (often rather informally established rights, and at small social scale levels, but rights nonetheless). The idea was that a Finnish private company, in partnership with some Nepali business houses, would be given responsibility for the regeneration of the entire forest, along with near-monopoly rights to its commercial exploitation. The aim, in line with the Washington consensus, was to introduce radical changes in a sector that had not hitherto been oriented to market-led approaches. From that point on it was all downhill, with the ambitious project eventually turning into a disaster so unmitigated and so universally acknowledged that the Finns swore never again to get involved in forestry in Nepal.

To their great credit, however, they commissioned a post-mortem (carried out by three Nepalis and a Finn) which arrived at the following conclusion:

> As a villager in Nepal is apt to say, it is the *dharma* of the bureaucracy to regulate, of the markets to innovate and of activist groups to advise caution. The case of Bara was one in which *dharma* had gone wrong—a situation characterized by the bureaucracy assuming the role of the market and vice versa. It further showed that the hierarchic order that has broken down is no instrument for the implementation of radical reforms, without first mending that order and restoring its legitimacy, or as the villager might say, restoring the *dharma*. (Sharma et al. 2004: 241–2)

This conclusion, we cannot help feeling, says it all. But it does need a little bit of unpacking.

*Dharma*, especially in the West, is often equated with fate, but it is much more than that. *Dharma* is "the law" or, more properly, "the righteousness that underlies the law". Hence the emphasis on legitimacy, and the accompanying consequence—breakdown (or "explosiveness")—when that legitimacy is eroded. Legitimacy, moreover, cannot spring from the interaction of just hierarchy and individualism—the only sort of interaction that is entertained in the Washington consensus (and, as we have seen, in most of social science). Legitimacy requires a third form of solidarity—egalitarianism—and it is manifested here (as it is, for instance, in the Movement for the Survival of the Ogoni People) as the "activist groups" whose *dharma* is to advise caution. Fatalism, however, *is* involved, in the sense that the more the *dharma* goes wrong the more likely it is that people (as in the Niger Delta, for instance, and in South Africa's platinum belt) will find themselves labouring under what has been called "the double burden": increasingly impoverished and increasingly subject to social exclusion (Tamang and Tulandhar 2014; Gupta and Thompson 2010).

So you cannot defuse a bomb by sitting on it—by imposing a nature-commoditizing hegemony of hierarchy and individualism and then silencing the dissenting voices. You can only do it by ensuring the *requisite* variety: by creating an institutional set-up in which each of the four voices is (a) able to make itself heard and (b) is the responsive to (rather than dismissive of) the others. And that is precisely what has now been achieved in Nepal, where there are now more than 19,000 thriving community forests, each managed by its villagers, with the state's new Forest Act enshrining their inalienable rights to its use. The previous Forest Act, by contrast, had empowered the Forestry Department officers, under certain circumstances, to shoot those villagers. The lesson from Nepal, of course, is that REDD+ as currently conceived and implemented, does not encompass the requisite variety; it is very much a case of *dharma* gone wrong.[7]

---

[7]At the regional level, however, it seems that the *dharma* is now being restored. The 37th Session of the SAARC (South Asian Association for Regional Cooperation) Council of Ministers, in March 2016, noted that, after years of decline, Nepal's forests are now moving strongly in the other direction, and went on to recommend community forestry for all of South Asia (around a quarter of mankind).

*The Hegemonic Discourse of REDD+*

The IPCC (Intergovernmental Panel on Climate Change) is the body of experts that, at the behest of the global bureaucracy—the United Nations with its Framework Convention on Climate Change (UNFCCC)—has now elaborated this discourse of control in terms of forests as "carbon sinks", and then gone on to factor Africa into the resulting governance framework. How, we will be asking, does this framework, and its legitimising discourse, handle those myriad local communities with their direct, but uncommoditised, claims to forest resources?

In their 2007 report, the IPCC's scientists estimated that forestry accounted for about 17% of global emissions of anthropogenic greenhouse gas: behind energy supply (at 25%) and ahead of that other great consumer of fossil fuels transport (at 14%) (Nabuurs et al. 2007). By 2010 the energy sector's contribution had risen to 35%, followed by 24% from AFOLU (Agriculture, Forestry and Other Land Uses) and then 21% from industry, 14% from transport, and 6.4% from the building sector, with the cumulative carbon dioxide emissions from forestry (and other land uses) having increased by about 40% since 1970 (IPCC 2014). Though re-vegetation acts as a "carbon sink", thereby offsetting the emissions from deforestation and degradation, when there is a negative balance (as has been happening), this leads to a diminution of global forest carbon stocks. And it is this type of degradation that has been responsible for 15% of total emissions from tropical forests (Baccini et al. 2012).

With the problem framed in this way—global-level and de-contextualized quantities of all the various stocks and flows—the solution—carefully planned and top-down managerial intervention—seems the obvious way to go. And that is precisely the way REDD+ is set on going; it is a scheme, under the ultimate control of a number of multi-lateral regime builders—the United Nations Framework Convention on Climate Change (UNFCCC), UN REDD itself, the Forest Carbon Partnership Facility (FCPF), the Global Environment Facility (GEF), the Forest Investment Program (FIP) and so on—that will pay countries (or, rather, the governments of those countries) not to chop down their forests, not to neglect their high conservation value forests, not to lessen biodiversity, and not to ride roughshod over the communities that live in and are reliant on those forests. Commoditizing things in this way, sceptics might object (indeed, have objected), will open the door to "moral hazard", encouraging a sort of thinly-veiled blackmail: "Give us the money or we'll remove this forest". And, those sceptics continue, relying on governments (especially those of countries that are placed near the top of Transparency International's corruption index, and that is where so many African countries are) is akin to putting King Herod in charge of childcare: a justified jibe, as we will see when we look at the havoc the Nigerian state has wreaked in the Niger Delta. But the architects of REDD+ insist that they have thought about these objections and have put the necessary mechanisms in place: careful monitoring and validation regimes will prevent moral hazard, and a strong emphasis on "safeguards" and "capacity-building", will ensure that governments always act in the best interests of their citizens and their environments. However, even if they are right, their commitment to a solution that relies on monetary transfers has denied any voice to those actors who are steadfastly opposed to the commoditization of their land resources. That common-pool goods might be part (perhaps almost all) of the solution has been ruled out from the word "go".

The perceived need to mobilize finance in order to preserve the world's remaining forests has been a central concern in the deliberations over global climate change talks, and this imperative's implementation, by way of REDD+, has been touted as a cost-effective way to keep emissions at bay and thereby stabilizing global temperatures: measures that are essential if we are to avoid "dangerous climate change". REDD+ is a policy mechanism, guided by the "polluter pays principle", that will create financial value for the carbon stored in forests, along with the other aforementioned *desiderata* that are covered by the plus sign. The trouble with all this, of course, is that that "perceived need" is not uncontested. There are, as we have seen, many actors—those who wish the land and its resources to be treated as common-pool goods—who see the placing of money values on nature as the problem, not the solution.

In other words, REDD+ is unremittingly hierarchical, as is evident from its accompanying slew of acronyms, together with its cumbersome multinational negotiation process, under the UNFCCC's Conference of Parties (COP), which began with the signing of the Kyoto Protocol in December 1997. It was conceived, from the start, as a strategy for dealing with climate change by working towards an agreement on how much to reduce the emissions of the participating countries and on financially compensating developing countries for reducing their forest sector emissions. Ten years later, REDD+ was endorsed, as a mitigation strategy, in the Bali Action Plan at the 2007 COP-13 in Bali, Indonesia (IPCC 2014).

At the national level, REDD+ is envisaged as occurring in three phases. Phase 1 is concerned with developing strategy and building capacity, Phase 2 with the implementation of the strategy and its accompanying policies and measures, and Phase 3 is focused on results-based payment activities and their measurement, together with the reporting and verification of emission reductions. Much is therefore expected of the national-level hierarchical actors who, in their turn, are subject to the controls exercised by their global-level counterparts. Moreover, a host of mechanisms and standards has had to be evolved for dealing with such thorny issues as the *leakage of activities* (to areas outside the project boundaries), *non-permanence* (as a result of natural disasters and other anthropogenic causes), *double-counting* (made possible by the intangible nature of carbon credits), *verification or independent auditing* (so as to ensure the credits are real and additional), and the institution of the *principle of conservative estimates* (so as to make sure that the methods used to measure the amount of carbon stored, the amount that has avoided being stored, and the overall rate of sequestration are accurate) (Estrada 2011: 49–59).

In the case of what are called "nested approaches", these mechanisms and standards then have to be applied to jurisdictional (or regional) level crediting, as well as at the project level, both being needed if there is to be an overall carbon accounting framework. To comply with these requirements, at least at the project level, costs may range from US$ 100,000 to US$ 300,000 (Olander and Ebeling 2011), that compliance being vital if a REDD+ project (and, in particular, a carbon quantification project design document: a PDD) is to be developed so as to lead to verified carbon credits. These carbon credits can then be sold, initially to mainly voluntary buyers but later, it is hoped, within an established "compliance market". The trouble is that, for small holders, there are high barriers to market entry, economies of scale cutting in only with the larger projects. And, even with these larger projects, there is a bottleneck, the implementation of the carbon quantification methodologies being often too expensive, in the absence of a regulatory driver or a strong enough demand, to justify the investment: a glaring instance, we would point out, of "*dharma* gone wrong". With the hierarchy taking on the role of the market actor (by striving to create compliance markets, for instance) market failure, it would appear, is not just possible; it is inevitable. Indeed, with the price for those carbon credits being insufficient of an incentive for the developers to continue their REDD+ projects sustainably, the market has failed before it has even come into existence![8]

On top of that, this regulated market-driven model is hyper-elegant, as is confirmed by the exclusion of those policy actors—the egalitarian advisors of caution—who are so implacably opposed to the commoditization of nature that is at the very heart of REDD+. The World Rainforest Movement (WRM), for instance, on the occasion of the 2014 United Nations climate change negotiations in Lima, Peru (known as COP20), has issued a passionate and uncompromising "Call to Action" that rejects both REDD+ and its reliance on payments for "environmental services". The latter, it holds:

> … involves the further financialization and commodification of nature, and signifies subjugating and enslaving it to capital. The carbon market, biodiversity offsets and water markets are part of this kind of capitalism (World Rainforest Movement 2014).

The fact that the WRM call's list of initial signatories (more are joining all the time) runs to 166 organizations from around the world, many of them representing indigenous communities that are directly dependent on forests for their livelihood, does tend to suggest that the proponents of REDD+ (despite those "safeguards") are not being attentive to all their legitimate stakeholders.

It is this—the fundamental wrongness of putting money values on the natural world—that is the central plank in the argument against REDD+ by those who speak with the excluded egalitarian voice. George Monbiot, in a speech at Sheffield University (Monbiot 2015), put this so eloquently and forcefully that it may well prove to be the final nail in REDD+'s coffin. Speaking about what the nature-commoditizers call "biodiversity offsets", he said "if you trash a piece of land here you can replace its value by creating some habitat elsewhere" (precisely that having been proposed, just a few miles from where he was speaking, by a developer who wanted to replace an ancient woodland with a motorway service area). "This", he continued, "is another outcome of the idea that nature is fungible and tradable, that it can be turned into something else: swapped either for money or for another place, which is said to have similar value." This speech went viral in a

[8]And even when they have come into existence, they are a failure. A 2009 investigation into France's Blue Next carbon exchange has revealed that as much as 90% of the trading was fraudulent, costing European tax payers 5 billion Euros over just 18 months (as reported by Jim Armitage, 15 August 2015, *The Independent*, London).

way no REDD+ publication is ever likely to, and it has evidently had some impact. One indigenous member of the REDD+ staff in the Democratic Republic of Congo, with whom we happened to be in contact, actually resigned after reading it.

George Monbiot, quite clearly, is speaking from the egalitarian apex: for the tabooing of the globe and against what he sees as the idiocies being propagated by those who are set on commoditizing it, who are usually the individualist actors: entrepreneurial types intent, as they say, on "turning an honest penny". But, with REDD+, those commoditizers, as we have seen, are the hierarchical actors. Yes, they are still intent on managing the globe but, in their embracing of the Washington consensus, they have taken on the role that, properly speaking, is that of the individualist actors. In other words, referring back to the Bara Forest fiasco, George Monbiot is following the egalitarian *dharma* (and to the letter) but when it comes to the proponents of REDD+, their *dharma* has gone seriously wrong. Only if the unsullied individualist and egalitarian voices can force their way into the current and hyper-elegant policy debate will the requisite variety be achieved and the hierarchical *dharma* restored. It is as simple, and as daunting, as that!

#### 4.3.2.2 Coal Mining's Legacy in South Africa: Acid Mine Drainage in the Upper Oliphants River Catchment

Water governance legislation in South Africa dates back to as early as 1903, when responsibility for mining impacts was placed on the owner of the mine until such time as a certificate releasing him/her was obtained (Hobbs et al. 2008). Increasingly, however, the environmental impacts came from mines that were abandoned, defunct and, in a sense, "ownerless". Eventually, in 1975, there were negotiations between the Minister of Water Affairs and the Chamber of Mines on the sharing of responsibility for abandoned mines, and these culminated in the Fanie Botha Accord (Hobbs et al. 2008). The state agreed to take sole responsibility for all mines closed before 1976. Mines closed from 1976 to 1986 would be 50% state responsibility and 50% owner responsibility, while all mines worked after 1986 would be 100% the responsibility of the owners (Hobbs et al. 2008: 422–423; Munnik et al. 2010: 8–10).

In 1991, new legislation was developed and implemented to regulate environmental management and mine closure processes, with all active mines being required to provide funds for environmental and social restoration in the event of their closure (Mineral Act 1991). Further legislation followed: notably in 1995, with the Policy and Strategy for Management of Water Quality Regarding the Mining Industry in South Africa being aimed at assisting mines in meeting their environmental responsibilities. Then, in 1998, came South Africa's National Water Act (Act 36), which was developed to protect water resources and was based on the "Polluter Pays Principle" (National Water Act (NWA) No. 36, 1998). In addition, the National Environmental Management Act (NEMA) (Act 107 of 1998) stipulates that pollution or degradation of the environment must be prevented or rectified such that if the person or company responsible for pollution fails to take the required action then further actions should be taken to recover the costs from the polluter (Hobbs et al. 2008).

It is clear from the foregoing that, prior to 1991, mining companies enjoyed leverage on mine closures in South Africa. What is not so clear, however, is why that leverage has been allowed to persist. Mines are still devastating watercourses and threatening communities, and those impacts are not being adequately addressed.

Mines, we should explain, inevitably alter the hydrology; they tap water that was previously untapped, depleting some watercourses and augmenting others. And they can dramatically alter the contents of that water, especially in the case of coal and iron mines. In the upper Oliphants River Catchment, Mpumalanga Province, the water that flows from the now-closed mines is strongly acidic (sulphuric acid, mostly) which, of course, was not the case prior to those mines being sunk. Since this water is relied upon by downstream communities, usually for both agriculture and drinking and cooking, there is inevitably an impact, and often a serious impact. Moreover, in the absence of any technological breakthrough, this impact is set to continue, unabated, into the foreseeable future. That is the legacy problem: common-pool goods, you could say, have been transformed into common-pool bads.

*Different Communities and Different Impacts*

Across Africa, with Botswana often being held up as the sole exception, mining brings serious social hazards for communities, health and the environment (Guichaoua 2009; Farrel et al. 2012; Marais 2013; Umejesi and Akpan 2013; Akiwumi 2014; Bolay 2014; Umejesi 2014). However, the direct consequences of mining activities vary from location to location. Much depends on the allocation of legal rights to land, on the nature of the compensation to the communities where exploitation sites are located, and on the distribution of benefits drawn from the industrial presence (which, according to Guichaoua (2009), often spark tensions, especially when some actors are excluded from those benefits). It should be noted that the country's change from apartheid to democracy, in 1994, has unsurprisingly had an impact on both the mining industry and the communities (Zimmerman 2000; Marais 2013). These impacts have revealed tensions between traditional and democratic structures, with disputes between traditional leaders sometimes resulting in community divisions (Farrell et al. 2012).

Caught up in all this, there is also a troublesome complexity of land rights, especially in the former 'homelands', where communal lands are held in trust by a traditional authority or the state, often with competing claims from multiple traditional leadership structures. For instance, Ramutsindela (2002) has noted that conflict arose when some Makuleke communities[9] in Mpumalanga Province were asked to relocate to other areas by a mining company without the company involving any social scientists (choosing, instead, to rely on its own engineering experts). The communities claimed that the company did not adequately understand the impacts on their livelihoods and accused their representatives of corruption and of sourcing labour from elsewhere. As a result, some households refused to relocate, protests

[9]Tsonga speakers living near the Kruger National Park (KNP), Mpumalanga Province, North East of South Africa.

erupted and the police shot and injured several of the protesters. Prior to this fracas, the government had remained silent while the company, for its part, had ignored community voices. There have also been cases where communities had been evicted from their ancestral lands with no compensation whatsoever (Ramutsindela 2002).

With acid mine drainage and its impacts on local communities being inevitably tied-into the whole issue of land rights, progress is often disappointing, even in situations where the mine-owners are acting with exemplary corporate social responsibility and adhering to all their voluntary principles. This is because the past racially-based land expropriations carried out by the apartheid government arestill very much in evidence in the democratic present. Land reform (as embedded, for instance, in the Land Restitution Act 22 of 1994) is seen as the solution, but it is not easily implemented. It is difficult to give land to some people without taking it away from others. And, even when this has been done in strict accordance with Act 22's principle that all people who were dispossessed of their land through racially discriminatory legislation are entitled to its restitution, this re-distribution has done little to alleviate rural poverty, and may indeed have aggravated rural inequalities (Zimmerman 2000). Unsurprisingly, therefore, this sort of trouble has now become commonplace in the Olifants River Catchment areas (Ramutsindela 2002; Munnik et al. 2010).

The mining industry is deeply entrenched in the South African economy and protected by historical precedent and legislation (Mccarthy and Pretorius 2009; Munnik et al. 2010). This has often been at the expense of the poor rural communities who reside in these mining regions, with those affected by mine-related pollution struggling to get companies and the government to mitigate these negative socio-ecological effects. Mathews Hlabane, a veteran community activist, has exposed the sort of tactics that companies use, whereby a mine's environmental manager will show you the good places and hide the bad. In some cases, they also buy off community leaders who then support the mines' practices (see Munnik et al. 2010: 16). Coal mining companies in the Olifants catchment, in pursuing these sorts of tactics, have created immediate poverty for the communities around the mines, displaced farming and associated livelihoods, caused air pollution with its accompanying health impacts, and brought about the long term destruction of water resources in the area.

It is, we can conclude, easy to see here another instance of the individualist/hierarchy alliance—the mining companies and the state (and sometimes the traditional rulers)—riding rough-shod over those who speak with the egalitarian voice: the local communities who have had their common-pool goods transformed into common-pool bads. The question then is: what, if anything, can be done about the legacy problem? The first essential, we suggest, is a move away from crap governance by the granting of "access" to the currently excluded egalitarian voice, together with measures that will ensure that the other voices become responsive to it. Extracting all that mineral wealth, while the poorest and most vulnerable of South Africa's citizens suffer as a result of the pollution of their environment, is simply not a democratic way of proceeding!

Ultimately, it comes down to devising ways of restoring those common-pool goods. Compensation (which, unfortunately entails commoditization) is the commonly resorted-to route, but ensuring that that compensation accrues to the community, and not to its individual members (still less, its traditional leaders), is not easy. Better, we feel, would be legislation (along the lines of Nepal's new Forest Act) that clearly defines each community's rights over its land and water resources. However, rights in what have become common-pool bads are hardly sufficient; there needs to be restitution: the land and the water must somehow be put back to the way they were pre-legacy. When it comes to the land, there is now a wealth of expertise—much of it acquired in efforts to remedy the environmental despoliation within the former Soviet Union (where hierarchy drowned out the other voices for 70 or so years)—that can be drawn on, but what, many will ask, can possibly be done about the water, now that it is laced with sulphuric acid and the dissolved salts of various heavy metals?

Well, in Sweden—a stable democracy with good (i.e. multi-voice) governance—they have for many years now resorted to the "liming" of lakes that have become over-acidic as a result of "acid rain". So that is one, rather low-tech, possibility. Beyond that, there is research (in Britain, at Bath University) on the acid mine drainage from the now-closed Wheal Jane tin mine in Cornwall. The idea is to breed up large quantities of the unusual bacteria that currently live in the water that is flowing out of the mine, so that they can be used to bio-concentrate the heavy metals *and* neutralise the sulphuric acid (Doward 2014: 1–4). It is early days yet but, if the method can be scaled up, then the heavy metals can be profitably extracted from the bacteria which can then be used (again profitably) to make biodiesel fuel. In this way, a burdensome waste is transformed into a valuable feedstock for a new and impressively "green" industry. If South Africa's governance were to become a little less crappy we could expect to see major investments in this line of research and its potential "new generation" industries, with the local communities then owning substantial stakes in them.

#### 4.3.2.3 Oil Extraction in the Niger Delta

When oil was discovered in the Niger Delta the oil companies (Shell predominantly, but there were others) were not free to just go in and exploit it; they had to negotiate terms with the Nigerian government, the ultimate owner of the land (that, along with sweeping expropriation rights, having been established by the preceding British colonial regime). This situation was thus quite similar to, but on a much larger scale than, Nepal's Bara Forest Management Plan: just two kinds of actors: hierarchical (the Nigerian government) and individualist (the oil companies). And, as with Bara, there was a massive "unknown known". The Delta was densely populated and was a mosaic of rights: small scale, informal and often shaped up in terms of common-pool goods: the kind of goods that, while supportive of the egalitarian solidarity, were alien and baffling to the two hegemons: the Nigerian government, casting it all in terms of public goods (for the use of which the oil companies would have to pay royalties) and the oil companies, casting it all in terms of private goods (concessions that would give them monopolistic rights, within designated areas and subject to royalty payments, over all the oil they could profitably extract). This is the classic markets-and-hierarchies settlement and, provided there were some clauses about environmental protection, what, many would ask, could possibly go wrong? The answer, as all the actors (especially the excluded residents of the Delta) now know, is everything!

Between 1970 and 2000 Nigeria earned around $350 billion in oil revenues, yet per capita income declined, poverty increased (from 36 to 70%) and inequality worsened sharply (Shaxson 2007). In other words, the money was creamed off by a corrupt elite: a practice that, sadly, is endemic across Africa. For instance, a recent report (led by the former United Nations Secretary General, Kofi Annan) observes that, in many African countries, multinational companies and political leaders collude to swindle the citizens out of their just revenue from their natural resources. Extractive industries, in consequence, "leave the poor behind and harm the environment" (Africa Progress Panel 2013). Sure enough, in the Delta itself, the mosaic of informal property rights that prior to the discovery of oil had left the trees in place (some species, having ancestors buried beneath them, being taboo), the farmland and orchards productive, and the waterways abundant with fish and other biota (and woe betide anyone who harmed a python: the tabooed "indicator species") was swept away. Now, with irresponsible oil spills by the oil companies (Shell has recently been found guilty in the Dutch courts) having been augmented by the deliberate acts of sabotage and piracy by alienated and displaced rebel groups, the environment has been laid waste and hitherto comfortable and environmentally benign livelihoods destroyed (Umejesi and Thompson 2015).

All this could have been avoided, Umejesi and Thompson (2015) point out, if the egalitarian solidarity had not been excluded from the policy process (and Kofi Annan, in pin-pointing the prevailing alliance between corporations and state officials, says much the same). All that was needed, at the start, was an acknowledgement of that matrix of largely common-pool rights (in the same sort of way that Nepal's new Forest Act enshrines the community forests as the inalienable property of the villagers). And even now, decades on and with the Delta and its inhabitants reduced to ruin, a clumsy solution of this kind—a shift towards better governance (i.e. towards a vibrant multi-vocality and away from the current unresponsive monologue)—is still feasible (as is outlined in Umejesi and Thompson 2015).

## 4.4 Conclusion: Towards the Light in Virunga

Virunga National Park—in many ways the most treasured and valuable portion of the Democratic Republic of Congo's vast (200 million hectares) expanse of tropical forest—has been poised to go the very same way as the Niger Delta and the Upper Oliphants River Catchment. SOCO—a British-based company—has been in negotiations with the government of the Democratic Republic of Congo for rights to explore and drill for oil in an area—Block 5, as it is designated—that takes in around half of the park, including Lake Edward (with its rich indigenous fishery that supports 27,000 livelihoods) (Gettleman 2014) and the habitats of the endangered mountain gorillas. Unfortunately, the area's abundant biodiversity—a key concern, of course, in REDD+—also includes several rampaging and well-armed militias that, as well as killing and looting indiscriminately, routinely use rape as a weapon.

Unsurprisingly, the excluded egalitarian actors, including the indigenous Batwa people (who, being hunters and gatherers, simply could not survive without their forest), have not remained silent. Grassroots resistance, stiffened and publicized with support from various international non-governmental organisations, has mushroomed, even as SOCO has been granted permits (from the Ministry of Environment, Nature Conservation and Tourism: the ministry that is charged with responsibility for REDD +) to explore for oil in Block 5 and, later, in an area that takes in 85% of the park. In particular, a documentary film denouncing SOCO's activities was released in November 2014 and, largely thanks to the furore that has created, the permissions have now been revoked.[10] Had REDD+ been able to continue, unchallenged, on its hierarchical way, the Virunga National Park would have become just one more depressing instance of that continent-wide swindle that, as Kofi Annan and his colleagues have warned, is being perpetrated through the collusion of multinational corporations and national political leaders.

So the prospect, in terms of governance, is now somewhat brighter in the Virunga National Park than it is in the Niger Delta or in the Upper Oliphants River Catchment. In Virunga, thanks to the egalitarian voice having managed to make itself heard, the policy terrain, unlike in the Niger Delta, is no longer uncontested. Indeed, with the cancellation of those SOCO permits (not to mention the likely imminent demise of REDD +), we have a quite dramatic transition from crap governance (hierarchy and individualism in club goods–creating cahoots) to good governance (with the hitherto excluded egalitarian voice being heard loud and clear).

## References


Achebe, C. (2009). 'My Regrets over Nigeria'. *Daily Sun* Newspaper Nigeria, 26 January, www.sunnewsonline.com/webpages/features/newsonthehour/2009/jan/26/newsbreak. [Accessed 9 March 2010].

ACMH. (2015). *African countries mineral industry handbook* (Vol. I). Washington DC: Strategic Information and Opportunities, International Business Publications.

Africa Progress Panel. (2013). Equity in Extractives: Stewarding Africa's natural resources for all. Retrieved from www.aricaprogresspanel.org [Accessed 21 January 2014].

Agukoronye, O. C. (2001). Landscape practices in traditional Igbo society. *Nigeria. Landscape Research, 26*(2), 85–98.

Akiwumi, F. A. (2014). Strangers and Sierra Leone mining: Cultural heritage and sustainable development challenges. *Journal of Cleaner Production, 84,* 773–782.

Ambedkar, B. R. (2009). *Buddha or Karl Marx*. Delhi: Siddhartha Books.

Amery, L. S. (1953). The Crown and Africa. *African Affairs, 52*(208), 179–185.

Baccini, A., Goetz, S. J., Walker, W. S., Laporte, M., Sun, M., Sulla-Menashe, D., et al. (2012). Estimated carbon emissions from tropical deforestation improved by carbon-density maps. *Nature Climate Change, 2*(3), 182–185.

Betts, R. F. (2006). *Decolonisation: Making of the Contemporary World*. New York: Routledge.


[10]The documentary is available at: http://virungamovie.com/blog/details/10824/team-virunga-update-take-action-to-protect-virunga [Accessed 1 March 2015].

Bolay, M. (2014). When miners become “foreigners”: Competing categorizations within gold mining spaces in Guinea. *Resources Policy, 40,* 117–127.
Davenport, T. H. R., & Saunders, S. (2000). *South Africa: A modern history*. London: Palgrave Macmillan.
Davidson, B., & Munslow, B. (1990). The crisis of the nation state in Africa. *Review of African Political Economy, 49,* 9–21.
de Jouvenel, B. (1952). *The Ethics of Redistribution*. Cambridge: Cambridge University Press.
Doward, J. (2014) Groundbreaking biofuel project brings new life to Cornish mine. https://www.theguardian.com/environment/2014/dec/27/algae-biofuel-wheal-jane-mine-cornwall [Accessed 21 June 2017].
Ebeku, K. S. A. (2001) Oil and Niger Delta People: The Injustice of the Land Use Act, *CEPMLP Online Journal*, 9 June 2007. www.dundee.ac.uk/cepmlp/journal/html/vol9article9–14.html [Accessed 20 April 2009].
Estrada, M. (2011). *Standards and methods available for estimating project-level REDD+ Carbon benefits reference guide for project developers*. Working Paper 52, Bogor Indonesia: CIFOR.
FAO. (2015). Nature and Faune: Enhancing natural resources management for food security in Africa, *29*(2), www.fao.og/3/a-i4872e.pdf [Accessed 9 February 2016].
Farrell, L. A., Hamann, R., & Mackres, E. (2012). A clash of cultures (and lawyers): Anglo Platinum and mine-affected communities in Limpopo Province. *South Africa Resources Policy, 37*(2), 194–204. https://doi.org/10.1016/j.resourpol.2011.05.003.
George, S. (1998). Preface to M. Goldman (ed) *Privatizing nature: Political struggles for the global commons*, London: Pluto Press.
Gettleman, J. (2014). Oil dispute takes a page from Congo’s bloody past. https://www.nytimes.com/2014/11/16/world/oil-dispute-takes-a-page-from-congos-bloody-past.html [Accessed 11 March 2016].
Ghazvinian, J. (2005). *The curse of oil in the Niger Delta*. The Virginia Quarterly Review: Nigeria.
Gilmour, D. A., & Fisher, R. J. (1991). *Villagers, forests and foresters*. Kathmandu, Nepal: Sahayogi Press.
Guichaoua, Y. (2009). Oil and political violence in nigeria. In J. Lesourne (Eds.), *Governance of Oil in Africa, Unfinished Business*, Les etudes, May 2009 (pp. 9–50), Paris: Institut Francais des Relations Internationales (IFRI).
Gupta, J., & Thompson, M. (2010). Development and development cooperation theory. In J. Gupta & N. van der Grijp (Eds.), *Mainstreaming climate change in development cooperation* (pp. 33–36). Cambridge: Cambridge University Press.
Hair, P. E. H. (1954). A Study on Enugu, unpublished manuscript, National Archives Enugu.
Hardin, G. (1968). The tragedy of the commons. *Science, 162,* 1248.
Hobbs, P., Oelofse, S. H. H., & Rascher, J. (2008). Management of environmental impacts from coal mining in the upper Olifants River catchment as a function of age and scale. *International Journal of Water Resources Development, 24*(3), 417–431.
Ikejiofor, C. U. (2006). Integrative strategies of functional interface? *Emerging trends in land administration in contemporary Enugu, Nigeria, IDPR, 22*(2), 137–158.
Ikejiofor, C. U. (2009). Planning within a context of informality: Issues and trends in land delivery in Enugu, Nigeria. A case study prepared for Revisiting Urban Planning: Global Report on Human Settlements, www.unhabitat.org/grhs/2009/ [Accessed 3 January 2009].
IPCC. (2014). R. K. Pauchauri, L. A. Meyer, & Core Writing Team (Eds.), *Synthesis report: Climate Change*, Geneva, Switzerland: Intergovernmental Panel on Climate Change.
Isichei, E. (1976). *The history of igbo people*. London: Longman.
Kropotkin, P., & Marshall, S. S. (1995). *Kropotkin: The conquest of bread and other writings*. Cambridge: Cambridge University Press.
Ladi, S. (2005). *Globalisation, policy transfer and policy research institutes*. Massachusetts: Edward Elgar Publishing Inc.
Lea, D. (1992). *Libertarian theory* (pp. 39–59). Reasons Papers, Fall: Customary Communal Ownership and Environmental Protection.

Lindblom, C. (1977). *Politics and markets: The world's political economic system*. New York: Basic Books.
Lukpata, V. I. (2013). Revenue allocation formulae in nigeria: A continuous search. *International Journal of Public Administration and Management Research, 2*(1), 32–38.
Mabogunje, A. L. (1979). Land and peoples of West Africa. In J. F. Ade Ajayi & M. Crowder (Eds.), *History of West Africa* (Vol. II). London: Longman.
Manson, A., & Mbenga, B. (2003). The richest tribe in Africa: Platinum-mining and the Bafokeng in South Africa's North West Province, 1965–1999. *Journal of Southern African Studies, 29*(1), 25–47.
Marais, L. (2013). Resources policy and mine closure in South Africa: The case of the Free State Goldfields. *Resources Policy, 38*(3), 363–372.
Mccarthy, T. S. & Pretorius, K. (2009). *Coal Mining on the Highveld and Its implications for future water quality in the vaal river system*, pp. 56–65.
Meek, C. K. (1946). A note on crown land in the colonies. *Journal of Comparative Legislation and International Law, 3*(28), 87–91.
Mineral Act. (1991). South African Minerals Act 50 of 1991, South Africa.
Mitee, L. (2002). *The centrality of self-identity in indigenous peoples' struggles: The struggle of the Ogoni people*. Paper presented at the Indigenous Peoples Rights in the Commonwealth Project, Africa Regional Expert Meeting, Cape Town, South Africa, pp. 16–18 October.
Mnwana, S. (2014). Mineral wealth-in the name of morafe? Community control in South Africa's Platinum valley. *Development Southern Africa, 31*(6), 826–842.
Monbiot, G. (2015). SPERI Annual Lecture, hosted by Sheffield University's Political Economy Research Institute. Transcript available at: http://theguardian.com/environment/georgemonbiot/2014/jul/24/price-nature-neoliberal-capital-road-ruin [Accessed February 27].
MPRDA. (2002). Mineral and Petroleum Resources Development Act 28 of 2002. Government Gazette, Republic of South Africa, No. 23922, Notice No. 1273.
Munnik, V., Hochmann, G., Hlabane, M., & Law, S. (2010). *The social and environmental consequences of coal mining in South Africa: A case study*. Cape Town: Environmental Monitoring Group.
Nabuurs, G. J., Masera, O., Andrasko, K., Benitez-Ponce, P., Boer, R., Dutschke, M., Elsiddig, E., Ford-Robertson, J., Frumhoff, P., Karjalainen, T., Krankina, O., Kurz, W. A., Matsumoto, M., Oyhantcabal, W., Ravindranath, N. H., Sanz Sanchez, M. J. & X. Zhang (2007). Forestry. In B. Metz, O. R. Davidson, P. R. Bosch, R. Dave & L. A. Meyer (Eds.), *Climate change 2007: Mitigation*. Contribution of Working Group III to the Fourth Assessment Report of the Intergovernmental Panel on Climate Change. New York: Cambridge University Press.
Ney, S. (2009). *Resolving messy policy problems*. London: Earthscan.
Nock, A. J. (1935). *Our enemy, the state*. New York: William Morrow & Company.
Obi, C. (2008). *Oil as a curse of conflict in Africa: Peering through the smoke and mirrors*. Paper presented at the XIV South African Sociological Association (SASA) Congress, Stellenbosch University, South Africa, 7–10 July 2008.
Ogon, P. (2006). Land and forest resources use in the Niger Delta: Issues in regulation and sustainable management. www.globetrotter.berkeley.edu/GreenGovernance/papers/Ogon2006.pdf [Accessed 27 April 2007].
Olander, J., & Ebeling, J. (Eds.). (2011). *Building carbon forest projects*. Washington, DC: Forest Trends.
Onoh, C. C. (1997). *Whose coal city?*. Enugu: Frontline Publishers.
Patt, A. (2015). *Transforming energy: Solving climate change with technology policy*. Cambridge: Cambridge University Press.
Prins, G., & Rayner, S. (2007). Time to ditch Kyoto. *Nature, 449*, 973–975.
Ramutsindela, M. F. (2002). The perfect way to ending a painful past? Makuleke land deal in South Africa. *Geoforum, 33*(1), 15–24. https://doi.org/10.1016/S0016-7185(01)00008-2.
Rebanks, J. (2015). *The Shepherd's Life: A Tale of the Lake District*. London: Allen Lane.
Samuelson, P. (1948). *Economics*. New York: McGraw Hill.
Scully, J. (1976). The treason of the clerks. *Minnesota Review, 6*, 5–9.

Sharma, S., Koponen, J., Gyawali, D., & Dixit, A. (2004). *Aid Under Stress: Water, Forests and Finnish Support in Nepal*. Lalitput, Nepal: Himal Books.

Shaxon, N. (2007). Oil, corruption and the resource curse. *International Affairs, 83,* 1123–1140.

Shipton, P. (1994). Land and culture in tropical Africa: Soils, symbols, and the metaphysics of the mundane. *Annual Review of Anthropology, 23,* 347–377.

Snidel, D. (1991). Relative gains and the pattern of international cooperation. *American Political Science Review, 85*(3), 701–726.

Stamp, L. Dudley, & Hoskins, W. R. (1963). *The common lands of England and Wales*. London: Collins.

Tamang, M. S., & Tuladhar, M. T. (Eds.). (2014). *Social inclusion research: A source book*. Kathmandu, Nepal: Mandala Book Point.

Thompson, M. (2000). Global networks and local cultures: What are the mismatches and what can be done about them? In C. Engel & K. H. Keller (Eds.), *Understanding the impact of global networks on local social* (pp. 113–130). Nomos: Political and Cultural Values, Baden-Baden.

Thompson, M. (2008). *Organising and disorganising*. Axminsters: Triarchy Press.

Thompson, M., Ellis, R., & Wildavsky, A. (1990). *Cultural theory*. Boulder, Colorado: Westview.

Uchendu, V. C. (1978). State, land and society in Nigeria: A critical assessment of land use decree. *Journal of African Studies, 6*(2), 62–74.

Umejesi, I. (2014). Amnesty, patriarchy and women: The 'missing gender' voice in post-conflict Niger Delta region of Nigeria. *Gender and Behaviour, 12*(1), 6223–6237.

Umejesi, I. (2016). Colonial era practices of eminent domain and state-community contestation for mineral-resource ownership in postcolonial Nigeria: The search for a sustainable synergy. *South African Journal of Environmental Law and Policy, 22,* 65–83.

Umejesi, I., & Akpan, W. (2013). Oil exploration and the character of local opposition in colonial Nigeria: Exploring the roots of state-community conflict in the Niger Delta region. *South African Review of Sociology, 44*(1), 111–130.

Umejesi, I., & Thompson, M. (2015). Fighting elephants, suffering grass: Towards a clumsy solution to the current violence and environmental degradation in the Niger Delta region of Nigeria. *Journal of Organisational Change Management, 28*(5), 791–811.

Verweij, M. (1999). Whose behaviour is affected by international anarchy? In M. Thompson, G. Grendstad, & P. Selle (Eds.), *Cultural theory as political science*. London: Routledge.

Verweij, M. (2006). Is the Kyoto protocol merely irrelevant, or positively harmful, for the efforts to curb climate change? In M. Verweij & M. Thompson (Eds.), *Clumsy solutions for a complex world* (pp. 31–60). Basingstoke: Palgrave Macmillan.

Verweij, M. (2011). *Clumsy solutions for a wicked world*. Basingstoke: Palgrave Macmillan.

Wa Thiong'o, N. (1972). *Homecoming: Essays on African and Caribbean Literature, Culture and Politics*. Cited in Hugh Web, Passion Spaces: African Literature and the Post-colonial Context, Perth: Postcolonial Press.

Williamson, O. (1975). *Markets and hierarchies, analysis and anti-trust implications*. New York: The Free Press.

Wolverkamp, P. (1999). Prologue. In P. Wolverkamp (Ed.), *Forests for the future: Local strategies for forest protection, economic welfare and social justice* (pp. 1–28). New York: Zed Books.

World Rainforest Movement. (2014). Call to action: To reject REDD+ and extractive industries. To confront capitalism and defend life and territories. Available at www.wrm.org.uy/wp-content/uploads/2014/11/Call-COP-Lima_NoREDD.pdf [Accessed 24 October 2015].

WRI. (2015). RELEASE: African countries launch AFR100 to restore 100 million hectares of Land. www.wri.org/news/2015/12/release-african-countries-launch-afr100-restore-restore-100-million-hectares-land [Accessed 12 February 2016].

Zimmerman, F. J. (2000). Barriers to participation of the poor in South Africa's land redistribution. *World Development, 28*(8), 1439–1460. https://doi.org/10.1016/S0305-750X(00)00029-2.

Zolberg, A. R. (1968). The structure of political conflict in the new states of tropical Africa. *American Political Science Review, 62*(1), 70–87.

# Part II
# Water-Energy-Food Nexus

# Chapter 5
# Introduction to Part II: Integrative Frameworks and Participatory Governance for Effective Water-Energy-Food Systems Management

**Shonali Pachauri**

**Abstract** The establishment of the United Nations Sustainable Development Goals has provided greater impetus for nexus thinking and integrated planning and policy between sectors, systems and resources. Water, energy and food (WEF) are three key resources that are fundamental to human survival. This section includes studies that examine important interdependencies and interactions between these resources and solutions for more integrated management of them.

Since the establishment of the United Nations Sustainable Development Goals (SDG) in 2015, nexus thinking has gained traction internationally. This has meant that interdependencies and interactions between different sectors, systems, and resources are increasingly acknowledged and being accounted for. Water, energy and food (WEF) are fundamental to human survival, but policies for managing the provision of these resources in South Africa and other parts of the world are routinely still sector-specifically determined. Such silo thinking is neither constructive to capturing important interdependencies between systems nor conducive to ensuring consistency across scales. How will future environmental and socioeconomic change affect the supply of WEF services, and how can these systems be made resilient to such change? How can policy coherence in WEF resource management to maximize synergies and avoid tradeoffs be achieved? The contributions in this section focus specifically on some of these key questions by adopting a systems perspective to specific WEF resources.

In the contribution made by Kanyerere et al. (2018), the focus is on solutions to advance water security in sub-Saharan Africa. They highlight the significance of water governance strategies, including increasing stakeholder engagement, for

S. Pachauri (✉)
Energy Program, International Institute for Applied Systems Analysis, Laxenburg, Austria
e-mail: pachauri@iiasa.ac.at

 
P. Mensah et al. (eds.), *Systems Analysis Approach for Complex Global Challenges*,
https://doi.org/10.1007/978-3-319-71486-8_5

improving water management and resolving conflict between competing needs and actors. Jarbandhan et al. (2018), also argue for greater participatory governance, but their focus is on energy rather than water resources. They conclude that the need to incorporate public participation within the project cycle and institutionalize it cannot be overemphasized, and is central to the success of renewable energy projects. A detailed analysis of existing energy, air quality, and climate change policies in South Africa is the focus of the contribution by Klausbruckner et al. (2018). They argue in favor of integrated assessment frameworks to resolve existing policy inconsistencies and design coherent new policies to achieve future targets and goals. Finally, Ncube et al. (2018), focus on precision agriculture and its impacts on food security in Africa. They conclude that by ensuring more efficient use of inputs such as fertilizer and water, and reduced detrimental outcomes such as environmental pollution and degradation, significant benefits can be accrued by commercial and small-scale farmers through the implementation of precision agriculture practices.

These contributions all highlight the need for integrative and participatory frameworks to better understand interactions between and to promote efficient, inclusive and sustainable management of WEF resources. This is also the conclusion of a recent report by the World Wide Fund for Nature—South Africa (Gulati 2014). The report makes recommendations for a resilience-based integrated framework with a focus specifically on the links between food and energy systems. Such findings have also led to the genesis of South Africa's new water-energy- food nexus lab, which aims to enhance inter, multi- and trans-disciplinary research on water, energy and food security within the South African science system. Such efforts, it is hoped, will also encourage the application of good governance and management principles to achieve the WEF related SDG, by placing special emphasis on nexus between these systems.

## References


Gulati, M. (2014). *Understanding the food energy water nexus: Through the food and energy lens*. South Africa: WWF-SA.

Jarbandhan, J., Komendantova, N., Xavier, R., & Nkoana, E. (2018). Transformation of the South African energy system: Towards participatory governance. In P. Mensah, S. Hachigonta, D. Katerere & A. Roodt (Eds.), *Systems Analysis Approach for Complex Global Challenges*. Heidelberg: Springer.

Kanyerere, T., Tramberend, S., Mokoena, P., Mensah, P., Chingombe, W., Goldin, J., et al. (2018). Water futures and solutions: Options to enhance water security in sub-Saharan Africa. In P. Mensah, S. Hachigonta, D. Katerere & A. Roodt (Eds.), *Systems Analysis Approach for Complex Global Challenges*. Heidelberg: Springer.

Klausbruckner, C., Henneman, L. R. F., Rafaj, P., & Annegarn, H. J. (2018). Energy policy, air quality, and climate mitigation in South Africa: The case for Integrated Assessment. In P. Mensah, S. Hachigonta, D. Katerere & A. Roodt (Eds.), *Systems Analysis Approach for Complex Global Challenges*. Heidelberg: Springer.

Ncube, B., Mupangwa, W., & French, A. (2018). Precision agriculture and food security in Africa. In P. Mensah, S. Hachigonta, D. Katerere & A. Roodt (Eds.), *Systems Analysis Approach for Complex Global Challenges*. Heidelberg: Springer.

# Chapter 6
# Water Futures and Solutions: Options to Enhance Water Security in Sub-Saharan Africa

**Thokozani Kanyerere, Sylvia Tramberend, Audrey D. Levine, Portia Mokoena, Paul Mensah, Wisemen Chingombe, Jacqueline Goldin, Sumbul Fatima and Mayank Prakash**

**Abstract** *Background and Significance of the topic*: Water security is one of the greatest health, ecological, environmental, and human rights challenges of our time. Africa sits at the epicenter of this quandary, with the need to build resilience into already over allocated water resources. This chapter focuses on Sub-Saharan Africa and stresses the inter-related physical and social dimensions that underpin water security. The chapter highlights the value of engaging stakeholders through meaningful dialogue towards outcome oriented and adaptable governance strategies. *Methodology*: A desktop review was conducted to provide an overview of the challenges and opportunities to advance water security in Africa. *Application/*

T. Kanyerere (✉) · A. D. Levine · J. Goldin
Environmental and Water Sciences Program, Department of Earth Sciences, University of the Western Cape, Private Bag X17, 7535 Bellville, South Africa
e-mail: tkanyerere@uwc.ac.za

S. Tramberend
Water Program, International Institute for Applied Systems Analysis, Schlossplatz 1, Laxenburg, Austria

A. D. Levine
University of California, Santa Cruz, CA, USA

P. Mokoena
Department of Water and Sanitation, Surface and Groundwater Information, Pretoria, South Africa

P. Mensah
Institute for Water Research, Rhodes University, Grahamstown, South Africa

W. Chingombe
School of Biology and Environmental Sciences, University of Mpumalanga, Private Bag X 11283, 1200 Mpumalanga, South Africa

S. Fatima
Aligarh Muslim University, Aligarh, India

M. Prakash
International Institute for Population Science, Mumbai, India

 
P. Mensah et al. (eds.), *Systems Analysis Approach for Complex Global Challenges*,
https://doi.org/10.1007/978-3-319-71486-8_6

*relevance to systems analysis*: While Integrated Water Resource Management (IWRM) has been adopted to various degrees around the world, it is still in its infancy in sub-Saharan Africa. Additional research, ground-truthing, and on-the-ground field experience are necessary for tailoring IWRM to meet the individual and collective water security challenges that confront Sub-Saharan African countries. *Policy and/or practice implications*: The feasibility of applying evidence-based decision-making is enhanced by technology developments and advances in data collection, validation, curation, and interoperability. *Discussion and conclusion*: Water security is a global imperative and sub-Saharan Africa can benefit from 'lessons learned' to implement short-term and long-term strategies.

## 6.1 Introduction

Water security is one of the greatest health, ecological, environmental, and human rights challenges of our time. Competition for scarce water resources is a key concern in many countries, including large emerging economies such as China, India, South Africa and Brazil. Depletion of groundwater resources has accelerated (Aeschbach-Hertig and Gleeson 2012), water quality continues to deteriorate in many regions, and threats to river biodiversity are pervasive (Vörösmarty et al. 2010). A UN-Water (2010) study noted that "the total usable freshwater supply for ecosystems and humankind is merely around 200,000 km$^3$ of water—less than 1% of all freshwater resources.

Africa sits at the epicenter of this quandary, with the need to build resilience into the already over allocated water resources to support a population that is anticipated to more than double to over 2 billion by 2050 (UNECA 2016). Adequate and reliable sources of safe drinking water, food, and energy are necessary to support the health and welfare of this growing population, promote public health, and mitigate environmental contamination, particularly under recurring droughts and future climate uncertainties (NASCAS 2014). In Sub-Saharan Africa, 319 million people are without access to reliable drinking water sources, while 695 million of a global 2.4 billion people living without improved sanitation facilities live in the sub-continent (WHO 2015). Water (in) security is a consequence of inadequate or inequitable access to safe drinking water either due to water scarcity, insufficient or unreliable water infrastructure, intense storms and flooding, poor water quality, lack of resources, or ineffective governance. Water (in)security is exacerbated by climate disruption.

This chapter focuses on Sub-Saharan Africa and stresses the inter-related physical and social dimensions that underpin water security. The chapter highlights the value of engaging stakeholders through meaningful dialogue towards outcome oriented and adaptable governance strategies. The application of Integrated Water Resource Management (IWRM) is introduced as an analytical framework for strengthening water security in sub-Saharan Africa and beyond. While IWRM has been adopted to various degrees around the world, it is still in its infancy in sub-Saharan Africa. Additional research, ground-truthing, and on-the-ground field experience are necessary for tailoring IWRM to meet the individual and collective water security challenges that confront Sub-Saharan African countries.

## 6.2 Social and Physical Dimensions of Water Security

Water systems involve the complex interplay among physical, social, economic, and political factors, including risks and uncertainties (Grey et al. 2013). The concept of water security refers to equitable access to freshwater that is of an acceptable quantity and quality to sustainably and reliably support health, sanitation, livelihoods, ecosystems, and economic stability (AMCOW 2012). Physical water scarcity may be attributed to climatic or geographical factors, unsustainable consumption or over exploitation, while economic water scarcity may be due to inadequate infrastructure or lack of equitable access to available water resources. In some cases, there may be more than enough water available, but the ability to access and use the water is compromised due to insufficient capacity for storing, conveying, and provisioning the water to communities, individuals, or industries. In other cases, there may be an imbalance in water availability to supply competing water-users. In addition, the health and environmental consequences of water contamination can exacerbate water insecurity. Africa, being home to many of the world's poorest nations, is especially vulnerable to climate- and weather-induced disruptions to water resources, such as prolonged and recurring droughts, flooding, infrastructure failures, or extreme precipitation events. Water scarcity negatively impacts on local and regional human populations as reflected by the Water Poverty Index (WPI)[1] where the Southern African Development Community (SADC) Member States do not currently score above 61.9.

The physical aspects of water security include

- technical and economically viable access to surface and groundwater resources;
- the infrastructure to convey, process, store, deliver, and manage the water for specific end-uses;
- safeguards to protect against the consequences of drought, flooding, or hazardous conditions.

While the physical dimensions of water security define the availability and quality of accessible water resources, complex social dimensions reflect the intersection between environments (non-human) and people. Water security requires agility and responsiveness as individuals, households, communities—including whole cities—financially, socially, and legally adjust and adapt to direct and indirect consequences and impacts. In addressing the social dimension—we move away from the emphasis

[1]Of the many water poverty indices developed, the most notable is that conceptualized by Lawrence et al. (2002). Their composite water poverty index has three components namely, water availability, access to safe water and sanitation, and time and effort to collect domestic water.

**Table 6.1** Comparison of physical and social dimensions of resilience

| Dimension | Adaptation examples | Capacity building |
|---|---|---|
| Physical | • Restoring infrastructure<br>• Remediating flood damage | • Adoption of dry crops or new crops that require less water<br>• More resilient and sustainable crop and seed choices |
| Social | • Collective action and participatory processes<br>• Co-production of knowledge<br>• Firsthand learning about what does and doesn't work within a given setting<br>• Direct involvement in planning for the future | • Diversifying livelihoods,<br>• Physically moving away from flood plains<br>• Fostering dignity, hope, pride, and other intangible goods that are, as essential to small scale farmers as fertiliser or seed (Goldin 2015) |

on material infrastructure towards the intangible or non-material goods that reflect the broader notions of human development and well-being.

Water security is intrinsically linked to resiliency, which is the ability of a social or ecological system to resist, absorb, accommodate and recover from the effects of a hazard, whether by persisting or by adapting and changing towards new pathways (Folke 2006). In essence, resiliency is a measure of the amount of change a system can undergo, the degree to which it can re-organise, and the extent to which it can build capacity to learn and adapt (AMCOW 2012; Folke et al. 2002; Berkes et al. 2008). The physical and social dimensions of how resiliency can influence adaptation and capacity building are outlined in Table 6.1.

It is evident that management and governance strategies must balance and address the technical and social dimensions to:

- safeguard water quality,
- promote public health,
- protect water resources and ecosystems,
- mitigate sources of contamination, and
- respond and remediate acute, transient, intermittent, or chronic water resource threats or hazards.

### 6.2.1 *The Capability Approach (CA) and Ideas of Social Justice*

The Capability Approach (CA), a human development paradigm, provides a useful framework for highlighting the freedoms people have reason to value, giving primacy to multi-dimensionality and the idea of capabilities, opportunity, social justice, and hope rather than solely the distribution of material goods. The main premise of Amartya Sen's (1999) Development as Freedom is that individuals either achieve or fail to realise important freedoms due to social, political and economic constraints or opportunities. Many of these freedoms are absent in environments where people lack water security and concurrently are 'victims' of known and unknown external stressors (such as climate change).

A secure water world is one which is just—and this implies that the water users are treated fairly not only in the equitable access to safe water supplies, but in the distribution of knowledge so that they are able to make informed decisions about water resource protection, use, control, management and conservation. This also means that citizens have the necessary information and decision-making authority to become agents of change and custodians of the water resources upon which they depend. The CA provides a diagnostic—evaluative tool for considering resilience in particular localities. Progress depends on participatory parity (Fraser 2009) and expanded notions of human wellbeing (Sen 1999, 2012; Atique 2014; Goldin 2015; Nussbaum 2001). The application of CA towards water security incorporates material and non-material principles along with attention to ideals of justice.

### 6.2.2 Justice and 'Good' and 'Bad' Resilience

While water security is global in nature, its impacts are not expected to be globally homogeneous but rather differentiated across regions, generations, age classes, income groups, occupations and gender. To some degree, all societies will be affected by climate change. However, the impacts will vary by location, exposure, and context-specific social characteristics, power relations, and socio-political factors (Nelson 2011). To develop practical solutions, it is important to consider local, regional, and global scales in terms of planning short-term and long-term responses to known and unknown threats.

The application of a gender lens adds an important dimension across the spectrum of water resources management. Important determinants are income, education and the role of women who often collect and manage water within households and communities. Integrating a gender-sensitive approach to development can have a positive impact on the effectiveness and sustainability of water interventions and on the conservation of water resources (Barnes 2013; UN-Water 2013). It is important to recognize that methods for engendering inclusivity can motivate all sectors of society to contribute their capabilities, knowledge and relevant skills to enhance local and regional water security. With appropriate resources and empowerment, women and men have the collective robust capabilities to develop complex adaptive strategies to build resilience to climate variability and change; conversely, a social context with inadequate knowledge flows is unlikely to provide those outcomes (Goldin 2015 and Goldin et al. 2017). When considering water security, the 'intangible' goods that people hold in their hearts and minds are just as significant as the 'tangible' components. Exclusion and social isolation are tangible attributes that include a range of non-material emotions that are apparent not only when water is unsafe or too far away but also in contexts where people are far removed from decision-making opportunities or/and are experiencing exclusion and social isolation (Goldin et al. 2017; FAO 2000).

### 6.2.3 *Building Water Security Resilience Towards Climate Variability*

The impacts of rapidly changing economies, populations, and climate on fresh water fluxes are still poorly understood, although it is clear that most of the impacts of climate change on society will be transmitted by water. For example, the average global temperature is projected to increase by 1.4–5.8 C and there would be a substantial reduction in freshwater resources by 2050 in Sub-Saharan Africa, causing an estimated 10% drop in rainfall and 17% decrease in drainage (Misra 2014; Schulze and Perks 2000). Various complex and interconnected factors that determine water security exist, ranging from socio–political through climate change, resource constrains, water quantity, water quality and biophysical, to financial, infrastructural and institutional aspects (UNESCO IHP 2006).

The seasonal and inter-annual variations in rainfall are some of the elements that lead to vulnerability in water systems in Southern Africa (Schulze and Perks 2000). The climate in most of southern Africa is considered to be semi-arid and climate disruption is anticipated to introduce more uncertainties in the frequency and variability of rainfall, intense storms, and drought (Batisani and Yarna 2010; Ramos and Martinez-Casasnovas 2006; Kozo 2008; Cook et al. 2004; Reason and Jagadheesha 2005; Reason and Mulenga 1999; Pahl-Wostl 2007). The impacts of drought are exacerbated by anthropogenic activities such as deforestation, land use, and water resource management decisions that alter the storage of water on the land. In Sub-Saharan Africa, drought can have direct effects on the 70% of the working population who are engaged in rain-dependent agricultural activities that contribute to about 25% of the average gross domestic product (GDP) (Dixon et al. 2001; Sheffield et al. 2014). It is widely acknowledged that water demand already exceeds supply in many parts of South Africa (Muller et al. 2009a, b). As a consequence of the imbalances between supply and demand, ensuring adequate and resilient water storage is critical (Keller et al. 2000; Muller 2000; Kiker 2000). At the global level, impacts of climate change on increasing irrigation water requirements could be nearly as large as the changes projected from socioeconomic development in the last century (Fischer et al. 2007).

The Water Futures and Solutions (WFaS[2]) initiative, launched in 2013, is a stakeholder-informed, scenario-based assessment of water resources and water demand. The WFaS takes a systems analysis approach to identifying water-related policies and management practices that work together consistently across scales and sectors to improve human well-being through water security. The emerging ensembles of state-of-the-art socioeconomic and hydrological models provide a promising approach

[2]http://www.iiasa.ac.at/web/home/research/water-futures.html.

to defining the necessary scaffolding for water security. For sub-Saharan Africa, the WFaS initiative plays a critical role in determining the specific data and information needs to support short-term and long-term decisions relevant to water security (Burek et al. 2016). The consistent, multi-model global water scenarios embedded in WFaS also consider the parallel (in)securities of food and energy. The importance of the water-food-energy-climate-environment nexus is emerging as a major research challenge at local, regional, and global scales (Tramberend et al. 2015; Wada et al. 2016).

## 6.3 Water Governance for Managing Complex Systems

The concept of water governance is evolving to incorporate greater levels of market involvement (privatization of water services), greater levels of civil society involvement (water user associations, more public participation), and more independent bodies (such as river basin committees). Governance, as defined by Pierre and Peters (2000), links the efforts by private and public actors to steer, control, or manage water resources with the relevant institutions[3] and the normative underpinnings of these efforts and institutions. Water is a prime example of shared natural-resource systems, potentially subject to the "tragedy of the commons" (Hardin 2009). In principle, it is in the interests of the users of a commons to manage it prudently, and they often devise complex social schemes for maintaining common resources efficiently (Ostrom et al. 1999, 2002).

Water governance strategies are progressing towards institutional cooperation across numerous governance units (e.g. agriculture, water, housing, energy, and environment), the private sector and non-governmental organizations. For example, stakeholders for improving water quality in river basins include decision-makers from: ministries for water (e.g. building a reservoir, monitoring), agriculture (run-off of fertilizer and pesticides), industry (point-source pollution), energy (water temperature). Potential conflicting interests between different water users requires careful mediation and management.

### *6.3.1 Spatial Dimension*

Water moves over large areas in the form of precipitation, surface run-off, subsurface flows and along river channels. The quality of water depends on the local watershed characteristics and land-uses. Water quality can deteriorate as a consequence of direct and indirect contamination, untreated sewage, industrial pollution, and changes in salinity due to evaporation/evapotranspiration. For example, dams

[3]The clean-up of the Rhine river (Malle 1996) is an example of a successful water management strategy.

in the South African part of the Inkomati River basin reduced freshwater flows into the Inkomati estuary in Mozambique and led to increased salt levels (Vaz and Pereira 2000).

Surface and groundwater, which cross national boundaries, present additional complexities evoked by institutional limitations and potential tensions in riparian relations. Globally, the almost 300 transboundary lake and river basins covering nearly half of the Earth's land surface, account for 60% of global freshwater flow and provide water to support about 40% of the world's population (Sadoff and Grey 2005; UN-Water 2008). However, many of the transboundary water-use agreements apply only to specific portions of basin, and frequently do not include all basin riparians (Giordano et al. 2014). The lack of deliberate co-operation may cause a deterioration in water quantity or quality, or both (Wolf 2007). The advent of transboundary watershed commissions holds promise for increased water management cooperation and collaboration. The global adoption of Geographic Information Systems (GIS) technology and availability of spatial data sets is providing unprecedented access to simulation models, visualization techniques, and decision-support systems. The interoperability of data is beginning to facilitate communication among distant water users through supporting evidence-based strategic water governance decisions that directly or indirectly encourage co-operation.

### 6.3.2 Temporal Variability

The extent of inter- and intra-annual rainfall variability has been dubbed the "curse of freshwater variability" which is manifested by floods, droughts, and chronic uncertainties (Hall et al. 2014). Coping with variable access to freshwater represents a major economic and social challenge that is exacerbated by poverty. Of the 35 river basins with a population greater than one million that are classified by the World Bank as "low income", 19 experience critical and intermittent water shortages, flooding, and other water (in)securities (Hall et al. 2014).

Water managers are entrusted with managing a highly variable resource to meet often rapidly changing demand. Institutions and governance need to engage in proactive planning and development to manage and share risks (e.g. land zoning, watershed protection, insurance, water allocation mechanisms, food trade liberalization). Investments in infrastructure help buffer variability and decrease risks (reservoir storage, dams, water transfers, wastewater treatments, efficiency improvements in water use).

### 6.3.3 Behavioral Factors

Water, extracted, treated and distributed at different spatial scales involves multiple stakeholders in its management. Household water management is determined by various factors including perception about water quality, type and material of water storage container used, knowledge of water treatment prior to consumption, and hygiene observance. Diseases that proliferate due to overcrowding, poor housing, and ineffective hygiene are exacerbated by poverty, where physical access to water and sanitation are lacking and there is a need for vigilance towards mitigation and prevention through promoting best practice behaviors (Goli et al. 2011; Karn et al. 2003; UN-Habitat 2004). Domestic household management is an important short-term water-related health intervention in many developing countries (Clasen and Cairncross 2004; WHO 2002).

### *6.3.4 Competition Across Water Use Sectors*

Typically, water withdrawals support the needs of agriculture (irrigation), industry (energy generation and manufacturing), the municipal sector (drinking water, sanitation) and 'environmental flows' to protect aquatic ecosystems and habitats (Smakhtin et al. 2004; Pastor et al. 2014). At the global level, irrigated agriculture produces as much as 44% of total crop production (FAOSTAT, AQUASTAT, Alexandratos and Bruinsma 2012). In Africa, irrigation development is highly concentrated and only 5% of total cultivated areas are equipped for irrigation due to lack of access to freshwater resources coupled with inadequate infrastructure and competing water demands. Scenario assessment under envisioned population growth and economic development indicates significant increases in future industrial and domestic water uses (Wada 2016).

While local and regional water governance can be complex, the increasing trade of agricultural commodities has spawned 'virtual water trade' which is the hidden flow of water when food and other commodities are traded from a region of production to regions of consumption (Hoekstra and Hung 2005; Hoekstra 2003; Allan and Allan 2002). For sub-Saharan Africa, the 'virtual water trade' can affect water security, particularly when commodities with embedded virtual water are exported from already water scarce regions.

## 6.4 Systematic Approach to Enhance Water Security

Since water security is a function of sustainable development, effective water management to achieve water security dictates that the resource is managed sustainably. Key components include water infrastructure and water quality as summarized in Table 6.2.

The Africa Water Vision 2025 aspires towards equitable and sustainable use and management of water resources for poverty alleviation, socioeconomic development, regional cooperation, and the environment and offers a context within which the resource may be sustainably managed (NASAC 2014). This calls for a paradigm shift from traditional approaches used to manage water resources (Rietveld et al. 2016). A systems approach (also holistic or ecosystem approach) has often been

**Table 6.2** Examples of infrastructure and water quality opportunities to enhance water security (adapted from NASAC 2014)

| Component | Governance | Examples |
|---|---|---|
| Water infrastructure for sustainable economic growth | • Invest in operations, maintenance, and water use efficiency<br>• Promote appropriate legislation<br>• Foster public-private partnerships<br>• Incentivize community involvement and economic development | • Enhance water storage by water harvesting<br>• Adopt conjunctive water uses<br>• Promote cost-effective wastewater treatment, reclamation, reuse, and recycling, and Reduce losses in the supply chain |
| Water quality protection and enhancement | • Commit to ensuring sustainable access to water resources<br>• Implement safeguards to mitigate health risks that can results from exposure to microorganisms (pathogens) and chemicals, especially for the poor | • Develop and maintain secure water resources to improve access to safe water and sanitation<br>• Ensure that wastewater treatment plants and recycling function efficiently before discharging into receiving water resources<br>• Develop appropriate and effective water quality treatment technologies and low-cost sanitation solutions |

advocated for water management in the form of integrated water resource management (IWRM) in many developing countries, including those in Sub-Saharan Africa (Bruce 2005), where governments play a key role in its implementation. Using systems approach to ensure water security in Sub-Saharan African will require governments to build efficiency into water resource use through investment in education and innovation, green technologies and multiple uses of recycled water.

Central to the success of IWRM is careful consideration of how to operationalize and sustain data collection, validation, curation, and interoperability. Currently, water resources assessment, information and communication management instruments do not produce data systematically to inform better decisions on water resources.

The Network of African Science Academies (NASAC) provides a cogent application of water security in the document "*The grand challenge of water security in Africa: recommendations to policy makers.*" On the climate change threat to water security, NASAC opined that mitigating disaster risks requires careful development planning, scientific knowledge, and early warning systems that are people centred, as wells as effective mechanisms for disaster responses, including future risks related to climate variability and change. Governments on the sub-continents are advised to put in place new strategies or to review existing strategies and policies to counteract the impact of global and climate change on water resources and to incorporate climate change adaptation strategies in their development plans and programmes (NASAC 2014).

## 6.5 Towards Sustainable Water Management

Globally, water management strategies are shifting towards social, institutional and environmental sustainability principles (UN-Water 2008; Grey and Sadoff 2007; UN-Malawi 2010). During the twentieth century, many Sub-Saharan countries developed constitutions that purport to promote people's welfare through sustainable management of natural resources such as water. In many countries, there is a significant time-lag for revising and implementing new regulatory policies. For example, although catchment management authorities (CMA) were initiated in South Africa over 20 years ago, to-date only two are currently functioning (e.g. the Breede-Gouritz and Inkomati-Usuthu CMAs). The 'lessons learned' from these CMAs can inform future policy directions. In addition, constitutional obligations can be stymied by limitations in finances, equipment, and human personnel (WWAP 2009).

In general, the role of IWRM is increasingly acknowledged as a pathway towards sustaining the availability and quality of water resources (UNDP 2010). The Rio Declaration on Environment and Development, Agenda 21 and the Statement for the Sustainable Management of Water (UN-Water 2008; WWAP 2009) propose an integrated approach to:

- poverty relief through community and stakeholder participation and
- sustainable resource management and development
  - by protection of quality and supply of freshwater resources and
  - application of the integrated approaches to the development, management and use of water resources.

Globally, IWRM has been partially or fully implemented in conjunction with socioeconomic and environmental development plans, as shown in Table 6.3 (UN-Water 2008). We now have the opportunity to learn from these vanguard efforts and develop the next-generation of tools that leverage advances in monitoring, analytical tools, data interoperability, and linking the social and physical dimensions of water security.

There are ongoing efforts to apply these basic concepts to Sub-Saharan Africa (Braune et al. 2008; Baumann and Danert 2008) that include:

- establishing a National Water Resources Board,
- implementing catchment management authorities,
- engaging stakeholders in water resources management, and
- promoting the use of various water resources management tools:
  - economic instruments by water utility institutions,
  - water quality regulatory guidelines,
  - water allocation rules,
  - water resources assessments and
  - water resources information management

**Table 6.3** Examples of local, provincial/state, transboundary, and international applications of IWRM

| Spatial coverage: Scale | Problem for IWRM | IWRM action | IWRM benefit |
|---|---|---|---|
| 1. Local level: Malawi (Dzimphutsi Village in Chikwawa District: Lower Shire valley) | – Water management<br>– Existing old water laws versus new water policy<br>– Appropriate institutions<br>– Coordination issues | – Established IWRM priority issues<br>– Set up coordination unit<br>– Piloted IWRM in Dzimphutsi village | – Piloted & assessed Dzimphutsi IWRM<br>– Started coordination meetings: national level<br>– Drafted water laws |
| 2. Transboundary level: Mozambique/Zimbabwe | – Floods<br>– Water quality | – Started studies within Pungwe project | – Strengthen interstate<br>– Gained knowledge |
| 3. National level: Uganda | – Water quality<br>– legal & institutions issues | – Set up coordination<br>– plan piloting IWRM | – Coordination was set<br>– Piloted IWRM in 2008 |
| 4. National level: Morocco | – Water scarcity (demand)<br>– Water reform not practiced | – involved NGO<br>– Pilotedwater projects<br>– Set up best practices | – Soussa Massa Agency implemented reform<br>– Provided water |
| 5. National Level: Sri Lanka | – Water policy not practiced<br>– Water related disasters<br>– inadequate water | – Baseline assessment<br>– set up institutions and disaster management | – Flood impact reduced<br>– Early warning given<br>– water storages set up |
| 6. National level: Chile | – Increasing water demand<br>– Increasing water use | – Water assessment<br>– New water laws | – Water use improved<br>– Clean environment |
| 7. National level: Kazakhstan | – Disputes<br>– Water shortage<br>– Water pollution | – Basin council set up<br>– IWRM & WE set up<br>– Legaltools developed | – Created basin council<br>– Created organisation<br>– Amended water law |
| 8. State Level: United States of America (New York City: Croton & Catskill/Delaware Watershed) | – Water quality (building new treatment water supply plant or improving the protection of water sources) | – Chose to protect the source of water<br>– Set up partnership & programmes with stakeholders | – 350 farms started best management practices with watershed<br>– Saved US$4400 million on water costs |

(continued)

**Table 6.3** (continued)

| Spatial coverage: Scale | Problem for IWRM | IWRM action | IWRM benefit |
|---|---|---|---|
| 9. Provincial level: China | – Water pollution<br>– Water shortage<br>– Deforestation upstream | – Set up coordination commission<br>– Enforced water policy | – Reduce conflicts<br>– Reduced pollution<br>– Stopped deforestation |
| 10. Local level: Colombia | – Deforestation<br>– Water diversion of water Amazon river to Pacific | – Set up partnership with stakeholders<br>– Set us committees to participate in decision making process | – Committees worked with ministry of environment to stop plans to divert water<br>– 387 families doubled their income and food |
| 11. International level: Fergana Valley (Central Asia) | – Disputes<br>– inefficient use of water<br>– Safe drinking water | – Set up commission for coordination<br>– Train stakeholders<br>– Set up committees<br>– Give service as pilot | – Partnership set up<br>– 28water committees<br>– 28 village got safe drinking water; 320 Ecosan toilets built |

*Source* UN-Water (2008), van Koppen et al. (2009)

- decentralizing community-based maintenance (CBM) functions:
  - cost sharing of financial resources
  - enabling beneficiaries to invest in projects of their choice.

However, policy harmonization remains a challenge. For example, ministries of environmental affairs, forestry, and water develop independent and sometimes contradictory policies to protect headwaters, wetlands and rivers banks. Yet, the Ministry of Agriculture can inadvertently foster agricultural activities on fragile lands such as riverbanks, hill-slope areas through various food-security initiatives (FAO 2008). By extension, cropping in mountainous and hill slopes (Healy 2010) can impair natural hydrological processes such as groundwater recharge and promote sediment erosion and deposition in surface water resources. However, appropriate policies, institutions and implementation instruments must be developed to ensure sustainable management of water resources especially under scarcity.

The IWRM approach provides a wealth of global experience that can help to reconcile inter- and intra-institutional disparities and contradictions (UNDP 2008).

However, it must be accompanied by appropriate policies, institutions and implementation instruments in order to ensure sustainable management of water resources especially under scarcity.

## 6.6 Conclusion

Water security in Sub-Saharan-Africa is at a critical cross-roads. Climate uncertainties, coupled with population growth and aspirations towards economic stability in an increasingly competitive global marketplace magnify the exigency of water security. Water security is also at the heart of well-being, health, food security, and energy reliability. This chapter has provided an overview of the challenges and opportunities to advance water security and draws four key conclusions:

- To build resilience for water security amongst communities, the intertwined social and physical dimensions can benefit from the capability approach. This requires a shift from traditional top-down approaches, towards community participation in defining the choices and opportunities that can strengthen water security.
- To manage water resource complexities, it is helpful to combine Ostrom and Hardins design principles to avoid the tragedy of the commons. However, it is also necessary to field-test these principles at local, regional, and national scales to develop location-specific feasible solutions tailored to short-term and long-term needs and aspirations of Sub-Saharan African countries.
- Historically, water resource management has been skewed towards engineering and scientific solutions. However, a more holistic approach can equalize the technical and engineering aspects with the social and political realities.
- The consensus among Sub-Saharan African countries is that IWRM has the potential to achieve socioeconomic development without compromising environmental integrity. However, the feasibility, implementation and progress of IWRM principles require a concerted effort to overcome the pervasive gaps between rhetoric and practice. The feasibility of applying evidence-based decision-making is enhanced by to technology developments and improved mechanisms for data collection, validation, curation, and interoperability.

**Acknowledgements** The authors would like to thank the National Research Foundation (NRF) of South Africa and the International Institute for Applied Systems Analysis (IIASA) of Laxenburg in Austria through the funded joint collaborative capacity building research program of the Southern African Young Scientists Summer Program (SA-YSSP) which was designed to develop capacity in systems analysis and expose scholars to an array of additional competencies and skills required to be successful in knowledge-driven societies. The Fulbright Specialist Program also provided collaboration opportunities. The authors appreciate the helpful and thoughtful comments provided by the reviewers.

## References

Inocencio, A. Sally, H. & Merry, D. J. (2003). *Innovative approaches to agricultural water use for improving food security in Sub-Saharan Africa*. Working Paper 55. International Water Management Institute. Colombo, Sri Lanka.

Aeschbach-Hertig, W., & Gleeson, T. (2012). Regional strategies for the accelerating global problem of groundwater depletion. *Nature Geoscience, 5,* 853–861.

Alexandratos, N., & Bruinsma, J. (2012). World Agriculture towards 2030/2050: The 2012 Revision. FAO, Rome. ESA Working Paper *12*(3).

Allan, J. A. & Allan, T. (2002). The middle east water question: Hydropolitics and the global economy, Ib Tauris.

AMCOW (Africa Ministrial Council on Water). (2012). Water security and climate resilient development. Technical background document.

Anyamba, A., Tucker, C. J., & Mahoney, R. (2002). From El Niño to La Niña: Vegetation response pattern over East and Southern Africa during the 1997–2000 Period. *Journal of Climate, 15,* 3096–3103.

Atique, A. (2014). Assessing capabilities approach as an evaluative framework for climate justice.

Banerjee, S. G., & Morella, E. (2011). Africa's Water and sanitation infrastructure. Access, affordability, and alternatives, World Bank, pp. 217–219.

Barnes, J. (2013). *Who is a water user? The politics of gender in Egypt's water use associations*. In: L. Harris, J. Goldin, & C. Sneddon, (Eds.), Contemporary water governance in the global South: Scarcity, marketization and participation. London, UK: Routledge.

Batisani, N., & Yarnal, B. (2010). Rainfall variability and trends in semi-arid Botswana: Implications for climate change adaptation policy. *Applied Geography, 30,* 483–489.

Baumann, E., & Danert, K. (2008). Operation and maintenance of rural water supplies in Malawi: study findings. Zurich: SKAT—Swiss Resource Centre and Consultancies for Development. Available at http://www.rural-watersupply.net/_ressources/documents/default/208.pdf.

Berkes, F., Colding, J. & Folke, C (2008) *Navigating social-ecological systems: Building resilience*. Cambridge University Press.

Berry, H. L., Bowen, K. & Kjellstrom, T. (2010). Climate change and mental health: A causal pathways. International Journal of Public Health, *55*, 123–132.

Bradley, D. J., & Bartram, J. K. (2013). Domestic water and sanitation as water security: Monitoring, concepts and strategy. *Philosophical Transactions of Royal Society, A*(371), 20120420.

Braune, E., Hollingworth, B., Xu, Y., Nel, M., Mahed, G., & Solomon, H. (2008). Protocol for the Assessment of the Status of Sustainable Utilization and Management of Groundwater Resources—With Special Reference to Southern Africa. WRC Report No. TT 318/08. Water Research Commission, Pretoria, South Africa.

Bruce, M. (2005). Integrated water resource management, institutional arrangements, and land—use planning. *Environment and Planning A, 37,* 1335–1352.

Burek, P., Satoh, Y., Fischer, G., Kahil, M., Scherzer, A., Tramberend, S., Nava, L., Wada, Y., Eisner, S. & Flörke, M. (2016). Water futures and solution-fast track initiative.

Cao, M., & Prince, S. D. (2005). Climate-induced regional and interannual variations in terrestrial carbon uptake. *Tellus, 57,* 210–217.

Carius, A., Tänzler, D., & Maas, A. (2008). *Climate change and security: Challenges for German development cooperation*. Germany: Deutsche Gesellschaft für Technische Zusammenarbei.t.

Clasen, T., & Cairncross, S. (2004). Household water treatment: refining the dominant paradigm. Tropical Medicine and International Health *9*:187–191.

Cook, C., Reason, C. J. C., & Hewitson, B. C. (2004). Wet and dry spells within particularly wet and dry summers in the South African summer rainfall region. *Climate Research, 26,* 3–17.

Davies, B., & Day, J. (1998). Vanishing waters. Cape Town, South Africa. University of Cape Town press.

Dixon, J., Gulliver, A. & Gibbon, D. (2001). Farming systems and Poverty: Improving farmers' livelihoods in a changing world. FAO and World Bank, 407 pp.

DWAF (Department of Water Affairs and Forest). (2012). Municipal biodiversity summary project. Department of water affairs and forest. economics research papers 2002/19.

FAO. (2000). The state of food and agriculture: Lessons from the past 50 years. Available at http://www.fao.org/docrep/017/x4400e/x4400e.pdf [Accessed 9 February 2016].

FAO. (2008). BIOFUELS: prospects, risks and opportunities. Available at http://www.fao.org/3/a-i0100e.pdf [Accessed 9 February 2016].

Fischer, G., Tubiello, F. N., van Velthuizen, H., & Wiberg, D. A. (2007). Climate change impacts on irrigation water requirements: Effects of mitigation, 1990–2080. *Technological Forecasting and Social Change, 74,* 1083–1107.

Folke, C., Carpenter, S., Elmqvist, T., Gunderson, L., Holling, S. & Walker, B. (2002). Resilience and sustainable development: Building adaptive capacity in a world of transformations. *Journal of Human Environment 31*(5), 437–440.

Folke, C. (2006). Resilience: The emergence of a perspective for social—ecological systems 725 analyses. Global Environmental Change, *16,* 253–267.

Fraser, N. (2009). Scales of justice: Reimagining political space in a globalizing world. New York: *Geography, a new radical journal 8*(3), 24–36.

Giordano, M., Drieschova, A., Duncan, J. A., Sayama, Y., De Stefano, L. & Wolf, A. T. (2014). A review of the evolution and state of transboundary freshwater treaties. *International Environmental Agreements: Politics, Law and Economics, 14,* 245–264. *Global Environmental Change 16*(3), 253–267.

Goldin, J. (2015). Hope as a critical resource for small scale farmers in Mpumalanga. *Human Geography, a new radical journal, 8*(3), 24–36.

Goldin, J., Botha J., Koatla T., Anderson, J., OWEN, G. & Lebese, A. (2017). Towards an ethnography of climate change variability: perceptions and coping mechanisms of women and men from Lambani Village, Limpopo Province. *Human Geography, A New Radical Journal, 10*(2), forthcoming.

Goli, S., Arokiasamy, P., & Chattopadhayay, A. (2011). Living and health conditions of selected cities in India: Setting priorities for the National Urban Health Mission. *Cities, 28,* 461–469.

Grey, D., Garrick, D., Blackmore, D., Kelman, J., Muller, M., & Sadoff, C. (2013). Water security in one blue planet: Twenty-first century policy challenges for science. *Philosophical Transactions of the Royal Society A, 371,* 2002.

Grey, D., & Sadoff, C. W. (2007). Sink or Swim? Water security for growth and development. *Water Policy., 9*(6), 545–571.

Hall, J., Grey, D., Garrick, D., Fung, F., Brown, C., Dadson, S., et al. (2014). Coping with the curse of freshwater variability. *Science, 346,* 429–430.

Hardin, G. (2009). The tragedy of the commons. *Journal of Natural Resources Policy Research, 1,* 243–253.

Healy, R. W. (2010). Estimating ground water recharge. New York: Cambridge University Press.

Hoekstra, A. Y. E. (2003). Virtual water trade—Proceedings of the International Expert Meeting on Virtual Water Trade. *IHE Delft. Human Environment The Netherlands, 31*(5), 437–440.

Hoekstra, A. Y., & Hung, P. Q. (2005). Globalisation of water resources: International virtual water flows in relation to crop trade. *Global Environmental Change, 15,* 45–56.

IPCC (International Panel on Climate Change). (2011). Summary for Policymakers. In: Intergovernmental panel on climate change special report on managing the risks of extreme events and disasters to advance climate change adaptation. In C. B., V. Barros, T. F. Stocker, D. Qin, D. Dokken, K. L. Ebi, M. D. Mastrandrea, K. J. Mach, G. K. Plattner, S. K. Allen, M. Tignor, & P. M. Midgley (Eds,). Cambridge University Press, Cambridge, UK and New York, USA. Available at: http://ipcc-wg2.gov/SREX/.

Kahl, C. (2005). *States, scarcity and civil strife in the developing world.* Princeton: Princeton University Press.

Karn, S. K., Shikura, S. & Harada, H. 2003. Living environment and health of urban poor: A study in Mumbai. *Economic and Political Weekly,* 3575–3586.

Katambara, Z., & Ndiritu, J. (2009). A fuzzy inference system for modelling stream flow: Case of Letaba River, South Africa. *Physics and Chemistry of the Earth, 34,* 688–700.

Keller, A., Sakthivadivel, R., & Seckler, D. (2000). Water scarcity and the role of storage in development. Research Report 39. International Water Management Institute, Colombo, Sri Lanka.

Kiker, G. A. (2000). *South African country study on climate change: Synthesis report for the vulnerability and adaptation assessment section.* Pretoria: South African Department of Environmental Affairs and Tourism.

Kozo, N. (2008). Similarities and differences among the South Indian Ocean convergence zone, North American convergence zone, and other subtropical convergence zones simulated using an AGCM. *Journal of the Meteorological Society of Japan, 86,* 141–165.

Lawrence, P., Meigh, J. & Sullivan, C. (2002). Water poverty index: An international comparison. Keele Economics Research Papers. Available at http://citeseerx.ist.psu.edu/viewdoc/download?doi=10.1.1.13.2349&rep=rep1&type=pdf.

Lenton, R., & Muller, M. (Eds.) (2009). Integrated water resources management in practice: Better water management for development. Stockholm: GWP & London: Earthscan. LLM Environmental Regulation and Sustainable Development Newcastle Law School.

Malle, K. -G. (1996). Cleaning up the river Rhine. *Scientific American, 274,* 70–75.

Midgley, G. F., Chapman, R. A., Hewitson, B, Johnston, P., De Wit, M., Ziervogel, G., Mukheibir, P., Van Niekerk, L., Tadross, M., Van Wilgen, B. W., Kgope, B., Morant, P., Theron, A., Scholes, R. J. & Forsyth, G. G. (2005). A status quo, vulnerability and adaptation assessment of the physical and socio-economic effects of climate change in the Western Cape, Report to the Western Cape Government, Cape Town, South Africa. Report No. ENV-S-C 2005-073. Stellenbosch, CSIR.

Midgley, S. J. E., Davies, R. A. G., & Chesterman, S. (2011). *Risk and vulnerability mapping in southern Africa: Status quo (2008) and future (2050) in Southern Africa.* Synthesis Report: Regional Climate Change Programme.

Misra, A. K. (2014). Climate change and challenges of water and food security. *International Journal of Sustainable Built Environment, 3,* 153–165.

Mostert, E. (2008). Managing water resources infrastructure in the face of different values. *Physics and Chemistry of the Earth, 33,* 22–27.

Mukheibir, P. & Sparks, D. (2002). Water resource management and climate change in South Africa: Visions, driving factors and sustainable development indicators. Report for Phase I of the Sustainable Development and Climate Change project. Energy & Development Research Centre: University of Cape Town.

Muller, M. (2000). *How national water policy is helping to achieve South Africa'sdevelopment vision.* In C. L. Abernethy (Ed.), Inter-sectorial management of river basins (pp. 3–10). Colombo, Sri Lanka: International Water Management Institute.

Muller, M. (2007). Parish pump politics: The politics of water supply in South Africa. *Progress in Development Studies, 7*(1), 33–45.

Muller, M., Schreiner, B., Smith, L., Sally. H., Aliber, M., Cousins, B., et al. (2009a). Water security in South Africa. Development Planning Division. Working Paper Series No.12, DBSA: Midrand.

Muller, M., Schreiner, B., Smith, L., Van Koppen, B., Sally, H., Aliber, M., et al. (2009b). Water security in South Africa. *Development planning division.* Working Paper Series, (12).

NASAC (Network of African Science Academies). (2014). The grand challenge of water security in Africa: Recommendations to policy makers. Network of African Science Academies.

Ndaruzaniye, V. (2009). *Water for conflict prevention.* Belgium: Global Water Institute. Brussels.

Nelson, V. (2011). Gender, generation, social protection and climate change, a thematic overview.

Nussbaum, M. (2001). Upheavals of thought. The intelligence of emotions, Cambridge.

OSTROM, E. (1990). Governing the commons: The evolution of institutions for collective action. Cambridge University Press Cambridge.

Ostrom, E., Burger, J., Field, C. B., Norgaard, R. B. & Policansky, D. (1999). Revisiting the commons: Local lessons, global challenges. science, (Vol. 284, pp. 278–282). Overseas Development Institute (ODI).

Ostrom, E. E., Dietz, T. E., Dolšak, N. E., Stern, P. C., Stonich, S. E. & Weber, E. U. (2002). *The drama of the commons*. National Academy Press.

Pahl-Wostl, C. (2007). Transitions towards adaptive management of water facing climate and global change. *Water Resources Management, 21,* 49–62.

Pastor, A. V., Ludwig, F., Biemans, H., Hoff, H., & Kabat, P. (2014). Accounting for environmental flow requirements in global water assessments. *Hydrology and Earth System Sciences, 18,* 5041–5059.

Perret, S. (2002). Water policies and small holding irrigation schemes in South Africa. Department of Agricultural economics, University of Pretoria, South Africa.

Pierre, J., Peters, B.G. (2000). Governance, politics, and the state. Basingstoke: Palgrave.

Pietz, D. A. (2015). The yellow river: The problem of water in modern China, Harvard University Press.

Ramos, M. C., & Martinez-Casasnovas, J. A. (2006). Trends in precipitation concentration and extremes in the Mediterranean Penedes-Anoia region. *NE. Climate Change, 74*(4), 457–474.

Reason, C. J. C., & Jagadheesha, D. (2005). A model investigation of recent ENSO impacts over southern Africa. *Meteorology and Atmospheric Physics, 89,* 181–205.

Reason, C. J. C., & Mulenga, H. M. (1999). Relationships between South African rainfall anomalies in the south-west Indian Ocean. *International Journal of Climatology, 19,* 1651–1673.

Rietveld, L. C., Siri, J. G., Chakravarty, I., Arsénio, A. M., Biswas, R. & Chatterjee, A. (2016). Improving health in cities through systems approaches for urban water management. *Environmental Health, 15*(Suppl 1), 31, 151–160.

Sadoff, C. W., & Grey, D. (2005). Cooperation on international rivers: A continuum for securing and sharing benefits. *Water International, 30,* 420–427.

Schulze, R., & Perks, L. (2000). Assessment of the Impact of climate change on hydrology and water resources in South Africa. ACRUcons report 33. Pietermaritzburg, School of Bioresources Engineering and Environmental Hydrology, University of Natal, South Africa.

Sen, A. (1999). *Development as freedom,* Knopf, New York, USA.

Sheffield, J., Wood, E. F., Chaney, N., Guan, K., Sadri, S., Yuan, X., Olang, L., Amani, A., Ali, A., Demuth, S., & Ogallo, L. (2014). A drought monitoring and forecasting system for Sub-Saharan African water resources and food security. *Bulletin of the American Meteorological Society*, *95*:861–882.

Smakhtin, V., Revenga, C. & Döll, P. (2004). A pilot global assessment of environmental water requirements and scarcity. *Water International, 29*, 307–317.

Stocker, T., Qin, D., Plattner, G., Tignor, M., Allen, S., Boschung, J., Nauels, A., Xia, Y., Bex, V. & Midgley, P. (2013). Summary for policymakers. In: IPCC (Ed.), Climate Change 2013: The physical science basis, contribution of working group I to the fifth assessment report of the intergovernmental panel on climate change. Cambridge, New York, USA.

Swatuk, L. A. (2007). Southern Africa, environmental change and regional security: An assessment. Externe Expertese für dasWBGU-hauptgutachten 'Welt im Wandel: Sicherheistsrisika Klimawandel'. Berlin: WBGU.

The World Bank East Asia and Pacific Region. (2002). China Country Water Resources Assistance Strategy. World Water Development Report No. 2. UNESCO/WWAP.

Tramberend, S., Wiberg, D., Wada, Y., Flörke, M., Fischer, G., Satoh, Y., Yillia, P., Van Vliet, M., Hizsnyik, E. & Nava, L. 2015. Building global water use scenarios.

Tronto, J. (1993). Moral boundaries: A political argument for an ethic of care. Routledge, New York.

UN (UNITED NATIONS). (2010). The millennium development goals report 2010. New York: United Nations. University Press, Cambridge, UK.

UNDP (United Nations Development Programme). (2006). Beyond scarcity: Power, poverty and the global water crisis. Human Development Report, www.undp.org. Retrieved 20 February, 2017.

UNECA (United Nations Economic Commission for Africa (2016). The Demographic Profile of African Countries. https://www.uneca.org/sites/default/files/PublicationFiles/demographic_profile_rev_april_25.pdf. Retrieved 20 February, 2017.

UNESCO/IHP. (2006). Urban water conflicts: An analysis of the origins and nature of water-related unrest and conflicts in the urban context. Paris: International hydrological Programme (IHP).

UN-HABITAT. (2004). The challenge of slums: Global report on human settlements 2003. *Management of Environmental Quality: An International Journal, 15*, 337–338.

UNICEF/WHO. (2012). Progress on drinking water and sanitation update. Report 4. Accessed from http://www.unicef.org/media/files/JMPreport2012.pdf. Retrieved 20 February, 2017.

United Nations Development Programme. (2008). UNDP climate change country profiles: Mozambique.

United Nations Environmental Programme. (2008). Sudan post-conflict environmental assessment. http://postconflict.unep.ch/publications.php?prog=sudan. Retrieved 20 February, 2017.

UN-WATER. (2008). Transboundary waters: Sharing benefits, sharing responsibilities. UN-WATER. Thematic Paper.

UN-WATER. (2013). UN water annual report 2012. United Nations. UN-Water TFIMR, 2009. Monitoring progress in the water sector: A selected set of indicators (Final Report). World Water Assessment Programme (WWAP).

van Koppen, B., Chisaka, J., & Shaba, S. S. (2009). Lessons learnt from the IWRM demonstration projects: innovations in local-level integrated water resource development in Malawi, Mozambique, Swaziland and Zambia. Pretoria, South Africa: SADC/Danida Water Sector Support Programme; Pretoria, South Africa: International Water Management Institute (IWMI).

Vaz, A. C., & Pereira, A. L. (2000). The Inkomati and Limpopo international river basins: A view from downstream. *Water Policy, 2,* 99–112.

Vörösmarty, C. J., McIntyre, P. B., Gessner, M. O., Dudgeon, D., Prusevich, A., Green, P., et al. (2010). Global threats to human water security and river biodiversity. *Nature, 467,* 555–561.

Wada, Y., Flörke, M., Hanasaki, N., Eisner, S., Fischer, G., Tramberend, S. Satoh, Y., Van Vliet, M., Yillia, P. & Ringler, C. (2016). Modeling global water use for the 21st century: Water Futures and Solutions (WFaS) initiative and its approaches. *Geoscientific Model Development, 9*, 175–222.

WEF (World Economic Forum). (2017). Global risks report 2017. In: W. E. Forum (Ed.).

WHO. (2002). The world health report 2002: reducing risks, promoting healthy life, World Health Organization.

WHO (World Health Organization). (2015). Key facts from 2015 JMP Report. Available from: http://www.who.int/water_sanitation_health/publications/JMP-2015-keyfacts-en-rev.pdf?ua=1.

WHO/UNICEF (United Nations International Children's Emergency Fund/ World Health Organization). (2015). Joint monitoring programme for water supply and sanitation "2015 Report and MDG Assessment" Available from: http://files.unicef.org/publications/files/Progress_on_Sanitation_and_Drinking_Water_2015_Update_.pdf.

Wolf, A. T. (2007). Shared waters: Conflict and cooperation. *Annual Review of Environment and Resources, 32,* 241–269.

Wright, J., Gundry, S. & Conroy, R. (2004). Household drinking water in developing countries: A systematic review of microbiological contamination between source and point-of-use. *Tropical Medicine & International Health, 9,* 106–117.

WWAP (United Nations World Water Assessment Programme). (2009). The United Nations World Water Development Report 3: Water in a Changing World. Paris: UNESCO, and London: Earthscan.

Zaveri, E., Grogan, D. S., Fisher-Vanden, K., Frolking, S., Lammers, R. B., Wrenn, D. H., et al. (2016). Invisible water, visible impact: groundwater use and Indian agriculture under climate change. *Environmental Research Letters, 11,* 084005.

Zhu, X., Zhao, A., Li, Y., & Liu, X. (2015). Agricultural irrigation requirements under future climate scenarios in China. *Journal of Arid Land, 7,* 224–237.

# Chapter 7
# Energy Policy, Air Quality, and Climate Mitigation in South Africa: The Case for Integrated Assessment

**Carmen Klausbruckner, Lucas R. F. Henneman, Peter Rafaj and Harold J. Annegarn**

**Abstract** *Background and significance*: South Africa reports approximately 20,000 premature deaths due to air pollution annually. Policy in South Africa has typically addressed greenhouse gas emissions, energy supply, and air quality separately. Integrated assessment provides a framework in which policies related to these topics can be evaluated simultaneously. *Methodology*: The present study provides an overview of legal and policy documents and reviews available literature concerning existing energy, climate, and air quality policies in South Africa to highlight inconsistencies of different policy approaches and identify possible co-benefits. Previous applications of integrated assessment in South Africa are discussed as approaches to provide evidence-based decision support. *Application/relevance to systems analysis*: The analysis and results demonstrate that a complete analysis of the energy and industry sectors can identify inefficiencies and opportunities. The system was analysed through both a policy lens and a technical application of an integrated assessment model. *Policy and/or practice implications*: Multiple potential policy options have been identified for South Africa to meet future energy demand and reduce air pollution and greenhouse gas emissions. Combining GHG mitigation policies with subsidies to encourage the use of electricity or liquefied petroleum gas (LPG) for cooking is most promising to avoid trade-offs. *Discussion and conclusion*: The goal of this work is to provide an argument for assessing energy, air quality, and climate change policies in an

C. Klausbruckner (✉)
University of Natural Resources and Life Sciences, Vienna, Austria
e-mail: carmen.kb@gmail.com

L. R. F. Henneman (✉)
Department of Biostatistics, Harvard University T.H. Chan School of Public Health, Boston, MA, USA

P. Rafaj
Air Quality and Greenhouse Gases Program, International Institute for Applied Systems Analysis, Laxenburg, Austria

H. J. Annegarn
Department of Geography and Environmental Sciences, North-West University, Potchefstroom Campus, Potchefstroom, South Africa

 
P. Mensah et al. (eds.), *Systems Analysis Approach for Complex Global Challenges*,
https://doi.org/10.1007/978-3-319-71486-8_7

integrated assessment framework. Examples of current policy inconsistencies have been presented and published work detailing policy options to attain defined climate-related goals discussed. Integrated assessment can help to identify co-benefits and is a useful tool to improve decisions in complex policy environments. It is therefore recommended that integrated assessment tools be used to gain useful information for decision-making concerning climate change and air quality policies.

## 7.1 Introduction

South Africa is the largest greenhouse gas (GHG) emitter in Africa,[1] and the 13th largest in the world (U.S. Energy Information Administration 2013). Coal, which accounts for 90% of South Africa's electricity production, is abundant in the country, and the resulting low energy prices have driven economic success by attracting energy-intensive industries (Baker et al. 2014; Edkins et al. 2010a; Pretorius et al. 2015). The reliance on coal has led to high per capita emissions of GHGs and conventional air pollutants such as $SO_2$ and $NO_X$.

Historically (~1970–2000), South Africa maintained low energy prices and an electricity supply excess. In 2008, however, demand surpassed supply, leading to rolling blackouts and increasing electricity prices. Eskom, the state-owned utility that generates the majority of South Africa's electricity, has since commenced construction of two new large coal-fired generating plants along with other smaller renewable energy projects to increase generation capacity (Edkins et al. 2010b). Many of the older coal-fired power plants constructed during the late 1970s and 1980s will reach the end of their extended operating lifetime and be decommissioned between 2013 and 2040 (Worthington 2008).

South Africa has entered into multiple international agreements to reduce its GHG emissions. South Africa ratified the Kyoto Protocol in 2002 but as a developing country did not have reduction targets. In 2009, the state committed itself at the Copenhagen Climate Change Conference (Larson 2015) to reducing its GHG emissions compared to the business-as-usual trajectory (UNFCCC 2011); in 2016 it signed and ratified the Paris Agreement.

A recent update to the Global Urban Ambient Air Pollution Database of the World Health Organisation (WHO) shows severe (and worsening) air pollution in South African urban areas (Kings 2016). The South African Air Quality Information System, operated by the Department of Environmental Affairs (DEA), found average concentrations at 13 sites between 2011 and 2014 to be between 33 and 119 $\mu g\ m^{-3}$ (Vegter 2016). The World Bank and Institute for Health Metrics and Evaluation (IHME) (2016) estimate that 20,000 premature deaths per year in South Africa are attributable to air pollution. Electricity generation, industrial

[1]See http://data.worldbank.org/indicator/EN.ATM.CO2E.KT/countries (accessed 25.11.2017).

processes, domestic energy use, and vehicular exhaust emissions contribute to the majority of ambient air pollution emissions in South Africa (Scorgie et al. 2003). Fossil fuel burning is the primary anthropogenic air pollution source, and city centres, which contain high concentrations of industry and vehicles, typically experience the highest ambient concentrations. Solid fuel combustion for domestic heating and cooking contribute to elevated levels of ambient and indoor air pollution, specifically particulate matter (PM), in low-income and informal residential areas (Language et al. 2016). Barnes and co-workers (Barnes et al. 2009) found that indoor air pollution associated with acute lower respiratory infections causes up to 1400 child deaths per year in South Africa. In 2013, 28% of the South African population in rural areas were using solid fuels as a primary domestic energy source, compared with less than 5% in urban areas (total 12%) (WHO 2015).

All of these issues—energy production and consumption, air pollution emissions, and climate change mitigation—are connected to the idea of environmental justice in South Africa. The high cost of electricity relative to solid and liquid fuels, and economic circumstances drive low-income households to use inferior fuels and appliances, which emit air pollutants associated with adverse health impacts (Matinga et al. 2014; Shirindeet al. 2014; National Department of Health 2014). Other significant sources of air pollution include industrial processes, electricity generation, and vehicular exhaust, each of which exacerbates locally generated pollution, specifically in poorly located population centres such as low-income townships and informal settlements (Scorgie et al. 2003).

South African policy documents and legislative acts have historically focused on energy supply, air pollution, and GHG emissions separately. This approach is not always coherent, especially in the context of other pressing issues, such as poverty alleviation and employment policy. Because these three areas are all related, the most efficient approach is to adopt a new policy advised by evidence-based integrated assessment tools that consider consequences of policies on the entire system, instead of one area at a time.

This chapter lays out an argument for assessing South Africa's energy, air pollution, and climate change related policy within an *Integrated Assessment Framework* by evaluating legislation and policies, and reviewing existing literature. In the first section, the history and current state of knowledge relating to each of these issues is addressed. This is followed by a discussion of existing policies regarding each of these domains, including their intersections and inconsistencies. The third section discusses the *Greenhouse gas and Air pollution Interaction and Synergies* (GAINS) model, with an emphasis on two previous successful applications and the potential for use in South Africa as an evidence-based support tool. The final section summarises the findings, including a strong argument for taking an integrated approach to decision-making in South Africa regarding climate change mitigations and air quality.

## 7.2 The South African Environmental Regulatory Framework

The current section discusses important policy documents and regulations in South Africa, beginning with general documents that define broad policy goals. Next, we address climate change policy—both GHG reduction and impacts mitigation policies—and air pollution policy, including ambient and indoor air pollution. In the last section we present the trade-offs and inconsistencies between different policy measures and co-benefits leading to an argument for applying an integrated assessment to the South African policies.

### *7.2.1 Broad Policy Background*

South Africa's climate change mitigation, energy, and air quality policies are embedded in South African constitutional principles and overall development policies. The South African Constitution guarantees the '*right to an environment that is not harmful to human health or well-being*' [Section 24 (a)] and the '*right to have the environment protected*' [Section 24 (b)]. This right is limited but commits the government to take measures over time to give effect to the constitutional right.

Overall, South African policies are based on the National Development Plan (NDP). The NDP, adopted by the Cabinet in 2012, provides a long-term perspective on the development of South Africa with the time horizon 2030 (South Africa National Planning Commission 2011). The NDP's broad scope considers employment, poverty reduction, equality, growth, health, environment and infrastructure to set priorities for the political agenda. While employment and education are mentioned as the top priorities for South Africa, the extreme pressure on natural resources is considered to be a challenge, and the need for a sufficient supply of electricity is regarded as a prerequisite for economic growth. The NDP states in this context that GHG mitigation and health protection must be taken into account when making decisions in other policy areas. Concerning climate change mitigation, the NDP reinforces ideas from South African climate change policies (Climate Change Response White Paper). Measures such as the introduction of carbon taxes are examined criticallly, as they could lead to increases in electricity prices with consequent aggravation of household energy poverty (South African Government 2011).

Electrification as a developmental need is a priority, and the NDP states the goal of reaching 95% electrification of the South African population by 2030—that requires an additional 40,000 MW capacity to be built (South Africa National Planning Commission 2011). In 1993, before the end of apartheid, only 36% of South African households were connected to the grid, with access divided across racial lines. In the NDP, South Africa set out to increase renewable energy production to 20,000 MW by 2030 (South Africa National Planning Commission

2011). New energy infrastructure development must support the industrial, transport, mining, and agriculture sectors (the top four energy-use sectors).

The Medium-Term Strategic Framework (MTSF) 2014–2019 was published to implement the NDP (Republic of South Africa 2014). One of the top priorities mentioned in the MTSF is the electricity supply in total, with attention on expanding electrification to a greater percentage of households. The report advocates for a diverse energy mix incorporating coal, hydro, nuclear, shale gas, off-shore oil and gas and renewables. Climate change mitigation measures receive minor attention in the MTSF considering the prominent emphasis in the Reconstruction and Development Programme (RDP) and international commitments made by the President in this regard. However, the objectives for $CO_2$ reduction (34% in 2020 and 42% by 2025 from business as usual) from the NDP are repeated in the MTSF and should be achieved through market-based mechanisms.

### *7.2.2 Climate Change Policy*

South Africa is a major contributor of GHG emissions on a global level, but also vulnerable to climate change. It ranks 24th in the Climate Risk Index 2015 and 105th in the Climate Risk Index 1996–2015 (Kreft et al. 2016). South African water resources, food security, health, infrastructure, and biodiversity are under threat (Ziervogel et al. 2014). Adverse effects of climate change disproportionally affect the poor, especially women and children (South Africa National Planning Commission 2011). Here, we discuss two types of climate change-relevant policies —broader government-driven policies set goals for country-wide GHG emission reductions, and specific actions and regulations designed to decrease GHG emissions and meet those targets.

### *7.2.3 GHG Emissions Reduction Goals*

South African mitigation policies are based on international commitments. South Africa signed the Kyoto protocol, but as a developing country was not bound by any reduction targets. However, a national climate change response strategy for South Africa was published in 2004 (Department of Environmental Affairs and Tourism 2004) and in 2009 South Africa committed itself at the COP 15 to reduce its GHG emissions under the condition of an international agreement and support from the international community (Larson 2015). As a signatory country of the Paris Agreement,[2] South Africa has obligations and has already submitted its Intended Nationally Determined Contribution (South African Government 2016).

[2]South Africa signed the Paris Climate Agreements on 22 April 2016.

The content of the Intended Nationally Determined Contributions is based on the commitments made at COP 15 and in accordance with the 2011 National Climate Change Response White Paper (South African Government 2011). South Africa pledged to reduce its emissions in comparison to a Business as Usual (BAU) by 34% in 2020, and by 42% in 2025, which initially results in an increase in $CO_2$ emissions. These plans are based on the peak-plateau-decline trajectory (DEA 2011, 2014). This strategy proposes a peak production to occur sometime between 2020 and 2025 (within the range of 398 and 583 Mt $CO_2$ eq. by 2020 and 614 Mt $CO_2$ eq. by 2025), plateau for ten years and a decline after that to 212–428 Mt $CO_2$ eq. by 2050.

### 7.2.4 GHG Mitigation Measures

Climate change mitigation is based on the Long Term Mitigation Scenarios (LTMS) process - 2005 to 2008 (Scenario Building Team 2007) that examined possible long-term strategies to mitigate climate change and was the basis for the peak-plateau-decline trajectory.

GHG reduction measures planned in South Africa include energy efficiency, the use of advanced clean coal technologies (in combination with carbon capture and storage), an increase in renewables and nuclear. Power generation from waste incineration, the use of biofuels and hydropower (e.g. Grand Inga Dam Project) are other options mentioned as possible mitigation strategies. Measures in the transport sector are essential to curtailing GHG emissions. Legal measures include emission standards and reduction plans for transportation emitters. Other options are market-based instruments such as carbon taxes.

### 7.2.5 GHG Reduction Agenda in the Energy Sector

Recent developments in South African energy policy include attempts to reduce South African dependence on coal. The Integrated Resource Plan for Electricity 2010 (IRP) proposes reducing coal-derived power from 90 to 65% by 2030, through improved combustion technologies and phasing in of alternative energy sources. Nevertheless, two new coal-fired power plants are under construction (2016) to satisfy energy needs.

Increasing the nuclear power generating capacity (5% of the total capacity in 2014) is part of the strategy to decrease the carbon intensity of energy generation. Cabinet approval was already granted for the construction of a new nuclear fleet. However, implementation of the nuclear programme has repeatedly been delayed up to 2017, amid vigorous discussions about the affordability and necessity of nuclear energy infrastructure. The Minister of Energy's decision to develop 9.6 GW nuclear power triggered controversy and is currently even contested in court by

NGOs (Groundup 2016). Instead of the new eight reactors generating 9.6 GW by 2029, the earliest commission of nuclear plants is expected to be in 2037. The IRP proposes a total of 20.8 GW nuclear power to be added to the grid by 2050 according to the base case scenario (Cohen and Vecchiatto 2016).

In recent years, efforts were made to increase the share of renewables in South Africa. Independent power producers (IPPs) generated 2% of the total electricity in September 2015. The goal for renewable energy production is 20GW by 2030 (Energy Research Centre 2015). The mentioned proposed base case scenario foresees additional 37.4 GW of power from the wind, 17.6 GW of solar plants, 35.3 GW from gas and 15 GW from coal by 2050.[3] Measures to raise the share of renewable energies include the Renewable Independent Power Producer Procurement Programme (REI4P) that lead to 79 renewable energy IPP projects (an approved total of 5.2 GW) financed by private investors by 2015 and 6 GW under consideration.

Shale gas extraction from the Karoo basin located in central and southern regions of South Africa is under consideration as a future source of energy (Cohen and Winkler 2014; Esterhuyse et al. 2016). This measure is considered both a climate change mitigation measure (as gas would lead to lower carbon emissions compared to coal) and a step toward energy independence since natural gas consumption is dependent mainly on imports from Mozambique. It is unlikely that licensing decisions will be made before 2019. Methods for extraction carry risks for water resources-related impacts. Under the pressure of intense environmental protests, in 2016 the government appointed an independent scientific review of the environmental and economic risks and benefits of shale gas extraction. Any net reduction of GHG emissions will depend on whether the use of the shale gas displaces coal, and on the avoidance of methane leaks during fracking (Winkler et al. 2016).

### *7.2.6 GHG Reduction Agenda in the Other Sectors and Across Sectors*

Measures to reduce GHG emissions are spread across different sectors of the economy. The introduction of carbon taxes has been on the political agenda for several years. Implementation was planned for 2014 but was repeatedly postponed to 2015, 2016 and 2017, but then postponed again with no fixed date.[4] The proposed carbon tax is designed as a flat tax that will be implemented at R120/tonne of $CO_2$ eq, with a planned escalation of 10% per year. The regulations will include provisions for tax-free thresholds and offsets. Agriculture, forestry and land use will

---

[3]http://mg.co.za/article/2016-11-22-sa-delays-nuclear-plant-plan-as-economy-stagnates (accessed: 25.11.2017).

[4]South Africa released a draft Carbon tax bill in November 2015 indicating 1 January as a starting date.

be exempt in the first phase (Energy Research Centre 2015). In the transport sector, developing South African policies target public transport and carbon taxes on automobiles. Investments in public transport infrastructure, for instance, were R5 billion in 2012, with an anticipated growth of 5% annually. In the automobile sector, South Africa in 2010 introduced a one-time vehicle environmental levy at the point of manufacture or import amounting to R75 per g/km $CO_2$ emissions exceeding a threshold 120 g/km. In 2017, this levy had increased to R100.00 per g/km $CO_2$ for emissions exceeding 120 g/km.

Other measures include a move to a green economy. The 'green economies' initiatives and the SA Green Fund are concrete steps towards this goal. Measures include promoting green buildings, extended public and non-motorised transport infrastructure, clean energy and energy efficiency (e.g. off-grid options in rural and urban, solar heating and optimisation for large scale renewables), resource conservation and management, waste minimisation and recycling.

### *7.2.7 Air Pollution Policy*

Energy production and consumption, mining (particularly open pit mining) and traffic are the major contributors to ambient air pollution. Numerous studies have established links between air pollution and adverse health effects, from short-term respiratory problems (Lin et al. 2013; Pope 1989; Wichmann and Voyi 2006) and cardiovascular distress (Peel et al. 2007; Peel e al. 2010), to long-term effects such as birth outcomes (Parker e al. 2008; Rich et al. 2015), increased mortality rate (Dockery et al. 1993; Laden et al. 2006; Pope et al. 1992; Schwartz and Marcus 1990) and cancer (Laden et al. 2006; Raaschou-Nielsen et al. 2013). Particulate matter (PM) and ozone ($O_3$) are the pollutants of primary concern to the public, and the most difficult to reduce to remain below the air quality limit values.

Several policies and laws have been implemented to improve air quality in South Africa. South African air quality mitigation measures are based on the National Environmental Management: Air Quality Act (NEM: AQA) that establishes a framework for emissions and ambient standards, monitoring, modelling, emissions licensing, offences, and penalties. Air quality is also addressed in other legislative acts such as the National Health Act (Republic of South Africa 2004), National Road Traffic Act (Republic of South Africa 1996), and the National Energy Act (Republic of South Africa 2008) within the respective regulatory framework. Here, we discuss specific South African legislative measures and air pollution reduction programmes related to two major exposure domains: ambient and indoor.

### 7.2.8 Ambient Air Pollution

National ambient air pollution standards exist for sulphur dioxide, nitrogen dioxide, PM, $O_3$, benzene, lead, carbon monoxide and PM with aerodynamic diameter less than or equal to 2.5 μm ($PM_{2.5}$). Air Quality Management measures consist of Minimum Emission Standards (MES), Pollution Prevention Plans and the definition of priority areas for high pollution zones (currently three in the northeast of the country: Waterberg Bojanala Priority Area, Vaal Priority Area and Highveld Priority Area). GHG are not directly targeted when it comes to defining priority areas. However, the major coal-fired power plants are located in these three priority areas. The reduction of GHG can be among the objectives and is one of the aims of the Highveld Priority Area management plan.[5]

Minimum Emission Standards were introduced for PM, $SO_2$ and $NO_X$ (as $NO_2$) in 2010 and amended in 2013 as a measure to improve air quality. Standards for existing power plants had to be met by 1 April, 2015 and stricter standards for new power plants but also old power plants by 1 April, 2020. However, 37 companies applied to postpone the compliance deadlines and received permissions to delay by a maximum of five years after submitting compliance implementation plans (Department of Environmental Affairs 2015). An alternative option to the implementation of the Minimum Emission Standards, under active evaluation (2016–2017), is to allow offsets through reductions in near surface particulate matter pollution (specifically domestic coal emissions) in return for leniency in retrofitting existing older power plants with $SO_2$ and $NO_X$ gas scrubbers. This scheme will be managed through the Atmospheric Emission Licence (AEL) regulations, rather than altering the governing Air Quality Act, which contains no explicit provisions for emission offsets.

Using cleaner fuels in the transport sector is another measure to improve air quality. In 2006, the Clean Fuels I standard was introduced that is comparable with the Euro II standard for passenger vehicles. Lead in fuel was banned, and for diesel, the maximum sulphur level was lowered to 500 ppm. The Clean Fuel 2 standard (equivalent to Euro V) was planned to be implemented on the 1st of July 2017 but has subsequently been postponed to 2019. Under Clean Fuel 2, sulphur levels are reduced for petrol and diesel to $<10$ ppm.

### 7.2.9 Indoor Air Pollution

Indoor air pollution is considered a severe public health problem, given high concentrations and extended exposure (Bruce et al. 2000; National Department of Health 2014; Smith 2002). Household $PM_{2.5}$ has been found in Sub-Saharan Africa

---

[5]http://www.airqualitylekgotla.co.za/assets/hpa-aqmp-executive-summary—with-coverpage.pdf (accessed: 25.11.2017).

to contribute to higher GDP losses than ambient $PM_{2.5}$. In South Africa, indoor air pollution has been linked to severe health effects. Indoor use of liquid fuelled stoves and candles for illumination caused fires that affected 200,000 South Africans through deaths, burn injuries and property damage (Kimemia et al. 2014; The World Bank and IHME 2016).

Indoor air pollution is not addressed explicitly in the Air Quality Act. However, several projects to reduce the use of solid fuels in the domestic sector have been undertaken. These include the Integrated Household Clean Energy Strategy (ICHES), the National Liquefied Petroleum Gas Strategy, the *Basa Njengo Magogo* (a stove ignition method) Project (Le Roux et al. 2009), and the Low-Smoke Fuel Programme. Standards for kerosene stove designs exist on the national level.[6] The Department of Environmental Affairs is currently working on a new strategy to tackle air pollution in Dense Low Income Settlements (Department of Environmental Affairs 2016). A major shortcoming in previous policies—a missing coherent strategy within different state departments—is planned to be overcome by establishing a National Coordinating Committee on Residential Air Pollution (NCC) that ensures the prioritising of appropriate interventions (Department of Environmental Affairs 2016).

### *7.2.10 Trade-offs and Co-benefits—Need for an Integrated Assessment*

The preceding discussion shows the disjointed nature of energy, air pollution, and climate change policy papers in South Africa. Links do exist at certain points; however, an integrated strategy is missing.

Air pollution and climate change are targeted in legislative acts and policies. The South African constitution contains an environmental right that requires action to mitigate local air pollution. Climate change mitigation strategies, therefore, should consider their effects on local (indoor) air pollution. The Climate Change Response White Paper addresses the link between ambient air pollution and GHG emissions, but the document lacks measures and goals to avoid trade-offs. (South African Government 2011). The Intended Nationally Determined Contribution under the Paris agreement clearly states the priority of other policies (such as poverty alleviation). Partially, this can be traced to the fact that climate change and air pollution are not main issues in South Africa's overall development planning; poverty eradication, job creation and education are the top priorities (South Africa National Planning Commission 2011; South African Government 2016). Same as in previous South African documents, the INDC refers to South Africa's top priorities: equity, economic and social development and poverty eradication. It is clearly stated that

[6]Compulsory specification for coal-burning stoves and heaters for use in a dwelling (1982); Compulsory specification for non-pressure paraffin stoves and heaters (2006).

the *Intended Nationally Determined Contribution* commitments have to be regarded in the context of these national priorities. The National Climate Change and Health Adaption Plan 2014–2019 mentions exposure to air pollution and respiratory disease as a climate change related health risk (National Department of Health 2014). Higher ambient air pollution due to higher energy demands caused by air conditioning and the use of solid fuels in dwellings are mentioned as potential risks. However, the plan does not include concrete measures to decrease these risks.

GHG are now also targeted within the air pollution control legislation, but only to a minor extent and not in a comprehensive and systematic way (Thambiran and Diab 2010). GHG[7] were recently added as priority air pollutants making it necessary for 'significant emitters of greenhouse gasses' (emissions exceeding 0.1 Mt/a of $CO_2$ eq. that undertake activities listed in Annexure I[8]) to submit a *Pollution Prevention Plan*. Another air quality measure that includes GHG is *atmospheric emission licences* that are a prerequisite to operating for listed activities (including large solid and liquid fuel combustion installations). With the declaration of GHG as priority air pollutants, GHG play a more important role in air quality legislation.

A (controversial) policy related to climate change is the mandatory blending of a defined fraction of biofuels into the petroleum supply.[9] A trade-off between climate change mitigation and air quality measures might be caused by the introduction of carbon taxes and Minimum Emission Standards—measures to mitigate climate change and reduce ambient air pollution—as they are expected to lead to higher electricity prices and an increased use of solid fuels by low-income households. Possibilities to reduce adverse effects on socioeconomically disadvantaged families are discussed to accompany the introduction of a carbon tax (e.g. free basic electricity) that will most likely lead to higher electricity prices. Nevertheless, the planned inclusion of offsets could counteract negative consequences. A similar problem might occur with the introduction of Minimum Emission Standards for air pollutants. Another trade-off from GHG abatement might be higher water use due to water-intensive technologies such as shale gas production or carbon capture and storage (CCS). The feasibility of CCS is being researched as a potential strategy for reducing future GHG emissions from coal plants under work funded by the South African Energy Development Institute (SANEDI).

---

[7]Carbon dioxide, methane, nitrous oxide, hydroflurocarbons, perfluorcarbons and sulfur hexafluoride.

[8]Emissions that are the result of activities from the following sources are included: Fuel combustion, Fugitive emissions from fuels, industrial processes and other product use, agriculture, forestry and other land use and waste management. Industrial processes include: Coal mining, Production and/or refining of crude oil, Production and/or processing of natural gas, Production of liquid fuels from coal or gas, Cement production, Glass production, Ammonia production, Nitric acid production, Carbon black production, Iron and steel production, Ferro-alloys production, Aluminium production, Polymers production, Pulp and paper production, Electricity Production.

[9]Regulations regarding the mandatory blending of biofuels with petrol and diesel (2012): 5% v/v minimum concentration for biodiesel blending, permitted range for bio-ethanol blending: 2–10% (s 3 (5) (a + b)).

Measures against local/indoor air pollution due to the use of solid fuels are not linked with climate change policies and are not targeted in air quality legislation but in special programmes such as the Low-Smoke Fuel Programme or a new draft strategy for addressing air pollution in dense low-income communities. The plans to establish a National Coordinating Committee on Residential Air Pollution aim to improve the situation.

Figure 7.1, which highlights interrelations between air quality legislation and policies and climate change mitigation policies, shows that an overlap between climate change policies and air quality management exists, but local air pollution is more related with poverty alleviation and not directly addressed in climate mitigation policies.

Most trade-offs occur due to increased electricity prices, as they could lead to a rise in the use of solid fuels in the domestic sector creating indoor/local air pollution (OECD 2012). However, there is potential for co-benefits between air pollution and climate change mitigation policies. For example, the control of GHG emissions usually leads also to a reduction in local/regional air pollutants (Bollen et al. 2010; McCollum et al. 2013; Rao et al. 2013). Benefits of climate change mitigation policies can also create benefits in other sectors such as employment (Scenario Building Team 2007; Walwyn and Brent 2015).

If not designed properly, policies implemented to mitigate climate change might increase adverse health effects due to unanticipated outcomes (e.g. increased local air pollution), and these indirect consequences must, therefore, be considered when devising mitigation strategies. South Africa's Intended Nationally Determined

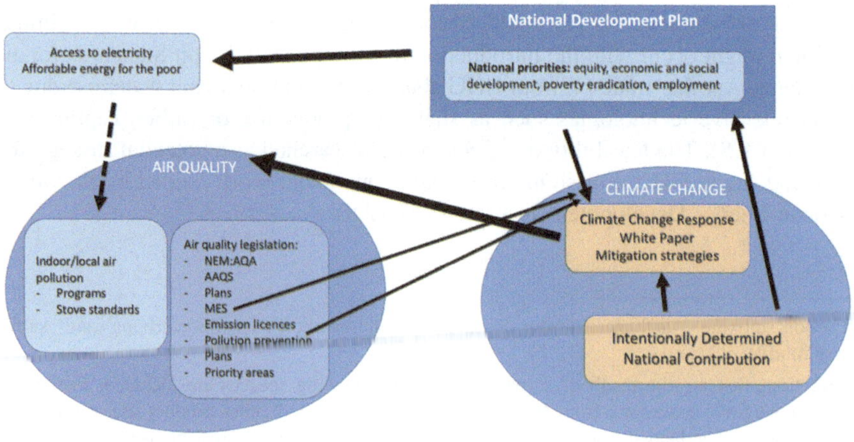


**Fig. 7.1** Interrelation between air quality legislation and climate change mitigation policies. Rectangles represent current laws or policies. Arrows of different sizes represent links to policies. The following abbreviations were used: NEM: AQA for National Environmental Management: Air Quality Act, AAQS for Ambient Air Quality Standards, and MES for Minimum Emission Standards [adopted from Klausbruckner et al. (2016)]

Contribution report does not directly mention trade-offs of climate change mitigation policies.

Lack of synergy in policy may be explained in part by the numerous institutional bodies that are involved, the influence of interest groups, the complexity of the topic, and the fact that an integrated approach to address GHG emissions and local air pollutants is missing. An evidence-based approach that takes into account GHG emissions and ambient air pollutants to tackle these challenges is needed. Appropriate consideration must be given to economic factors (affordability, cost optimisation), social factors (poverty alleviations, public health benefits) and political acceptability. An integration of GHG and air pollution policies is required along with set of criteria identified as requirements for an evidence-based decision support tool to resolve these contradictions in line with other national priorities (poverty alleviation, job creation, public health). A recent study indicated that the South African public is not supportive of climate change mitigation measures (Obradovich and Zimmerman 2016). Evidence-based approaches will become even more important to have sufficient backing from the public.

The GAINS model provides an evidence-based approach to investigate policy alternatives. It has already been used successfully as a policy support tool in Europe and Asia (Amann et al. 2011; Rafaj et al. 2011) and aims to support informed decision making that maximises synergies between different measures (co-benefits). It links GHG emissions and air pollution and can optimise on least cost, lowest GHG emissions, or health impacts (lowest DALYs[10]). Therefore, the implementation of the GAINS model would assist South Africa in the development of GHG and air quality policies and would be in line with the overall national development goals.

## 7.3 GAINS as an Integrated Assessment Tool

GAINS is a techno-economic optimising model that has been used to model air pollution and GHG policies throughout the world at different scales. Implementations vary from global (Rafaj et al. 2013), to continental (Europe) (Amann et al. 2011; Wagner et al. 2013; Winiwarter 2005), and to country (China, Pakistan and South Africa) (Amann et al. 2008; Henneman et al. 2015; Purohit et al. 2013). The model has been applied to evaluate costs and mutual benefits of climate mitigation policy impacts on air pollutant and GHG emissions, public and environmental health, and control costs (Rafaj et al. 2013). The GAINS model is accessible on an open-access basis through a web interface (http://gains.iiasa.ac.at/models/) (IIASA 2012).

[10]DALY—Disability Adjusted Life Years; see further Murray, 1994; Gao et al. 2015.

### 7.3.1 How Does GAINS Work?

The core of GAINS emissions model is a single equation (Winiwarter 2005):

$$E_p = \sum_{j,a,t} A_{j,a} ef_{j,a,p} \left(1 - eff_{t,p}\right) X_{j,a,t} \tag{7.1}$$

where $E_p$ is the emissions of pollutant $p$, and $j$, $a$, $t$, $p$ denote sector, activity, abatement technology, and pollutant, respectively. $ef_{j,a,p}$ (the uncontrolled emissions factor of pollutant $p$ for sector $j$ with activity $a$) and $eff_{t,p}$ (the reduction efficiency of abatement technology $t$ on pollutant $p$) are specific to the technologies being used. Of the variables in Eq. 7.1, $A_{j,a}$ (the activity level in sector $j$ with activity $a$) and $X_{j,a,t}$ (the implementation rate of technology $t$ sector $j$ and activity $a$) are the primary nodes for developing emissions control scenarios. GAINS can estimate both GHG (including $CO_2$, $CH_4$, $N_2O$, and fluorinated gases) and local air pollutant (including $SO_2$, $NO_x$, volatile organic compounds–VOCs, $NH_3$, and PM) emissions.

A base case activity level ($A_{j,a}$) is typically derived outside of the model and projected using an energy systems planning model, such as the *Model for Energy Supply Strategy Alternatives and their General Environmental Impact for the Energy System* (MESSAGE) (Amann et al. 2011; Messner and Strubegger 1995).

GAINS estimates control costs of emission abatement (for end-of-pipe measures) and considers an interest rate defined by the user. In the GAINS cost optimisation module, these costs are estimated per unit of activity and avoided emissions, and the model estimates cost curves ranking options for pollution control measures available to each country (Amann et al. 2011; Wagner et al. 2013).

Previous applications have linked GAINS outputs with atmospheric concentration estimation tools, such as the EMEP Eulerian model (Amann et al. 2011) and the Fast Scenario Screening Tool for Global Air Quality and Instantaneous Radiative Forcing, paired with TM-5 (TM5-FASST) (Chafe et al. 2015; Zhang et al. 2016). Both these approaches use reduced form alternatives to the full models to produce results for long-term runs quickly.

### 7.3.2 GAINS Applications in South Africa

Two studies have completed integrated assessments of future air pollution and climate policies in South Africa using GAINS. The first study projected emissions levels and control costs for multiple control and activity scenarios until 2050. The second study extended the analysis to exposure and health impacts.

### 7.3.3 Assessing Emissions Levels and Costs Associated with Climate and Air Pollution Policies in South Africa

In the first GAINS application in South Africa, Henneman and colleagues projected impacts of multiple energy production mixes and control policies on air pollution and $CO_2$ emissions until 2050 (Henneman et al. 2016). The researchers applied the base GAINS model using energy projections developed outside of the model but considering recent developments and policies within the energy sector. Energy industries are represented explicitly based on process and fuel types. These include specifically the energy consumption in South Africa's large coal-to-liquid (CTL) and gas-to-liquid (GTL) industries. Emissions factors are based on the information about adoption of emission control measures as well as on the information about fuel quality (for instance, South African coal has high ash content). Readers interested in more details of the GAINS model application in South Africa are referred to Henneman et al. (2016).

The authors devised several scenarios, which are defined by energy production mixes and controls on different sectors to estimate the effects of different policies. The baseline (termed *Business As Usual*–BAU) scenario was implemented using projections from the International Energy Agency's Energy Technology Perspectives (2012) report (International Energy Agency 2012).

Henneman et al. (2016) applied the GAINS model to estimate how selected policies (for instance, adopting aggressive GHG emissions mitigation strategies) affect emissions across various sectors. The authors used GAINS to identify policy contradictions that may arise because of energy policies that target only one part of the energy system (e.g. electricity generation, transport, industrial production and households).

Input parameters in GAINS were updated to ensure they matched the current policy situation in South Africa using data available to the public. For instance, coal power plants must begin following tighter $SO_2$ and $NO_x$ emission standards in 2020. Adjustments needed to be made for fuel inputs used in South Africa, e.g., South African coal is low in both sulphur ($\sim 1.5\%$) and calorific value ($\sim 20.4$ GJ/ tonne), but high in ash content (up to 40% used in power plant feedstock). References for this portion of the work included industry documents (e.g. utility reports documenting the quality of fuels used in their production), South African legislation (e.g. the 2004 NEM: AQA), peer-reviewed literature, and industry analyses.

Six policy scenarios were developed and implemented in the GAINS model besides the Business As Usual (BAU) scenario (Table 7.1). Two control scenarios were defined, i.e., scenarios that assume current projections for energy production, but assign control technologies to various emissions sectors: (i) installing *Maximum Feasible Controls* on all emitters by 2030 (MFC), and (ii) maintaining controls at the current (2015) levels (*No Further Controls*–NFC). Fuel input scenarios (termed activities scenarios) were based on scenarios implemented in prior modelling efforts of the energy policies in South Africa. These include: (iii) clean fuels in the *DOMestic* sector (DOM), (iv) installing *Clean Coal Technologies* for new generating capacity (CCT), (v) implementing changes to achieve the *2 Degrees Scenario* laid out in the International Energy Agency (IEA) Energy Technology Perspectives report (2DS) (International Energy Agency 2012), (vi) using *RENewables instead of coal* in all new generating capacity (REN). The group was chosen to give a representative group of scenarios that affect a range of sectors, including electricity production, industry, and domestic use. They were developed to reflect policies—both in the current regulatory discussion and undergoing feasibility studies by government agencies to show the range of potential options—targeting local/regional air pollutants (e.g. $SO_2$, $PM_{2.5}$, and $NO_X$) and GHG emissions.

Model outputs (in this case, emissions of the above pollutants and costs of air pollution controls) were first compared with countrywide emissions estimates from two sources in the literature (European Commission 2011; Witi et al. 2013). Results from the GAINS model were not outside of the reasonable range of previous estimates. Next, estimated emissions and costs were compared for the seven scenarios. The analysis focused on the ability of different policies to achieve a range of benefits, including reducing air pollution emissions, mitigation of GHG emissions, and lower cost of air pollution controls.

The results showed co-benefits and inherent contradictions in different policies (Fig. 7.2). For instance, the aggressive 2DS scenario reduces $CO_2$ by greater than 60% by 2050 and would yield substantial decreases in $SO_2$, $PM_{2.5}$, and $NO_x$. The CCT and REN scenarios, which target coal use for power production, shows

**Table 7.1** Control and activity scenarios employed in GAINS modelling (Henneman et al. 2015)

| Scenario | Acronym | Notes |
|---|---|---|
| *Baseline scenario* | | |
| Business As usual | BAU | Based on IEA 6 °C warming scenario and current legislation |
| *Control scenarios* | | |
| No further controls[a] | NFC | Freeze current control levels after 2015 |
| Maximum feasible controls[a] | MFC | Invest in best available technologies on all sources by 2030 |
| *Activity scenarios* | | |
| Clean fuels in DOMestic sector[a] | DOM | Replace solid fuels with LPG |
| Clean coal technologies[a] | CCT | Construct IGCC power plants |
| Coordinated global mitigation strategy (2 °C)[b] | 2DS | Based on IEA 2 °C warming scenario (International Energy Agency 2012) |
| RENewables only in electricity generation[b] | REN | No coal for electricity generation by 2050 |
| RENewables trade off[b] | RENT | Increasing domestic solid fuels with higher electricity prices |

[a]Air pollution reduction scenario
[b]Climate change mitigation scenario

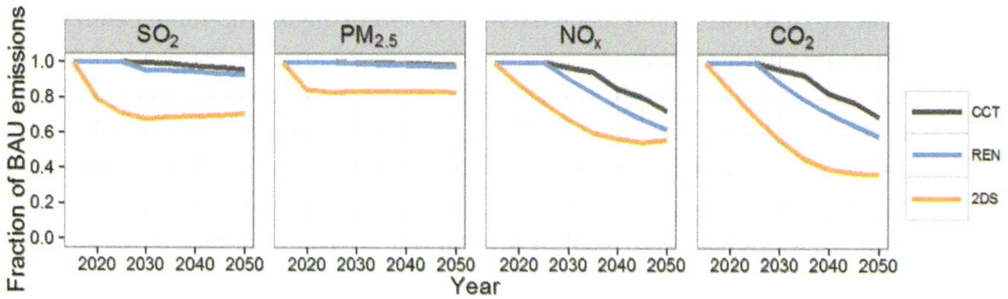


**Fig. 7.2** Estimated reductions in air pollutant emissions compared to the BAU scenario for three activity scenarios

substantial reductions in $NO_x$ and $CO_2$ emissions, but slight reductions in $SO_2$ and $PM_{2.5}$, as they are already controlled under current legislation. On the other hand, the 2DS scenario that assumes decarbonisation in all sectors achieves the largest reduction in each of the pollutants.

BAU emissions of $PM_{2.5}$ and $NO_x$ are close to the NFC case, while $SO_2$ emissions are nearly reduced as much as in the MFC case. This is because of the sources of the pollutants and the current legislation in place: $SO_2$ is emitted from the electricity generation sector, which will be required to implement strict controls, while $PM_{2.5}$ and $NO_x$ are emitted mainly from the domestic and transport sectors, which lag the electricity sector in required controls.

Altering the fuel consumption patterns in critical areas can have a substantial effect on emissions. For instance, replacing solid fuels in the domestic sector with cleaner burning liquid petroleum gas (LPG) reduces total $PM_{2.5}$ emissions by 40% compared to the BAU scenario. This reduction is already 50% of the reduction achieved by the MFC scenario in which industrial emissions are controlled at maximum feasible levels.

A contradiction would arise if large amounts of resources were put into replacing coal with renewables and nuclear in electricity generation and this action led to an abandonment of policies in the domestic sector aimed at curtailing solid fuels burning. This policy imbalance could cause a shift back to solid fuels triggered by price increases in electricity, which would increase local air pollution in this sector. Recent research, however, shows that renewables such as wind and solar are beginning to surpass the cost efficiency of even the newest coal plants in South Africa.[11] Emissions in the domestic sector are associated with the highest exposure and adverse health effects.

Costs of controls in the 2DS scenario are 50 and 35% of the expenses in the BAU scenario in 2030 and 2050, respectively. These costs are even less than the expenditure associated with the NFC scenario. These benefits are achievable

[11]http://reneweconomy.com.au/wind-solar-almost-half-the-cost-of-new-coal-generators-in-south-africa-75194/ (accessed: 25.11.2017).

because fuel mix changes towards cleaner fuels will decrease the reliance on costly controls to meet emissions targets.

The key findings of this project are the range of future emission levels and potential co-benefits that can be achieved with selected policies. For instance, the climate change mitigation strategies yield benefits across all air pollutants estimated and costs of air pollution controls. GAINS has the potential to improve the development of evidence-based policies in South Africa, and this work is a significant step toward a complete investigation of policy options available to the country. It shows how even a high-level analysis of countrywide emissions and costs can quantify co-benefits and inconsistencies inherent in the examined policies. Future implementation that includes quantification of environmental and health impacts, combined with scenario optimisation, can help refine the policies that exploit co-benefits and avoid policies with inherent contradictions.

### 7.3.4 Special Report on Energy and Air Pollution by the International Energy Agency 2016

Application of the capabilities of the GAINS model for South Africa was recently expanded to estimate population exposure and health impacts related to both ambient and household air pollution (International Energy Agency 2016). The authors then applied the global burden of disease methodology from Lim et al. (2013) to estimate premature mortality (i.e., numbers of premature deaths) and loss of life expectancy (estimated for the total population) attributable to air pollution exposure. This work was published as part of the International Energy Agency 2016 Energy and Air Pollution report, in which they laid out scenarios for investment in energy technologies to improve air quality and public health.

Exposure to ambient air quality was estimated by inputting GAINS emissions outputs into emissions-concentration sensitivities (transfer coefficients) developed using the EMEP MSC-W chemical transport model (Simpson et al. 2012). The sensitivity coefficients estimate the impacts of primary PM, $SO_2$, $NO_X$, ammonia ($NH_{3-}$), and volatile organic compounds on ambient $PM_{2.5}$ concentrations (Kiesewetter et al. 2015). Grid-based modelling was performed separately for urban and background portions of each grid cell. $PM_{2.5}$ concentrations from this method were used to estimate ambient pollution exposure.

For household air pollution exposure, the authors assumed typical exposures of 300 μg $m^{-3}$ for families that used traditional stoves and 70 μg $m^{-3}$ for clean stoves, based on results in Balakrishnan et al. (2013) and Larson (2015). IEA projections for numbers of people using each technology were used to calibrate the model.

The assessment was performed for ischemic heart disease, chronic obstructive pulmonary disease, stroke, lung cancer, and acute lower respiratory infections using the integrated exposure response (IER) functions described by Burnett et al. (2014). Using separate analyses for children under 10 than for persons over 30 years old, the authors adjusted the baseline mortality rates using IER functions applied to ambient air pollution, household air pollution, and smoking. The approach assumed that the ageing population becomes more vulnerable to air pollution-related diseases over time.

IEA projected air quality under two scenarios and compared air quality and health results to findings in 2015. The *New Policies Scenario* (NPS) assumes all policies that are currently in effect or planned will be implemented to their full capacities. These policies include those that require emissions controls and impact energy efficiency and technologies, such as those supporting renewable energy and those putting a price on carbon emissions. The *Clean Air scenario* (CAS) prescribes policies that seek to reduce air quality-degrading emissions aggressively. In both scenarios, policies are diagnosed by country and sector.

The findings show that under the Clean Air Scenario, the fraction of the South African population exposed to the WHO interim target 1 (25–35 $\mu g\ m^{-3}$) and above falls to zero, while under the policy-based scenario (which considers only implemented and planned policies), the fraction of population exposed to $PM_{2.5}$ above WHO interim target 1 is above 20%. Under the CAS, the 2040 premature deaths attributable to outdoor air pollution is 4000 (range: 2600–5600) compared to 16,900 (9000–23,700) under the scenario without additional measures. The authors note that such projections carry considerable uncertainty.

Of note, the number of premature deaths does not scale linearly with the fraction of utility sector emissions reductions but rather depends on the evolution of the energy demand, fuel mix choice and technological improvements in the domestic sector. Also, the linkage between emission cuts and health benefits is influenced by the following factors. First, the ageing population is expected to become more susceptible to adverse health effects. Second, large swaths of the population are moving to cities, and populations are increasing rapidly in developing countries, both trends of which will lead to increased total exposure. Third, the health models employ a nonlinear response curve, with declining impacts at higher concentrations.

The main conclusions of this work relating to South Africa are that using proven policy tools and technological measures, it is possible to reduce air pollution-related mortality significantly by 2040. This, however, will require new standards and investments in cleaner technologies.

### *7.3.5 Alternative Integrative Assessment Models*

GAINS is a useful tool, but it is not the only integrated assessment model available, and other approaches may be more appropriate depending on the specific questions or policies that need interrogation. For instance, previous integrated assessments in South Africa include the "Study to Examine the Potential Socioeconomic Impact of Measures to Reduce Air Pollution from Combustion" (Scorgie et al. 2003) and the

Long Term Mitigation Strategies (LTMS) report (Scenario Building Team 2007). Both studies estimated emissions and costs using independent frameworks to answer similar questions as the later GAINS studies. Slight differences in approaches, however, make it difficult to compare results directly. Since GAINS provides the ability to use the inputs from various models available to the public in the online framework, future efforts may decide to apply it using detailed data as they become available.

Other integrated assessment models that seek to investigate energy, air quality, and climate policy include the *Integrated Model to Assess the Global Environment* (IMAGE—Braspenning Radu et al. 2016) and the Massachusetts Institute of Technology Integrated Global System Model (MIT IGSM-CAM—Garcia-Menendez et al. 2015).

## 7.4 Conclusion

This chapter reviewed inconsistencies between various laws, reports, and regulations related to South African energy production and use, air quality, and climate change mitigation. The policy analysis showed that there is a need for a coordinated approach to managing these three critical areas. Integrated assessment models such as the GAINS model are used to analyse complex problems and also take into account social and economic factors. The GAINS model has been applied previously in many countries—including South Africa—to investigate problems of this nature, and further application has the potential to elucidate areas that could benefit from an integrated management approach.

The discussion has shown how a systems analysis approach can identify inefficiencies in environmental management and test various policy solutions. GAINS was able to estimate emissions for multiple scenarios on controls in different emissions sectors. For instance, replacing solid fuels with liquefied petroleum gas (LPG) in the domestic sector was shown to have the same effect on $PM_{2.5}$ emissions as costly controls installed in the industrial and power sectors. The 2DS scenario that aimed at contributing to the global 2 °C warming scenario (International Energy Agency 2012) substantially reduce air pollutants by targeting all sectors. It was shown that combining GHG mitigation policies with subsidies to encourage the use of electricity or LPG for cooking is most promising to avoid trade offs. While scientific analysis cannot determine the best policy by itself (politicians are charged with this duty), systems analyses such as this can offer a range of positive and negative impacts of various policies, allowing for direct comparisons between policy approaches.

## References

Amann, M., Bertok, I., Borken-Kleefeld, J., Cofala, J., Heyes, C., Höglund-Isaksson, L., et al. (2011). Cost-effective control of air quality and greenhouse gases in Europe: Modeling and policy applications. *Environmental Modelling & Software, 26*(12), 1489–1501. https://doi.org/10.1016/j.envsoft.2011.07.012.

Amann, M., Kejun, J., Jiming, H., Wang, S., Wei, W., Jia, X., et al. (2008). *Scenarios for cost-effective control of air pollution and greenhouse gasses in China*. Retrieved from http://gains.iiasa.ac.at/gains/download/GAINS-Asia-China-v8_MA.pdf.

Baker, L., Newell, P., & Phillips, J. (2014). The Political Economy of energy transitions: The case of South Africa. *New Political Economy, 19*(6), 791–818. http://www.tandfonline.com/doi/abs/10.1080/13563467.2013.849674.

Balakrishnan, K., Ghosh, S., Ganguli, B., Sambandam, S., Bruce, N., Barnes, D. F., et al. (2013). State and national household concentrations of PM2.5 from solid cookfuel use: Results from measurements and modeling in India for estimation of the global burden of disease. *Environmental Health: A Global Access Science Source, 12*(1), 77. https://doi.org/10.1186/1476-069X-12-77.

Barnes, B., Mathee, A., & Thomas, E. (2009). Household energy, indoor air pollution and child respiratory health in South Africa. *Journal of Energy in South Africa, 20*(1).

Bollen, J., Hers, S., & van der Zwaan, B. (2010). An integrated assessment of climate change, air pollution, and energy security policy. *Energy Policy, 38*(8), 4021–4030. https://doi.org/10.1016/j.enpol.2010.03.026.

Braspenning Radu, O., van den Berg, M., Klimont, Z., Deetman, S., Janssens-Maenhout, G., Muntean, M., et al. (2016). Exploring synergies between climate and air quality policies using long-term global and regional emission scenarios. *Atmospheric Environment, 140*, 577–591. https://doi.org/10.1016/j.atmosenv.2016.05.021.

Bruce, N., Perez-Padilla, R., & Albalak, R. (2000). Indoor air pollution in developing countries: a major environmental and public health challenge for the new millennium. *Bulletin of the World Health Organization, 78*, 1078–1092. https://www.ncbi.nlm.nih.gov/pmc/articles/PMC2560841/pdf/11019457.pdf.

Burnett, R. T., Arden Pope, C., Ezzati, M., Olives, C., Lim, S. S., Mehta, S., et al. (2014). An integrated risk function for estimating the global burden of disease attributable to ambient fine particulate matter exposure. *Environmental Health Perspectives, 122*(4), 397–403. https://doi.org/10.1289/ehp.1307049.

Chafe, Z. A., Brauer, M., Klimont, Z., Van Dingenen, R., Mehta, S., Rao, S., et al. (2015). Household cooking with solid fuels contributes to ambient PM2.5 air pollution and the burden of disease. *Environmental Health Perspectives, 122*(12), 1314–1320. https://doi.org/10.1289/ehp.1206340.

Cohen, M., & Vecchiatto, P. (2016, November). South Africa slows nuclear plans as rating assessments loom. *Bloomberg*. Retrieved from http://mg.co.za/article/2016-11-22-sa-delays-nuclear-plant-plan-as-economy-stagnates).

Cohen, B., & Winkler, H. (2014). Greenhouse gas emissions from shale gas and coal for electricity generation in South Africa. *South African Journal of Science, 110*(3/4). Retrieved from https://www.sajs.co.za/article/view/3731.

DEA. (2011). *Defining South Africa's desired greenhouse gas mitigation outcomes—research, concerns, issues and proposals*. Pretoria, SA. Retrieved from http://pmg-assets.s3-website-eu-west-1.amazonaws.com/docs/110329defining-edit.pdf.

DEA. (2014). *South Africa's Greenhouse Gas (GHG) Mitigation Potential Analysis*. South Africa: Pretoria.

Department of Environmental Affairs. (2015). Media Statement for Minister's announcement of decisions for applications for postponement of compliance time-frames for minimum Air Quality Emission Standard. Retrieved from https://www.environment.gov.za/mediarelease/molewa_airqualityemissionstandards.

Department of Environmental Affairs. (2016). *Draft strategy to address air pollution in dense low-income settlements*. Retrieved from https://www.environment.gov.za/sites/default/files/gazetted_notices/airpollution_strategy_g40088_gen356.pdf.

Department of Environmental Affairs and Tourism. (2004). *South African national climate change response strategy*. Pretoria, South Africa. Retrieved from https://www.environment.gov.za/sites/default/files/docs/climate_change_governance.pdf.

Dockery, D. W., Pope, C. A., Xu, X., Spengler, J. D., Ware, J. H., Fay, M. E., et al. (1993). An association between air pollution and mortality in six U.S. cities. *The New England Journal of Medicine, 329*(24), 1753–1759.

Edkins, M., Marquard, A., & Winkler, H. (2010a). Assessing the effectiveness of national solar and wind energy policies in South Africa, (June). Retrieved from https://open.uct.ac.za/bitstream/item/19472/Edkins_Assessing_effectiveness_2010.pdf?sequence=1.

Edkins, M., Marquard, A., & Winkler, H. (2010b). *South Africa's renewable energy policy roadmaps*. Retrieved from https://open.uct.ac.za/bitstream/item/19473/Edkins_South_Africa_039_s_renewable_2010.pdf?sequence=1.

Energy Research Centre. (2015). *Technical background information to support the development of the mitigation component of South Africa's intended nationally determined contribution, including supported required for mitigation*. Cape Town, South Africa. Retrieved from http://www.erc.uct.ac.za/sites/default/files/image_tool/images/119/Papers-2015/15-ERC-Technical_background_INDC_0.pdf.

Esterhuyse, S., Redelinghuys, N., & Kemp, M. (2016). Unconventional oil and gas extraction in South Africa: Water linkages within the population—environment—development nexus and its policy implications Unconventional oil and gas extraction in South Africa: Water. *Water International, 41*(3), 409–425. https://doi.org/10.1080/02508060.2016.1129725.

European Commission. (2011). Emission Database for Global Atmospheric Research. https://doi.org/10.2904/EDGARv4.2.

Gao, T., Wang, X. C., Chen, R., Ngo, H. H., Guo, W. (2015). Disability adjusted life year (DALY): A useful tool for quantitative assessment of environmental pollution. *Science of the Total Environment, 511*, 268–287.

Garcia-Menendez, F., Saari, R. K., Monier, E., & Selin, N. E. (2015). U.S. Air Quality and Health Benefits from Avoided Climate Change under Greenhouse Gas Mitigation. *Environmental Science and Technology, 49*(13), 7580–7588. https://doi.org/10.1021/acs.est.5b01324.

Groundup. (2016, November). South Africa: Understanding the Court Challenge to the Nuclear Deal. *AllAfrica.com*. Retrieved from http://allafrica.com/stories/201611250810.html.

Henneman, L. R. F., Liu, C., Mulholland, J. A., & Russell, A. G. (2016). Evaluating the effectiveness of air quality regulations: A review of accountability studies and frameworks. *Journal of the Air and Waste Management Association, 67*(2), 144–172. https://doi.org/10.1080/10962247.2016.1242518.

Henneman, L. R. F., Rafaj, P., Annegarn, H. J., & Klausbruckner, C. (2015). Assesing emission levels and costs associated with climate and air pollution policies in South Africa. *Energy Policy, 89*, 160–170. https://doi.org/10.1016/j.enpol.2015.11.026.

IIASA. (2012). GAINS Global. Laxenburg, Austria. Retrieved from http://gains.iiasa.ac.at/models/index.html.

International Energy Agency. (2012). *Energy Technology Perspectives 2012*.

International Energy Agency. (2016). Energy and Air Pollution. *World Energy Outlook—Special Report*, 266. Retrieved from https://www.iea.org/publications/freepublications/publication/WorldEnergyOutlookSpecialReport2016EnergyandAirPollution.pdf.

Kiesewetter, G., Borken-Kleefeld, J., Schöpp, W., Heyes, C., Thunis, P., Bessagnet, B., et al. (2015). Modelling street level PM10 concentrations across Europe: Source apportionment and possible futures. *Atmospheric Chemistry and Physics, 15*(3), 1539–1553. https://doi.org/10.5194/acp-15-1539-2015.

Kimemia, D., Vermaak, C., Pachauri, S., & Rhodes, B. (2014). Burns, scalds and poisonings from household energy use in South Africa: Are the energy poor at greater risk? *Energy for Sustainable Development, 18,* 1–8. https://doi.org/10.1016/j.esd.2013.11.011.

Kings, S. (2016, September). Air pollution kills 20 000 per year in South Africa—as many as in traffic. *Mail & Guardian*. Retrieved from http://mg.co.za/article/2016-09-12-00-air-pollution-kills-20-000-per-year-in-south-africa-as-many-as-in-traffic.

Klausbruckner, C., Annegarn, H. J., Henneman, L. R. F., & Rafaj, P. (2016). A policy review of synergies and trade-offs in South African climate change mitigation and air pollution control strategies. *Environmental Science & Policy, 57,* 70–78. https://doi.org/10.1016/j.envsci.2015.12.001.

Kreft, S., Eckstein, D., & Melchior, I. (2016). *Global Climate Risk Index 2017*. Bonn. Retrieved from https://germanwatch.org/de/download/16411.pdf.

Laden, F., Schwartz, J., Speizer, F. E., & Dockery, D. W. (2006). Reduction in fine particulate air pollution and mortality: Extended follow-up of the Harvard Six Cities study. *American Journal of Respiratory and Critical Care Medicine, 173*(6), 667–672. https://doi.org/10.1164/rccm.200503-443OC.

Language, B., Piketh, S. J., Wernecke, B., & Burger, R. (2016). Household air pollution in South African low-income settlements: a case study. In *WIT Transactions on Ecology and The Environment,* (pp. 227–236). WIT Press. https://doi.org/10.2495/AIR160211.

Larson, B. (2015). *Benefits and Costs of the Air Pollution Targets for the Post-2015 Development Agenda*. In *Post-2015 Consensus* (p. Conference Procedings). Copenhagen.

Le Roux, L. J., Zunckel, M., & McCormick, S. (2009). Reduction in air pollution using the "basa njengo magogo" method and the applicability to low-smoke fuels. *Journal of Energy in Southern Africa, 20*(3), 3–10.

Lim, S. S., Vos, T., Flaxman, A. D., Danaei, G., Shibuya, K., Adair-Rohani, H., et al. (2013). A comparative risk assessment of burden of disease and injury attributable to 67 risk factors and risk factor clusters in 21 regions, 1990–2010: a systematic analysis for the Global Burden of Disease Study 2010. *Lancet, 380*(9859), 2224–2260. https://doi.org/10.1016/S0140-6736(12)61766-8.

Lin, S., Jones, R., Pantea, C., Özkaynak, H., Rao, S. T. T., Hwang, S.-A., et al. (2013). Impact of NOx emissions reduction policy on hospitalizations for respiratory disease in New York State. *Journal of Exposure Science and Environmental Epidemiology, 23*(1), 73–80. https://doi.org/10.1038/jes.2012.69.

Matinga, M. N., Clancy, J. S., & Annegarn, H. J. (2014). Explaining the non-implementation of health-improving policies related to solid fuels use in South Africa. *Energy Policy, 68,* 53–59. https://doi.org/10.1016/j.enpol.2013.10.040.

McCollum, D., Krey, V., Riahi, K., Kolp, P., Grubler, A., Makowski, M., et al. (2013). Climate policies can help resolve energy security and air pollution challenges. *Climatic Change, 119*(2), 479–494. https://doi.org/10.1007/s10584-013-0710-y.

Messner, S., & Strubegger, M. (1995). *User's Guide for MESSAGE III* (IIASA Working Paper No. WP-95-069). Laxenburg, Austria.

National Department of Health. (2014). *National climate change & health adaptation plan 2014–2019*. Pretoria, SA. Retrieved from http://www.health.gov.za/index.php/shortcodes/2015-03-29-10-42-47/2015-04-30-08-29-27/2015-04-30-08-32-49?download=1776:national-climate-change-and-health-adaptation-plan-a4.

Obradovich, N., & Zimmerman, B. (2016). African voters indicate lack of support for climate change policies. *Environmental Science & Policy, 66,* 292–298. https://doi.org/10.1016/j.envsci.2016.06.013.

OECD. (2012). *OECD Environmental Outlook to 2050: The Consequences of Inaction*. Paris. Retrieved from http://www.keepeek.com/Digital-Asset-Management/oecd/environment/oecd-environmental-outlook-to-2050_9789264122246-en#page297.

Parker, J. D., Mendola, P., & Woodruff, T. J. (2008). Preterm birth after the Utah Valley Steel Mill closure: A natural experiment. *Epidemiology (Cambridge, Mass.)*, *19*(6), 820–823. https://doi.org/10.1097/EDE.0b013e3181883d5d.

Peel, J. L., Klein, M., Flanders, W. D., Mulholland, J. A., & Tolbert, P. E. (2010). *Impact of Improved Air Quality During the 1996 Summer Olympic Games in Atlanta on Multiple Cardiovascular and Respiratory Outcomes*. Retrieved from http://www.ncbi.nlm.nih.gov/pubmed/20575278.

Peel, J. L., Metzger, K. B., Klein, M., Flanders, W. D., Mulholland, J. A., & Tolbert, P. E. (2007). Ambient air pollution and cardiovascular emergency department visits in potentially sensitive groups. *American Journal of Epidemiology, 165*(6), 625–633. https://doi.org/10.1093/aje/kwk051.

Pope, C. (1989). Respiratory disease associated with community air pollution and a steel mill, Utah Valley. *American Journal of Public Health*, *79*(5), 623–8. Retrieved from http://www.pubmedcentral.nih.gov/articlerender.fcgi?artid=1349506&tool=pmcentrez&rendertype=abstract.

Pope, C. A., Schwartz, J., & Ransom, M. R. (1992). Daily mortality and PM10 pollution in Utah Valley. *Archives of Environmental Health: An International Journal, 47*(3), 211–217. https://doi.org/10.1080/00039896.1992.9938351.

Pretorius, I., Piketh, S., Burger, R., & Neomagus, H. (2015). A perspective on South African coal fired power station emissions. *Journal of Energy in Southern Africa, 26*(3), 27–40. Retrieved from http://www.scielo.org.za/scielo.php?script=sci_arttext&pid=S1021-447X2015000300004&nrm=iso.

Purohit, P., Munir, T., & Rafaj, P. (2013). Scenario analysis of strategies to control air pollution in Pakistan. *Journal of Integrative Environmental Sciences, 10*(2), 77–91. https://doi.org/10.1080/1943815X.2013.782877.

Raaschou-Nielsen, O., Andersen, Z. J., Beelen, R., Samoli, E., Stafoggia, M., Weinmayr, G., et al. (2013). Air pollution and lung cancer incidence in 17 European cohorts: Prospective analyses from the European Study of Cohorts for Air Pollution Effects (ESCAPE). *The Lancet Oncology*, *14*(9), 813–822. https://doi.org/10.1016/S1470-2045(13)70279-1.

Rafaj, P., Schöpp, W., Russ, P., Heyes, C., & Amann, M. (2013). Co-benefits of post-2012 global climate mitigation policies. *Mitigation and Adaptation Strategies for Global Change, 18*(6), 801–824. https://doi.org/10.1007/s11027-012-9390-6.

Rao, S., Pachauri, S., Dentener, F., Kinney, P., Klimont, Z., Riahi, K., et al. (2013). Better air for better health: Forging synergies in policies for energy access, climate change and air pollution. *Global Environmental Change, 23*(5), 1122–1130. https://doi.org/10.1016/j.gloenvcha.2013.05.003.

Republic of South Africa. (1996). National Road Traffic Act. Retrieved from https://www.gov.za/sites/www.gov.za/files/a93-96.pdf

Republic of South Africa. (2004). National Health Act, 2004. Retrieved from http://www.gov.za/sites/www.gov.za/files/a61-03.pdf.

Republic of South Africa. (2008). National Energy Act. Retrieved from http://www.energy.gov.za/files/policies/NationalEnergyAct_34of2008.pdf.

Republic of South Africa. (2014). Medium Term Strategic Framework 2014–2019.

Rich, D. Q., Liu, K., Zhang, J., Thurston, S. W., Stevens, T. P., Pan, Y., et al. (2015). Differences in birth weight associated with the 2008 Beijing Olympic air pollution reduction: Results from a natural experiment. *Environmental Health Perspectives*, *117*(11), 1713–1717. https://doi.org/10.1289/ehp.1408795.

Scenario Building Team. (2007). *Lont term mitigation scenarios strategic options for South Africa*. Long Term Mitigation Scenarios: Scenario Document, Department of Environment and Tourism.

Schwartz, J. & Marcus, A. (1990). Mortality and air pollution in London: a time series analysis. *American Journal of Epidemiology*, *131*(1), 185–194. Retrieved from http://www.ncbi.nlm.nih.gov/pubmed/2403468.

Scorgie, Y., Kneen, M., Annegarn, H. J., & Burger, L. (2003). Air pollution in the Vaal Triangle—quantifying source contributions and identifying cost-effective solutions. *Clean Air Journal, 13*(2), 5–18.

Shirinde, J., Wichmann, J., & Voyi, K. (2014). Association between wheeze and selected air pollution sources in an air pollution priority area in South Africa: a cross-sectional study. *Environmental Health: A Global Access Science Source, 13*(1), 32. https://doi.org/10.1186/1476-069X-13-32.

Simpson, D., Benedictow, A., Berge, H., Bergstrom, R., Emberson, L. D., Fagerli, H., et al. (2012). The EMEP MSC-W chemical transport model—technical description. *Atmospheric Chemistry and Physics, 12*(16), 7825–7865. https://doi.org/10.5194/acp-12-7825-2012.

Smith, K. R. (2002). Indoor air pollution in developing countries: recommendations for research. *Indoor Air, 12*(3), 198–207. https://doi.org/10.1034/j.1600-0668.2002.01137.x.

South Africa National Planning Commission. (2011). *Our future—make it work; National Development Plan 2030*. http://www.dac.gov.za/sites/default/files/NDP%202030%20-%20Our%20future%20-%20make%20it%20work_0.pdf.

South African Government. (2011). *National Climate Change Response White Paper*. Pretoria, SA. Retrieved from http://www.climateresponse.co.za/.

South African Government. (2016). South Africa's Intended Nationally Determined Contribution (INDC). Retrieved from https://www.environment.gov.za/sites/default/files/docs/sanational_determinedcontribution.pdf.

Thambiran, T., & Diab, R. (2010). A review of scientific linkages and interactions between climate change and air quality with implications for air quality management in South Africa. *South African Journal of Science, 106*.

The World Bank and IHME. (2016). *The Cost of Air Pollution—Strengthening the Economic Case for Action*. Washington DC. Retrieved from http://documents.worldbank.org/curated/en/781521473177013155/pdf/108141-REVISED-Cost-of-PollutionWebCORRECTEDfile.pdf.

UNFCCC. (2011). *Compilation of information on nationally appropriate mitigation actions to be implemented by Parties not included in Annex I to the Convention*. Retrieved from http://unfccc.int/resource/docs/2011/awglca14/eng/inf01.pdf.

U.S. Energy Information Administration. (2013). South Africa. Retrieved March 8, 2015, from http://www.eia.gov/countries/country-data.cfm?fips=SF#cde.

Vegter, I. (2016). *Air Quality: Missing the wood for the trees*. Johannesburg, South Africa. Retrieved from http://irr.org.za/reports-and-publications/occasional-reports/files/irr-air-quality.pdf.

Wagner, F., Heyes, C., Klimont, Z., & Schöpp, W. (2013). *The GAINS optimization module: Identifying cost-effective measures for improving air quality and short-term climate forcing*.

Wagner, F., Schöpp, W., & Amann, M. (2013b). Dealing with fixed emissions ceilings in an uncertain future: Offsetting under environmental integrity. *Journal of Environmental Management, 129*, 25–32. https://doi.org/10.1016/j.jenvman.2013.05.054.

Walwyn, D. R., & Brent, A. C. (2015). Renewable energy gathers steam in South Africa. *Renewable and Sustainable Energy Reviews, 41*(0), 390–401. https://doi.org/10.1016/j.rser.2014.08.049.

WHO. (2015). Global health observatory data repository. Retrieved from http://apps.who.int/gho/data/view.main.1701?lang=en.

Wichmann, J., & Voyi, K. V. V. (2006). Impact of cooking and heating fuel use on acute respiratory health of preschool children in South Africa. *The Southern African Journal of Epidemiology and Infection, 21*(2), 48–54.

Winiwarter, W. (2005). *The GAINS Model for Greenhouse Gases—Nitrous Oxide*.

Winkler, H., Altieri, K., Clarke, S., Garland, R. M., Kornelius, G., & Meas, M. H. (2016). Air quality and greenhouse gas emissions. In P. Scholes Lochner Schreiner, G., Snyman-Van der Walt, L. and de Jager, M., R (Ed.), *Shale Gas Development in the Central Karoo: A Scientific Assessment of the Opportunities and Risks*. Pretoria, SA. Retrieved from http://seasgd.csir.co.za/scientific-assessment-chapters/.

Witi, J., Stevens, L., Manzini, L., & Musee, N. (2013). *GHG Inventory for South Africa*.

Worthington, R. (2008). Cheap at half the cost: Coal and electricity in South Africa. In D. A. McDonald (Ed.), *Electric Capitalism* (pp. 109–148). HSRC Press.

Zhang, S., Worrell, E., Crijns-Graus, W., Krol, M., de Bruine, M., Geng, G., et al. (2016). Modeling energy efficiency to improve air quality and health effects of China's cement industry. *Applied Energy*, *184*, 574–593. https://doi.org/10.1016/j.apenergy.2016.10.030.

Ziervogel, G., New, M., Archer van Garderen, E., Midgley, G., Taylor, A., Hamann, R., et al. (2014). Climate change impacts and adaptation in South Africa. *Wiley Interdisciplinary Reviews: Climate Change*, *5*(5), 605–620. https://doi.org/10.1002/wcc.295.

# Chapter 8
# Transformation of the South African Energy System: Towards Participatory Governance

**Vain D. B. Jarbandhan, Nadejda Komendantova, Romao Xavier and Elvis Nkoana**

**Abstract** *Background & Significance of the topic*: Approximately 10% of South Africa's population has no access to electricity. Responding to this need for affordable and sustainable energy requires solutions that are environmentally friendly and not detrimental to human health. It has been demonstrated in countries such as Germany, Denmark, Canada and Wales that public participation contributes to the social acceptance of renewable energy. This study proposes that Arnstein's Ladder of Citizen Participation can be used to underpin the concept of stakeholder participation in emerging economies like South Africa, and that participation in renewable energy projects is dependent on leadership that is 'ecologically' attuned. From the onset, a renewable energy project that is geared for success must include opportunities for the public to participate in decision- making and to feel part of the success of the project. *Methodology*: A meta-analysis of the literature was conducted. *Application/Relevance to systems analysis*: This Chapter demonstrates how promoting public participation through applications such as climate modeling; assessment of impacts, vulnerability, mitigation, and adaptation options; and policy analysis can contribute to transforming a country's energy sector. *Policy and/or practice implications*: This study has special relevance for policy making in the energy sector in South Africa as it assists with long term projections for good governance and transformation of the energy sector. Incorporating public participation as part of the public policy process is essential to the success of transforming South Africa's energy system. Additionally, investing in building a cadet of

V. D. B. Jarbandhan (✉)
Department of Public Management and Governance, University of Johannesburg, Johannesburg, South Africa
e-mail: vainj@uj.ac.za

N. Komendantova
Risk and Resilience Program, International Institute of Applied Systems Analysis, Laxenburg, Austria

R. Xavier
University of Witwatersrand, Johannesburg, South Africa

E. Nkoana
University of South Africa, Pretoria, South Africa

 
P. Mensah et al. (eds.), *Systems Analysis Approach for Complex Global Challenges*,
https://doi.org/10.1007/978-3-319-71486-8_8

environmental leaders, especially in the public sector, would mitigate environmental degradation and embrace a transition to cleaner energy. *Discussion and conclusion*: The need to incorporate public participation within the project cycle and institutionalise it as part of the whole process is an important success feature, along with investing in the development of environmental leadership and monitoring and evaluation initiatives.

## 8.1 Introduction

Affordable access to electricity is crucial to alleviating poverty and attaining the Sustainable Development Goals (SDG) that are underlined by *inter alia* equality, access to health care, education, and clean water. Energy distribution in Southern Africa is skewed compared to other regions of Africa, with South Africa reaching targets of 85% electrification, while a country like Mozambique has a 40% electrification rate and Tanzania has an electrification rate of 24% as at 2012 (Africa Energy Outlook 2014; Statistics South Africa 2015). Figure 8.1 captures Africa's unequal energy landscape as at 2013. Half the population of West African countries, three quarters of East African countries and most of the Southern African countries (excluding South Africa) lack access to electricity. Making the case for adopting affordable and sustainable renewable energy options is therefore an important one.

In response to the need for affordable, sustainable clean energy solutions that have minimum negative impact on the environment and human health, the global energy sector has undergone enormous energy reform and transformation. Furthermore, energy production using coal in particular, through conventional techniques, has been identified as a major contributor to anthropogenic greenhouse

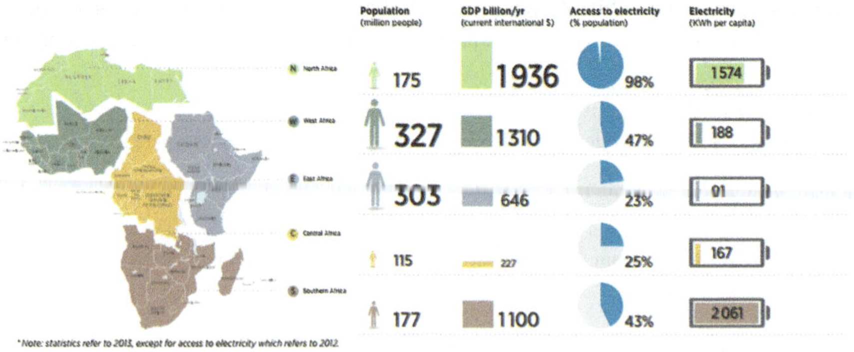


**Fig. 8.1** Africa's energy access in 2013. *Source* International Renewable Energy Agency (IRENA) (2015)

gas (GHG) emissions that lead to global warming and climate change as well as other direct health complications through pollution (Intergovernmental Panel on Climate Change [IPCC]-Climate Change Synthesis Report 2014). In order to minimise some of these negative effects, it is widely accepted that there is need to transform the way the world produces, distributes and uses energy globally.

South Africa is one of the major greenhouse gas emitters on the African continent. This is mainly due to its relatively high energy intensity per gross domestic product (GDP) on the continent and its heavy reliance on coal for electricity. South Africa accounts for 85% of the electricity produced in the region. Approximately, 82% of this electricity is generated by coal fired power stations with the remainder attributed to renewable (5%), 'peakers' (5%), nuclear (4%) and hydro electricity generation. South Africa's public utility, ESKOM generates 93% of the electricity in South Africa with independent energy producers contributing a minimal 7%. South Africa is responsible for 1% of the global greenhouse gas emissions and 18% of the sub-Saharan share (Pegels 2010). Additional data from the World Resources Institute Climate Analysis Indicator Tool (WRI CAIT) indicates that South Africa's GHG profile is dominated by emissions from the energy sector, which accounted for 84% of South Africa's total emissions in 2012. Moreover, indications are that South Africa's GHG emissions grew by 142 MtCO2e from 1990 to 2012, averaging 1.7% annually, while GDP grew by 77%, averaging 2.7% annually. In its National Climate Change Response Strategy, the South African government indicated that the country was a significant contributor to global climate change, with significant GHG emissions from its energy intensive, fossil-fuel powered economy (Department of Environment 2011). With its relatively high emissions and the carbon intensity of its economy being almost four times the world average, there is a need to examine ways of reducing South Africa's GHG emissions relative to GDP (USAID 2016).

In order to minimise negative environmental impact while producing energy, the South African government had found it prudent to adopt policies aimed at transforming the energy sector through diversification of the energy production portfolio in order to meet the electricity needs over a twenty year planning horizon towards 2030 (Integrated Resource Plan [IRP] 2010). The policies that have been adopted by the South African government in this regard is presented in Fig. 8.2.

Besides socio-political transformation, the transformation of the energy sector is paramount to promoting reconstruction and development in South Africa. Government policies, programmes, and projects are mainly geared towards poverty alleviation through job creation, especially for the youth. The unemployment rate in South Africa stood at 25.1% in 2015, and it is even higher for young people (OECD 2015; Statistics South Africa 2015). Unsurprisingly, South Africa's National Development Plan (NDP), is geared towards economic prosperity through employment creation, poverty alleviation, reducing inequality, and raising living standards while protecting the environment (National Planning Commission 2012).

In order to promote energy security, the NDP focuses on specific key areas including, economic infrastructure and environmental sustainability. The NDP identifies the need for South Africa to invest in a strong network of economic

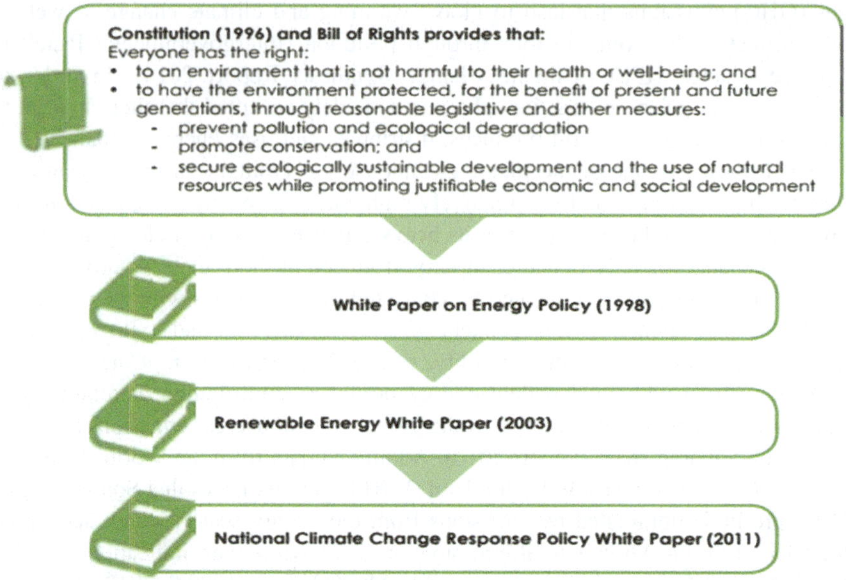


**Fig. 8.2** Key enabling policies for renewable energy in South Africa. *Source* Department of Energy (2015)

infrastructure designed to support the country's medium and long-term economic and social objectives, whilst protecting the environment (Mandela Institute & Konrad Adenauer Stiftung 2014). Additionally, energy infrastructure has to be robust and extensive enough to support commercial, households, and industrial needs.

Consequently, a robust energy infrastructure is important in the context of the black-outs that affected commercial and industrial establishments and households alike in 2008. Furthermore, the urgent need for considering an extensive energy infrastructure is driven by the political imperative to connect residents in remote rural areas to the electricity grid. Pressure from existing and new political parties threatens to deplete the ruling party's voting membership, particularly among the black youth and the black middle class that originate from these remote rural areas. It has also been realised that the need to connect residents to the national electricity grid is even more important given the escalation of service delivery protests in recent years and the loss of support experienced by the ruling party during the 2016 local government elections (Götz et al. 2016).

The deployment of renewable energy resources is one of the possible options for low carbon electricity generation. The costs of renewable energy is decreasing rapidly making it economically attractive and cost competitive with fossil fuels—something that has attracted the attention of stakeholders, including the government.

The government introduced the Renewable Energy Policy in 2003, and outlined a long-term vision for sustainable energy generation, with generous targets aimed at

10,000 GWh of renewable energy contribution to be achieved over a period of ten years (Department of Minerals and Energy 2003). The renewable energy sources were to be derived from biomass, wind, solar and small-scale hydro generation.

In 2011 the Department of Energy Affairs introduced the third policy paper on renewable energy. The paper was entitled The National Climate Change Response Policy White Paper whose main objective was to support South Africa's renewable energy focus (Department of Environment 2011). The debate around renewable energy received a significant boost with the government of South Africa committing to enhance greener energy at the COP15 meeting in Copenhagen, followed by South Africa hosting COP17 in Durban, where the government's Green Accord was signed with business and other stakeholders. South Africa's commitment to GHG reduction was further demonstrated in its carbon emission cap as indicated in the Integrated Resource Plan (IRP).

Given the background above, this chapter attempts to address the following key themes:

- Public participation in renewable energy project/s;
- Barriers to public participation; and
- Leadership to promote good governance in renewable energy projects.

## 8.2 Participatory Governance in Renewable Energy Projects: An International Perspective

Global renewable energy production grew by 85% over the past 10 years and reached 1700 GW in 2013 and is presently over 30% of all installed capacity (International Renewable Energy Agency [IRENA] 2014). The Renewable Policy Networks for the 21st Century (REN21) estimated that globally renewable energy accounted for more than 19% of the total human energy consumption and for around 23.7% of the electricity generation between 2014 and 2015. Later estimates also points to the fact that in 2015 renewables were the biggest contributor of mainstream energy production (REN21 2017). The volume of the overall investment into renewable energy capacity in 2015 was more than US$286 billion. Currently, the renewable energy sector accounts for more than 80 million jobs (REN21 2016). In fact, in 2015 the top countries leading deployment of renewable energies, including hydro generation were China, the United States of America, Brazil, Germany and Canada. The leaders at all non-hydro renewable power capacity were again China, the United States, and Germany, however new countries came to the top, such as Japan, India, Italy and Spain (REN21 2017).

China is the global leader in electricity production from renewable energy. For the total installed electricity capacity, China alone accounts for more than 25% of the total global renewable power capacity of more than 495 GW, where electricity is mainly generated from hydropower (excluding large scale hydropower) and wind. China also has the largest installation of hydro (excluding large scale hydropower), solar and wind capacity, which accounts for over 36% of its power generation. China features as a global leader in the production of solar photovoltaic (PV), and in 2016 became the largest manufacturer of PV, contributing to 63% of the market (Stamm et al. 2009).

Denmark is the world leader in wind energy as well as in manufacturing of wind turbines. In 2014 more than 57% of net electricity generation in Denmark came from renewable energy sources. The country envisages that 100% of its energy needs will be covered by renewable energy by 2050.

Renewable energy generation for electrification in Germany has been growing steadily. In the period from 2000 to 2016 the net-generation from renewable energy sources in the electricity sector in Germany increased from 6.3 to 34%. The renewable energy sector in Germany is one of the most innovative in the world and currently Germany is being called "the world's first major renewable energy economy" (Fraunhofer Institute for Solar Energy Systems ISE 2014).

One of the international best practices for participatory governance of energy transition was implemented in Austria as part of the climate and energy model (CEM) process. The goal of the CEM process focused on energy transition in rural and semi-rural regions of Austria, with the intention of renewable energy in these regions being a driver of socioeconomic development (Riegler et al. 2017).

In further studies by Hagget (2009) on public participation in renewable energy (wind) in Wales and England, it was reported that residents 'nimbyism' (opposition to the erection of wind turbines in the backyard of participants) was greater than the national interests of their respective countries. The study went on to report that allowing for effective public participation in renewable energy projects across Europe went a long way in enhancing the adoption of renewable energy.

Furthermore, studies by Jami and Walsh (2016) in renewable energy projects in Ontario, Canada indicate that 'knowledge-brokers' could be used to facilitate public participation. These 'knowledge-brokers' facilitate constant communication between the implementing authority and the community—constant communication and involvement of communities assists in keeping the project to its deadlines.

A further review of existing participatory governance practices indicated that the majority of practices are at the middle level of Arnstein's ladder, mostly informing inhabitants about the need for energy transition. However, some participatory practices are at the level of delegated authority. Inhabitants have an opportunity to discuss national funding for energy transition and to decide which measures they would like to have and how the funding should be spent in their region. The organisational frame for this discussion is given by energy groups, which meet four times per year. This practice proved to be successful as inhabitants were made aware of the energy transition needs and could suggest measures with the most positive impact on socioeconomic and sustainable development in their respective regions (Riegler et al. 2017).

The above literature review on public participation in renewable energy transition is supported by the words of Rennkamp and Bhuyan (2016) who wrote that the "concept of social shaping of technology suggests that the interplay of social,

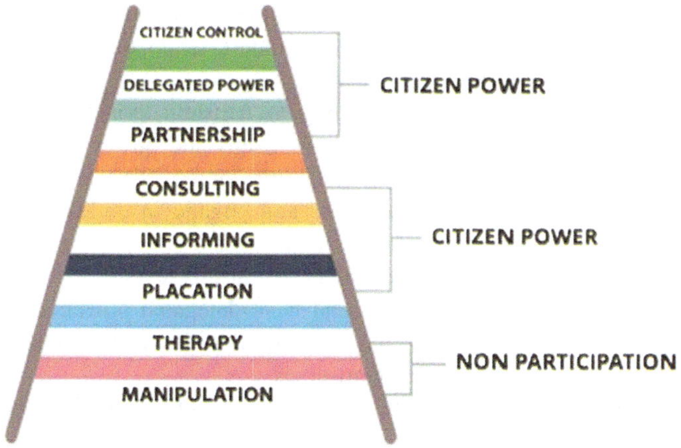


**Fig. 8.3** Arnstein's ladder of citizen participation. *Source* Adapted from Arnstein (1969)

political, economic, and cultural factors in a society shape the design and implementation of a technology". Part of shaping and designing the acceptance of renewable energy projects would require stakeholders to carefully plan the communication and consultation processes to ensure the success of communities accepting such projects.

The theoretical aspects related to public participation are highlighted in the work of Arnstein's 'ladder of citizen participation' (See Fig. 8.3). Arnstein's theory focuses on the meaning and purpose of public participation, and consists of eight rungs, with two levels of non-participation (Manipulation and Therapy), three degrees of tokenism, (Informing, Consultation, and Placation) and three degrees of citizen power (Partnership, Delegated Power and Citizen Control). Arnstein illustrated the characteristics of each type with examples from well-known federal programme (Arnstein 1969).

A further analysis of Arnstein's Ladder is encapsulated in the following discussion on the levels of participation:

i. **Manipulation**: is non-participation by the less powerful who are used by the powerful to achieve their own ends.
ii. **Therapy**: the aim is to cure or educate the participants. The proposed plan is put forward as the best, and participation's role is only to achieve public support through public relations rather than contributing.
iii. **Informing**: this is vital for legitimate participation, but all too frequently the emphasis is on a one-way flow of information, as there is no channel for feedback.
iv. **Consultation**: this is also a legitimate step in utilising apparatus such as attitude surveys, neighbourhood meetings, and public enquiries. However, Arnstein argues that this is just a window dressing ritual.

v. **Placation**: an example of this level is co-option. It allows citizens to play an advisory role or to plan, but power holders retain the right to judge the legitimacy or feasibility of the advice.

vi. **Partnership**: in this level, according to Arnstein, power is in fact redistributed through negotiation between citizens and power holders. Planning and decision-making responsibilities are shared. The have-not, poor or underpowered citizens can negotiate and engage in trade-offs with power holders, for example, through joint committees.

vii. **Delegated power**: Arnstein considers that at this level, citizens hold a clear majority of seats on committees and have delegated powers to make decisions. Thus, the public now has the power to assure accountability of the policies and programmes for themselves.

viii. **Citizen control**: the have-not citizens handle the entire job of planning, policy-making, and managing a programme, for example, neighbourhood cooperation, with no intermediaries between it and the source of funds or the state itself. Furthermore, citizens formerly without power obtain the majority of decision-making seats in the committees or full managerial power.

The first five levels represent what Arnstein called tokenism, where people can hear and be heard, but still lack the power to have their points considered by the powerful agents and decision makers. Therefore, at these levels of participation, there is no assurance that the *status quo* of policies, laws, or programmes will be changed.

Participation that is restricted to these five levels does not lead to a change in the *status quo* and hence would not bring about meaningful participation. As such, the powerless can achieve a certain level of advising (placation), but the power holders remain the ones who decide what can be done and how. The last three levels of participation represent the most balanced and effective concept of genuine participation. As Batel and Devine-Wright (2015) point out, "[a] change in *status quo* will be needed to implement local knowledge into the process of deployment of infrastructure projects. As evidence shows, local knowledge can contribute to a more sustainable deployment of infrastructure, with less impact on the environment and human health".

Arnstein (1969) points out some limitations to her conceptual framework. First, although this definition of participation shows that citizen participation is uneven and follows steps/levels or gradations, it is clearly limited because it is a simplification. Second, this conceptual framework considers have-not citizens and power holders as two homogeneous groups, undermining all the differences, cleavages, and competing interests that each group encompasses, including subgroups that emerge in the process. Finally, in the real world, people and programmes are more heterogeneous, there will thus be much more than eight levels of participation without clear sharp distinctions among them. Many of the characteristics used by Arnstein to illustrate each of the eight levels could be applicable to other levels.

Studies conducted by Xavier, Komendantova, Jarbandhan and Nel as part of the Southern African Young Scientist Summer Programme (SA-YSSP), which was published in the Journal of Cleaner Production (2017), concluded that the level of public participation in ten infrastructure projects analysed in South Africa was almost uneven, with some projects showing complete lack of public involvement and some being only at the first stage of participation as espoused in Arnstein's ladder. Moreover, major concentration of public involvement is at the information stage, consultation and placation, known as tokenism and few were at the initial stage of partnership and very few at the stage of citizen control. Thus, most of the cases analysed show a weak approach to participation, giving it less attention over the technical and management aspects of the project. Most of the project reports had no consideration for public participation at all, focusing their analysis on the economic gains and results of the project—without consideration of the governance aspect of the project and the socio-political contribution to empowering citizens.

## 8.3 Community Inclusivity in Transitioning Towards Renewable Energy: A South African Perspective

The South Africa Bill of Rights recognises that promoting economic development, attaining energy security, and achieving environmental sustainability will require investment in renewable energy (Constitution of the Republic of South Africa 1996). The Constitution has set a firm foundation for succeeding laws and strategies including the White Paper on Energy Policy (1998), Renewable Energy White Paper (2003), the National Climate Change Response Policy White Paper (2011), and the Integrated Resources Plan (2010) that ushered in the transition to renewable energy in South Africa.

A number of strategic policies including the NDP and the Integrated Resource Plan give clear guidance on South Africa's future energy position regarding diversification of electricity production over a 20 years planning horizon to 2030 as well as targeted production of 10,000 MW of additional electricity capacity by 2025 against the 2013 baseline of 44,000 MW. In line with the national commitment to a transition to a low carbon economy, 17,800 MW of the 2030 target are expected to be from renewable energy sources, with 5000 MW to be delivered online by 2019 and further 2000 MW (i.e. combined 7500 MW) online by 2020. These renewable energy sources will come from onshore wind, concentrated solar power, solar photovoltaic (PV), biomass, biogas, landfill gas, and small hydro (Mandela Institute & Konrad Adenauer Stiftung 2014).

As part of the energy mix diversification process, the South African government has put in place policies that encourage the contribution of independent power producers to the national grid through the Renewable Energy Independent Power Producers Procurement Program (REIPPPP) through a competitive tender process. The Preferential Procurement Policy Framework Act of 2000 and the Broad-Based Black Economic Empowerment Act 53 of 2003 strongly recommend that independent power producers (IPPs) be awarded tenders to develop renewable energy based on a 70/30 allocation of points to price factors and non-price factors (socioeconomic development) respectively. In order to enhance community inclusion and job creation, the Department of Energy (DoE) promotes inclusive projects that incorporate socioeconomic needs of the surrounding communities. Despite the potential benefits of the 'community inclusive process', a number of challenges exists including the lack of guidance on how surrounding community empowerment targets are prepared and evaluated, the incorporation of local stakeholder's concerns are not based on socioeconomic analysis, and short timeframes to prepare proposals between the bidding periods (*See windows 1 and 2*).

*Window 1*

Local community empowerment targets refer to specific community-based initiatives that directly benefit the community instead of "economic development" that may include domestic industrialisation, local content, and preferential procurement.

*Window 2*

Environmental leadership is the "ability to influence individuals and mobilise organisations to realise a vision of long-term ecological sustainability" (Boiral et al. 2007).

To address some of these challenges, the IPPs have put in place mechanism that focuses on community empowerment (job creation and socioeconomic development). For instance, the government appointed independent institutions to monitor and evaluate IPPs reporting and confirm their compliance with economic development (surrounding community empowerment) targets. Furthermore, economic development interventions are designed to incentivise bidders to promote job growth, domestic industrialisation, community development, and black economic empowerment through job creation, local content, local ownership, management control, preferential procurement, enterprise development, and socioeconomic development (i.e. through a community trust).

The 30% apportionment of economic development factors in the bid evaluation process is used as a vehicle to include different stakeholders during the REIPPPP process and beyond. These stakeholders are mainly emerging black (Africans, Indians, and Coloureds) entrepreneurs and surrounding communities. This criterion is critical considering that black emerging entrepreneurs are not able to compete with multinational corporations that are bidding for renewable energy projects on price, experience, expertise and other related factors. A component of the 30% apportionment is on local community empowerment targets that are intended to benefit community within a 50 km radius of the awarded renewable energy project.

During the bidding process, the IPPs are expected to submit a plan detailing how they will deliver the local community empowerment targets during the operation of their renewable energy project. Local community development targets include job creation for locals, promoting local content, supporting local businesses and community-based organisations, and setting up a community trust that will disburse around 1% of IPP profits to surrounding communities. To comply with this, IPPs meet with surrounding communities and conduct a community needs analysis and the final report forms part of their bidding documents.

Through this process, it is expected that the successful IPP will follow-up and deliver on its economic empowerment and local community empowerment promises. IPPs make lofty promises during the REIPPPP bidding process and once they are awarded the tender, challenges in the implementation of their commitments start to emerge. These challenges are related to the participation of various stakeholders (particularly surrounding community members) in the renewable energy infrastructure development and management processes.

## 8.4 Participatory Governance in the Renewable Energy Sector

A number of studies have stressed that participatory governance, involves not only state but economic and social actors, community-based institutions and unstructured groups and the media at the local, national, regional and global levels (Weiss 2010). Furthermore, participatory governance denotes forms of governance in which non-governmental actors, usually citizens, are empowered to use the resources of the state to make decisions about matters that directly concern them (Lukensmeyer and Torres 2006; Marsh 2002; OECD 2003). An effective participatory governance approach is capable of facilitating involvement of citizens in decision-making on issues that directly impact their livelihoods. If designed and implemented appropriately, this approach is capable of addressing specific needs and priorities relevant to citizens and at the same time assisting in their empowerment to fully participate (Lukensmeyer and Torres 2006; Marsh 2002; OECD 2003; Reddel and Woolcock 2004).

However, successfully designing and implementing participatory approaches is demanding, requiring buy-in and expert skills and resources among others. It has been noted that when these forms of participation emerge spontaneously, they are mostly informal and ephemeral or ad hoc and hence unsustainable, whereas when they are promoted and driven by public decision-makers, questions around their sustainability and effectiveness arise as they rely on active strategies to sustain them over time (Keast and Brown 2006).

A study by Weiss (2000) highlighted the importance of participatory governance in improving the outcomes of development projects as well as good governance.

Participatory governance leads to improvement of governance systems and provides an impetus to development, which increases circulation of information, transparency and accountability (Coelho and Favareto 2006). Collective participation is considered beneficial due to its potential to enhance learning processes and improve the quality of decisions as well as its potential to contribute to empowerment and to promote democratic citizenship (Turnhout et al. 2010).

Despite the potential benefits, "participatory governance has some degree of restrictions on who should be involved; the space for negotiation, assumptions on what the issue at stake is, and a level of expectations on the outcome and how the participants are expected to behave" (Turnhout et al. 2010).

## 8.5 Environmental Leadership in Transitioning Towards Renewable Energy Technologies

The long term sustainability of the environment is dependent on the stakeholder (which includes government, NGOs, industry, etc.) commitment to adapt to an environmentally-friendly leadership style. It is evident that the adopted leadership behaviour/s (style/s) could either be transactional or transformational, depending on the situation that the leaders find themselves in (Jarbandhan 2014).

An environmentally aware transactional leader may use contingent rewards to elicit immediate responses from followers, whereas a transformational leader can use this style of leadership to establish a long-term environmental vision for that organisation. In addition, a transformational leader can bring about change in followers to adapt to an environmentally friendly conscience. The transactional environmental leader can however, reward followers for adapting to an environmentally sustainable outlook.

Furthermore, ecologically sustainable public sector leadership requires that public organisations be transformed to promote a participatory form of governance so that the relevant stakeholders in the environmental sector are consulted in order to better manage the environment. In addition, ecological sustainability requires South African public sector leaders to have the requisite skills (which is not the focus of this article), such as, technical skills and monitoring and evaluation skills to roll-out environmentally friendly development and reconstruction initiatives. Finally, if South Africa is to promote environmentally sustainable development, it has to promote a form of leadership that fosters social, economic and environmental development. Currently, leaders focus on social and economic transformation at the behest of environmental custodianship. There should be a move by government to inculcate in public sector leaders the need to embrace environmental stewardship through its training and development programmes (Jarbandhan 2014).

## 8.6 Barriers to Transforming the Energy Sector

There are a number of barriers which prevent transformation of the energy system. Some of these barriers are of a technical nature, such as intermittence of renewable energy supply or the lack of efficient storage. The need to transform community systems and behavioral patterns also adds to these barriers and includes issues such as public and social acceptance of renewable energies, willingness to pay for consumption, participation and engagement in deployment of renewable energies in order to make their impact more sustainable and beneficial for society.

Evidence shows that transformation of the energy system is often faced with risks and boundaries for implementation of national strategies and policies on climate change mitigation. These boundaries exist along the entire deployment process of renewable energy sources and in relation to different private and public stakeholders as well as their perceptions of existing risks (Schinko and Komendantova 2016; Battaglini et al. 2012). This includes barriers and risks in the project planning and financing phases but also during the implementation phase of new technologies, which may be confronted with a lack of social acceptance of these innovations. There is an urgent need to understand these boundaries in order to be able to adequately support the development of decision-making processes and governance policy. This chapter advances knowledge about human factors by focusing on economic, social, political and participatory governance issues of energy transition and how these issues are addressed in South Africa. It is also necessary to assess social feasibility of deployment of renewable electricity by taking into account stakeholders' perceptions of risks, preferences for different energy transition options and willingness to engage in the decision-making process.

International legislation, such as the Aarhus Convention, requires dissemination of information for public comment prior to decision-making on infrastructure projects (Barthes and Mays 1998). The Aarhus Convention also requires that the people whose environment will be affected by infrastructure deployment have an opportunity to express their views in time, as this can still influence decisions related to the implementation of such projects (Rydin and Pennington 2000). In spite of this, the manner in which participation is practiced is perceived as insufficient and additional research is required to shed more light on how deployment of renewable energy and energy transitions that place stronger emphasis on technological and economic approaches, hampers effective public participation (Devine-Wright 2005). The need to understand individual positions, concerns and views of different stakeholders in decision-making processes is crucial (Webler and Tuler 2010). A perusal of research undertaken in the deployment of renewable energy indicates that only 3% of all existing research in the area addresses the topic of human factors (social issues) of energy transition and transformation of the energy system (Sovacool et al. 2014).

Public interest in energy infrastructure projects differs today from half a century ago when infrastructure projects represented technological progress necessary for society's progress (Klaas 2012). The traditional energy system aimed to provide electricity at the least possible costs. Aspects such as impact on the environment, interests of separate communities and people living in the vicinity of the projects were of secondary importance. Nowadays the views of citizens on the architecture of energy systems are changing, with a strong emphasis on sustainable development. The influence of non-governmental organisations in energy deployment is also growing, given their increased ability to organise, mobilise and articulate public opinion regarding infrastructure projects. Public participation and acceptance is also linked to the democratic principle of inclusion to enable people to share in decision-making that will affect their lives (Beierle and Cayford 2002).

Different views exist about how the public should be involved in decision-making on energy transition. Some argue that traditional decision-making processes on energy should be left to experts and scientists (Perhac 1998) while others suggest that the limited capacity and adequacy of stakeholder knowledge could hinder complex decision-making. For instance, it is questionable whether the public can understand concepts such as "uncertainty" and "probability". Others argue that public participation is needed to supplement the available knowledge of experts, which may be limited, for example the understanding of experts on local areas (Jasanoff 1998). Furthermore, evidence shows that integrating the views of 'lay people' and public values with the opinion of 'knowledgeable experts', can enhance trust and the legitimacy of decision-making processes (Renn 2008).

The issue of public resistance and concerns to the implementation of renewable energy technology close to residential communities remains a pressing concern. Such public resistance has been recorded in Australia (D'Souza and Yiridoe 2014), China (Liu et al. 2013), Colombia (Rosso-Ceron and Kafarov 2015), Germany (Musall and Kuik 2011), Japan (Maruyama et al. 2007), the Netherlands (van Os et al. 2014), Britain (Lock et al. 2014), Scotland (Shamsuzzoha et al. 2012) and even in South Africa (Pegels 2010). Public resistance to energy technologies can even be traced back to the contested decisions on the establishment of nuclear power plants, nuclear waste storage facilities, or large hydropower dams (Wustenhagen et al. 2007). The success of protest action by the Treasure Karoo Action Group (TKAG) has compelled the South African government to put a moratorium on Shale Gas exploration in the pristine Karoo region of the Northern Cape. The public's resistance to large scale renewable energy deployment is generally driven by environmental (i.e. noise, visual, and birds route impacts), cultural (i.e. ancestral burial land), financial (i.e. investors, consumers), and socio-political (i.e. job creation, trust, social justice) factors (Wustenhagen et al. 2007; de Araujo and de Freitas 2008). Scientific literature also provides examples of how these concerns have been addressed by *inter alia* designing wind turbines that minimise noise, recognising the aesthetics and tourism potential of wind turbine sites, contributing to an increase in quality of life through increased income of the local population, availability of electricity, and upgrading of infrastructure (de Araujo and de Freitas 2008; Yazdanpanah et al. 2015). Reviewing the case for public participation in energy transition and the acceptance of renewable energy deployment in the European context will provide valuable insights for similar projects in emerging economies like South Africa.

In most European countries the responsibility to decide, plan and impose renewable energy infrastructure remains with the central government. This may often result in conflict at local government level and with communities where the infrastructure is to be constructed. For example at the national level, government may be interested in welfare and the development of new infrastructure, whereas local government may be more concerned about human health and environmental impact. The central government is often in support of regulatory siting procedures and the utilitarian view of fairness which is channeled through an expert-government dominated siting process (Kunreuther et al. 1994).

The public participation in energy transition and the acceptance of renewable energy deployment in the European context is typically discussed in relation to the so-called "decide-announce-defend" (DAD) model and "Not-in-My-Backyard" (NIMBY) model (Komendantova et al. 2015). In the DAD model, decisions are made by educated experts, project developers and government and they are simply communicated to the public. Evidence shows that DAD often leads to social conflict, delays and even cancellation of the projects (Wolfsink 2000). The NIMBY concept is best illustrated when there is a high level of public support for renewable energy deployment at the national level, but the implementation of the actual projects faces hostility at the local level (Bell et al. 2005). Some scientists highlight the need for caution when using NIMBY to understand local objections and concerns as it may be misleading (Burningham et al. 2006; Devine-Wright 2009).

Numerous research has been published on public response to renewable energy in Europe. One example of micro-generation by Sauter and Watson (2007) demonstrated how to transform passive acceptance of energy installations to a more active willingness to participate in energy transition. A waste to energy power plant near the city of Thessaloniki received severe opposition from local residents (Achillas et al. 2011). Furthermore, severe opposition to the construction of the wind park in Welshpool stemmed from a perception that the community would not benefit from the electricity produced by the wind turbines (Pidgeon 2012). The introduction of large-scale wind turbines in Greece provoked serious reaction from the local population, resulting in cancellation of parts of the projects. These protests were directed against new installations, even though the acceptance of existing wind parks was high (Kaldellis 2005). A study of attitudes of inhabitants towards hydrogen vehicles in Norway showed that even though public acceptance was high at the beginning, it declined during the three-year span of the project (Tarigan and Bayer 2012). Finally, in the Netherlands, most of the population accepted the need for deployment of renewable energy sources but did not accept deployment of electricity transmission grids (Wolfsink 2010).

According to Grubler et al. (2016) social factors are critical for turning passive acceptance into active willingness to use renewable energies in Europe. In the European examples presented, social factors were critical for turning passive acceptance into active willingness to use renewable energies. The social factors included perceptions of benefits, perceived relative advantages of new technologies, social visibility and awareness of the technology as well as its compatibility with existing social norms and practices.

## 8.7 Conclusion

This chapter presented studies that show that a more participatory approach to the policy-making process is key to integrating renewable energy into South Africa's energy planning and aligning renewable energy policy with its broader climate and development policies (Masullo and Brown 2014). The opportunity for South Africa to advance the transformation of its energy system must include mechanisms that strengthen public participation and the development of environmental leadership. To enable genuine participation of previously disadvantaged communities (in particular black Africans, Coloureds, and Indians), the government should put in place mechanisms that address and monitor the supply of the IPPs value chain and community engagement processes. This approach, supported by social acceptance literature, makes the argument for participatory governance that includes the public in decision-making in order to raise awareness of the benefits of renewable energy technologies, maximise acceptance and improve the likelihood of success of such projects (Devine-Wright 2008; Evans et al. 2011; Strazzera et al. 2012; Cohen et al. 2014). Arnstein's ladder of participation was used to deepen understanding of public participation in energy infrastructure projects and highlight the critical role of environmental leadership in the governance of large scale renewable energy infrastructure projects. Developing countries such as South Africa would find it wise to invest in improving the mechanisms for public participation in social innovation programmes.

## References


Achillas, C., Vlachokostas, C., Moussiopoulos, N., Banias, G., Kafetzopoulos, G., & Karagiannidis, A. (2011). Social acceptance for the development of a waste-to-energy plant in an urban area. *Resources, Conservation and Recycling, 55*(9–10), 857–863.

Africa Progress Report. (2015). *Africa Progress Panel.* http://www.africaprogresspanel.org/publications/policy-papers/2015-africa-progress-report. Accessed 2 May 2016.

Arnstein, S. (1969). A ladder of citizen participation. *Journal of the American Planning Association, 35*(4), 216–224.

Barthes, Y., & Mays, C. (1998). *High profile and deep strategy: Communication and information practices in France's underground laboratory siting process*. Technical Note SEGR/98, 18, Institute De Protection Et De Surete Nucleaire.

Batel, S., & Devine-Wright, P. (2015). A critical and empirical analysis of the national-local 'gap' in public responses to large-scale energy infrastructures. *Journal of Environmental Planning and Management, 58*(6), 2015.

Battaglini, A., Komendantova, N., Brtnik, P., & Patt, A. (2012). Perception of barriers for expansion of electricity grids in the European Union. *Energy Policy, 47,* 254–259.

Beierle, T., & Cayford, J. (2002). *Democracy in practice: Public participation in environmental decisions.* RFF Press: An Imprint of Routledge. Washington, D.C.

Bell, D., Gray, T., & Haggett, C. (2005). Policy, participation, and the "social gap" in wind farm siting decisions. *Environmental Politics, 14,* 460–477.

Boiral, O., Cayer, M., & Baron, C. M. (2007). The action logics of environmental leadership: A developmental perspective. *Journal of Business Ethics, 85*(4), 479–499.

Burningham, K., Barnett, J., & Thrush, D. (2006). The limitations of the NIMBY concept for understanding public engagement with renewable energy technologies: A literature review. Beyond Nimbyism research project Working Paper. http://geography.exeter.ac.uk/beyond_nimbyism/deliverables/bn_wp1_3.pdf. Accessed 16 May 2016.

Coelho, V., & Favareto, A. (2006). *Participatory governance and development: In search of a casual nexus*. Hoboken: Blackwell Publishing Ltd.

Cohen, J. J., Reichl, J., & Schmidthaler, M. (2014). Re-focussing research efforts on the public acceptance of energy infrastructure: A critical review. *Energy, 76*, 4–9.

Conference of the Parties, Twenty-first session. (COP 21). Paris, 30 November to 11 December (2015). ADOPTION OF THE PARIS AGREEMENT: Proposal by the President, Draft decision -/COP.21.

D'Souza, D., & Yiridoe, E. K. (2014). Social acceptance of wind energy development and planning in rural communities of Australia: A consumer analysis. *Energy Policy, 74*, 263–270.

De Araujo, M. S. M., & De Freitas, M. A. V. (2008). Acceptance of renewable energy innovation in Brazil-case study of wind energy. *Renewable and Sustainable Energy Reviews, 12*, 584–591.

Department of Energy. (1998). *White paper on the energy policy of the Republic of South Africa*. Pretoria: Government Printers.

Department of Energy. (2009). *Digest of South African energy statistics*. Pretoria: Government Printers.

Department of Energy. (2013). *Integrated Resource Plan for Electricity (IRP) 2010-2030 updated report*. Pretoria: Government Printers.

Department of Energy. (2015). *State of renewable energy in South Africa*. Pretoria: Government Printers.

Department of Environment. (2011). *National climate change response policy white paper*. Pretoria: Government Printers.

Department of Minerals and Energy. (2003). *Republic of South Africa white paper on renewable energy*. Pretoria: Government Printers.

Devine-Wright, P. (2005). Beyond NIMBYism: Towards an integrated framework for understanding public perceptions of wind energy. *Wind Energy, 8*(2), 125–139.

Devine-Wright, P. (2008). Reconsidering public acceptance of renewable energy technologies: A critical review. In M. Grubb, T. Jamasb, & M. Pollitt (Eds.), *Delivering a low carbon electricity system*. London: Cambridge University Press.

Devine-Wright, P. (2009). Rethinking NIMBYism: The role of place attachment and place identity in explaining place-protective action. *Journal of Community and Applied Social Psychology, 19*(6), 426–441.

Evans, B., Parks, J., & Theobald, K. (2011). Urban wind power and the private sector: Community benefits, social acceptance and public engagement. *Journal of Environmental Planning and Management, 54*(2), 227–244.

Fraunhofer Institute for Solar Energy Systems ISE. (2014). Solar energy systems. Internet source. www.ise.fraunhofer.de/content/dam/ise/en/. Accessed on 12 Apr 2017.

Götz, G., Khanyile, S., & Katumba, S. (2016). Voting patterns in the 2016 local government elections. *Gauteng City Region Observatory*. Retrieved on 12 April 2017 from http://www.gcro.ac.za/outputs/map-of-the-month/detail/voting-patterns-in-the-2016-local-government-elections.

Gruebler, A., Wilson, C., & Nemet, G. (2016). Apples, oranges, and consistent comparisons of the temporal dynamics of energy transitions. *Energy Research and Social Science, 22*, 18–25.

Haggett, C. (2009). Public Engagement in Planning for Renewable Energy. In Davoudi, S., Crawford, J. & Mehmood, A. (Eds.). *Planning for Climate Change: Strategies for Mitigation and Adaptation for Spatial Planners*. Routledge, 297–307.

Intergovernmental Panel on Climate Change. (2014). Summary for policymakers. In: O. Edenhofer, Pichs-Madruga R., Sokona Y., Farahani E., Kadner S., Seyboth K., et al. (Eds.), *Climate Change 2014, Mitigation of Climate Change. Contribution of Working Group III to the Fifth Assessment Report of the Intergovernmental Panel on Climate Change*. Cambridge, United Kingdom and New York, NY, USA: Cambridge University Press.

Integrated Resource Plan. (2010). Executive Summary and Report. Department of Energy. Republic of South Africa. Government Printer. Pretoria.

International Energy Association. (2014). Africa Energy Outlook. A Focus on Energy in Sub-Saharan Africa. Paris, France.

International Renewable Energy Agency. (2014). Renewable Power Generation Costs in 2014. International Renewable Energy Agency.

International Renewable Energy Agency. (2015). Africa 2030: Roadmap for renewable energy future. International Renewable Energy Agency. www.irena.org/remap. Accessed on 20 Aug 2017.

Jarbandhan, D. B. (2014). Adopting an environmentally—Friendly leadership style for the South African Public Sector: A literature review. *Journal of Public Administration, 49*, 1.

Jami, A. A., & Walsh, P. R. (2016). Wind power deployment: The role of public participation in decision-making in Ontario, Canada. Sustainability. Available at www.sustainability08-0073. Accessed on 15 Aug 2017.

Jasanoff, S. (1998). The political science of risk perception. *Reliability Engineering and System Safety, 59*(1), 91–99.

Kaldellis, J. K. (2005). Social attitude towards wind energy applications in Greece. *Energy Policy, 33*(5), 595–602.

Keast, R., & Brown, K. (2006). Adjusting to new ways of working: Experiments with service delivery in the public sector. *Australian Journal of Public Administration, 65*(4), 41–53.

Klass, A. B. (2012). Renewable Energy and the Public Trust Doctrine, 45 U.C. Davis L. Rev. 1021 (2012). Available at http://scholarship.law.umn.edu/faculty_articles/34. Accessed 15 Aug 2017.

Komendantova, N., Voccciante, M., & Battaglini, A. (2015). Can the BestGrid process improve stakeholder involvement in electricity transmission projects? *Energy, 2015*(8), 9407–9433. https://doi.org/10.3390/en8099407.

Kunreuther, H., Linnerooth-Bayer, J., & Fitzgerald, K. (1994). Siting hazardous facilities: Lessons from Europe and America. Wharton Risk Management and Decision Processes Center.

Liu, W., Wang, C., & Mol, A. P. J. (2013). Rural public acceptance of renewable energy deployment: The case of Shandong in China. *Applied Energy, 102*, 1187–1196.

Lock, S. J., Smallman, M., Lee, M., & Rydin, Y. (2014). "Nuclear energy sounded wonderful 40 years ago": UK citizens views on CCS. *Energy Policy, 66*, 428–435.

Lukensmeyer, C., & Torres, L. (2006). Public Deliberation: A manager's guide to citizen engagement, IBM Centre for the Business of Government, http://www.businessofgovernment.org/sites/default/files/LukensmeyerReport.pdf. Accessed 22 July 2017.

Mandela Institute (MI) and Konrad Adenauer Stiftung (KAS) Foundation. (2014). Report of the scoping workshop on climate change, energy law and the environment, November 2014, Johannesburg, South Africa.

Marsh, I. (2002). Governance in Australia: Emerging issues and achievements. *Australian Journal of Public Administration, 61*(2), 3–9.

Maruyama, Y., Nishikido, M., & Iida, T. (2007). The rise of community wind power in Japan: Enhanced acceptance through social innovation. *Energy Policy, 35*, 2761–2769.

Masullo, I., & Brown, H. (2014). Lessons from South Africa: Mobilizing investment in renewable energy. http://www.wri.org/blog/2014/05/lessons-south-africa-mobilizing-investment-renewable-energy. Accessed on 4 May 2016.

Musall, F. D., & Kuik, O. (2011). Local acceptance of renewable energy: A case study from South East Germany. *Energy Policy, 39*(6), 3252.

National Planning Commission. (2012). National Development Plan 2030: Our future—Make it work. Pretoria: The Presidency.

Organisation for Economic Cooperation and Development (OECD). (2003). Open Government: Fostering dialogue with civil society, Organisation for Economic Cooperation and Development, Paris.

Organisation for Economic Cooperation and Development (OECD). (2015). Education at a Glance: Global Launch. Available at: https://www.slideshare.net/OECDEDU/education-at-a-glance-2015-global-launch.

Pegels, A. (2010). Renewable energy in South Africa: Potentials, barriers and options for support. *Energy Policy, 38,* 4945–4954.

Perhac, R. (1998). Comparative risk assessment: Where does the public fit in. Science, technology and human value. http://journals.sagepub.com/doi/abs/10.1177/016224399802300204. Accessed on 12 Aug 2017.

Pidgeon, N. (2012). Climate change risk perception and communication: Addressing a critical moment? *Risk Analysis, 32*(6), 951–956.

Reddel, T., & Woolcock, G. (2004). From consultation to participatory governance? A critical review of citizen engagement strategies in Queensland. *Australian Journal of Public Administration, 63*(3), 75–85.

Renn, O. (2008). Risk governance: Coping with uncertainty in a complex world. *Earthscan, 2008,* 455.

Rennkamp, B. (2017). Handbook in social and political research. [in press] Chapter 3 Out of Sync: Innovation and policy and theory in unequal societies. [in press]. Accessed at https://www.elgaronline.com/view/9781783471904.00009.xml. Accessed on 15 Aug 2017.

Rennkamp, B., & Bhuyan, R. (2016). *The social shaping of nuclear energy technology in South Africa*. United Nations: United Nations University World Institute for Development Economics Research Unit.

Renewable Policy Networks 21. (2016). Annual Report 2016—Connecting the DOTS: Convening Multi-Stakeholders on Renewable Energy. C/o UN Environment. Economy Division. (Paris: REN21 Secretariat).

Renewable Policy Networks 21 (REN21). (2016). Renewables 2016 Global Status Report (Paris: REN21 Secretariat). ISBN 978-3-9818107-0-7.

Renewable Policy Networks 21 (REN21). (2017). Renewables 2017 Global Status Report (Paris: REN21 Secretariat). ISBN 978-3-9818107-6-9.

Republic of South Africa. (2000). Preferential Procurement Policy Framework Act 05 of 2000. Pretoria: Government Printers.

Republic of South Africa. (1996). Constitution of the Republic of South Africa Act 108 of 1996. Pretoria: Government Printer.

Republic of South Africa. (2003). *Broad-Based Black Economic Empowerment Act 53 of 2003*. Pretoria: Government Printers.

Riegler, M., Vogler, C., Neumueller, S., & Komendantova, N. (2017). Engaging inhabitants into energy transition in climate and energy model (CEM) regions: Case studies of Freistadt, Ebreichsdorf and Baden. IIASA Working Paper. IIASA, Laxenburg, Austria: WP-17-003.

Rosso-Ceron, A., & Kafarov, V. (2015). Barriers to social acceptance of renewable energy systems in Colombia. *Current Opinions in Chemical Engineering, 10,* 103–110.

Rydin, Y., & Pennington, M. (2000). Public participation and environmental planning: The collective action problem and the potential of social capital. *Local Environment, 5*(2), 153–169.

Sauter, R., & Watson, J. (2007). Strategies for the deployment of micro-generation: Implications for social acceptance. *Energy Policy, 35*(5), 2770–2779.

Schinko, T., & Komendantova, N. (2016). De-risking investment into concentrated solar power in North Africa: Impacts on the costs of electricity generation. *Renewable Energy, 92,* 262–292.

Shamsuzzoha, A. H. M., Grant, A., & Clarke, J. (2012). Implementation of renewable energy in Scottish rural area: A social study. *Renewable and Sustainable Energy Reviews, 16,* 185–191.

Sovacool, B. K. (2014). The importance of open and closed styles of energy research. *Social Studies of Science, 40*(6), 903–930. https://doi.org/10.1177/0306312710373842.

Stamm, A., Dantas, E., Fischer, S., Ganguly, R., & Rennkamp, B. (2009). Discussion paper: Sustainability-oriented innovation systems. German Development Institute. Bonn: Deutsches Institut für Entwicklungspolitik gGmbH.

Statistics South Africa. (2015). 2014–2015 Annual Report. Pretioria: Governament Printer.

Strazzera, E., Mura, M., & Contu, D. (2012). Combining choice experiments with psychometric scales to assess the social acceptability of wind energy projects: A latent class approach. *Energy Policy, 48,* 334–347.

Tarigan, A., & Bayer, S. (2012). Temporal change analysis of public attitude, knowledge and acceptance of hydrogen vehicles in Greater Stavanger, 2006–2009. *Renewable and Sustainable Energy Review, 16*(8), 5535–5544.

Turnhout, E., Van Bommel S., & Aarts, N. (2010). How participation creates citizens: Participatory governance as performative practice. *Ecology and Society, 15*(4), 26. [online] URL: http://www.ecologyandsociety.org/vol15/iss4/art26/. Accessed on 12 May 2016.

USAID. (2016). Greenhouse Gas Emissions in South Africa. Retrieved from http://pdf.usaid.gov/pdf_docs/pa00msrg.pdf.

Van Os, H. W. A., Herber, R., & Scholtens, B. (2014). Not under our back yards? A case study of social acceptance of the Northern Netherlands CCS initiative. *Renewable and Sustainable Energy Reviews, 30,* 923–942.

Webler, T., & Tuler, S. (2010). Fairness and competence in citizen participation: Theoretical reflections from a case study. *Administration & Society, 32*(5), 566–595.

Weiss, T. G. (2010). Governance, good governance and global governance: Conceptual and actual challenges. *Third World Quarterly, 21*(5), 795–814.

Wolsink, M. (2000). Wind power and the NIMBY-myth: Institutional capacity and the limited significance of public support. *Renewable Energy, 21*(1), 49–64.

Wolsink, M. (2010). Contested environmental policy infrastructure: Socio-political acceptance of renewable energy, water, and waste facilities. *Environmental Impact Assessment Review, 30*(5), 302–311.

Wustenhagen, R., Wolsink, M., & Burer, M. J. (2007). Social acceptance of renewable energy innovation: An introduction to the concept. *Energy Policy, 35,* 2683–2691.

Xavier, R., Komendantova, N., Jarbandhan, D. B., & Nel, D. (2017). Participatory governance in the transformation of the South African energy sector: Critical success factors for environmental leadership. [SA-YSSP]. *Journal of Cleaner Production, 154,* 621–632.

Yazdanpanah, M., Komendantova, N., & Ardestani, R. S. (2015). Governance of energy transition in Iran: Investigating public acceptance and willingness to use renewable energy sources through socio-psychological model. *Renewable and Sustainable Energy Review, 45,* 566–673.

# Chapter 9
# Precision Agriculture and Food Security in Africa

**Bongani Ncube, Walter Mupangwa and Adam French**

**Abstract** *Background and Significance of the topic*: The chapter gives an overview of precision agriculture and its impacts on food security in Africa. *Methodology*: Methods and concepts of precision agriculture are described including crop, soil and position sensors; which include global positioning and remote sensing applications in detection of crop stress, monitoring variability, soils, weeds, and diseases. Machine controls and computer based systems are also briefly described. *Application/Relevance to systems analysis*: There are a number of operations that can benefit from precision agriculture at field level, including soil preparation, fertilisation, irrigation and weed management. In Africa, the benefits of precision agriculture include improved food security through increases in water and nutrient use efficiency, and timely management of activities such as weed control. Precision agriculture has saved costs of inputs in both commercial and smallholder farming in Africa. Pollution control of ground and surface water sources has slowed down where fertiliser and agrochemical applications are now more efficient. *Policy and/or practice implications*: Two examples of precision agriculture application in Africa are presented; FruitLook which is used by farmers in the Western Cape in South Africa as a state-of the art information technology that helps deciduous fruit and grape farmers to be water efficient and climate-smart. The Chameleon and Wetting Front Detector Sensors have enabled small scale farmers in Mozambique, Tanzania, and Zimbabwe to cut down irrigation frequency fifty times and double productivity. *Discussion and conclusion*: It is clear that precision agriculture has played a major

B. Ncube (✉)
Centre for Water and Sanitation Research, Cape Peninsula University of Technology, Bellville, South Africa
e-mail: NcubeB@cput.ac.za

W. Mupangwa
International Maize and Wheat Improvement Center (CIMMYT), ILRI Sholla Campus, Addis Ababa, Ethiopia

A. French
Risk and Resilience (RISK) and Advanced Systems Analysis (ASA) Programs, International Institute for Applied Systems Analysis, Laxenburg, Austria

 
P. Mensah et al. (eds.), *Systems Analysis Approach for Complex Global Challenges*, https://doi.org/10.1007/978-3-319-71486-8_9

role in improving food security in Africa through the efficient use of inputs such as fertiliser and water, while also reducing environmental pollution and degradation.

## 9.1 Defining Precision Agriculture

Precision agriculture (PA) refers to an integrated crop management system that attempts to match the kind and amount of inputs with actual crop needs for small areas within a field (Davis et al. 1998). The PA concept recognises that variations occur within agricultural fields, and therefore seeks to identify the spatial location and extent of such variations (Rilwani and Ikhuoria 2006). PA then matches resource application and agronomic practices with soil and crop requirements as they vary in space and time within a field (Whelan and MacBratney 2000). PA has also been defined as all those agricultural production practices that use information technology either to tailor input use to achieve desired outcomes or to monitor those outcomes (Bongiovanni and Lowenberg-DeBoer 2004).

PA may also be defined as a management system that is information and technology based, is site specific and uses one or more of the following sources of data: soils, crops, nutrients, pests, moisture, or yield, for optimum profitability, sustainability, and protection of the environment (McLoud et al. 2007). PA increases the number of correct decisions per unit area of land associated with net benefits. The PA definition has evolved over time as technologies and the understanding of what is achievable changes (MacBratney et al. 2005).

PA is also explained in terms of other concepts. Bongiovanni and Lowenberg-DeBoer (2004) refer to site specific management (SSM) which they define as doing the right things, at the right place and the right time. SSM enables the automation of PA, which increases the number of (correct) decisions per unit area of land per unit time, with associated net benefits. The decisions can be made by electronic sensors or humans. PA should also consider the economic and environmental components, effort and risk associated with management (MacBratney et al. 2005). PA is also seen as a way of thinking, a systems approach, and not a specific technology (Smith and Baillie 2009).

Within precision agriculture, other concepts such as precision irrigation have evolved. Precision irrigation focuses on 'differential irrigation' treatment of field variation as opposed to the 'uniform irrigation' treatment that underlies traditional irrigation management. Precision irrigation saves water and reduces costs by applying only the optimum amount of irrigation to individual plants or small areas within a field, while the traditional practice takes a 'whole-field' approach (Smith and Baillie 2009).

McLoud and co-workers (McLoud et al. 2007) see the main goal of PA as the optimization of inputs for agricultural production according to the capability of the land. This is also echoed by Smith and Baile (2009) who see precision irrigation as a tool to potentially alter on-farm decision-making, and simultaneously achieve the multiple objectives of enhancing input use efficiency, reducing environmental impacts, and increasing farm profits and product quality.

## 9.2 Precision Agriculture: Methods and Concepts

Precision agriculture is defined by the technologies that are employed, and the use of information is a key ingredient to the success of precision farming (Davis et al. 1998).

### 9.2.1 *Crop, Soil, and Positioning Sensors*

PA is supported by technologies such as global positioning systems (GPS) and information platforms such as geographical information systems (GIS). Precision agriculture is a rapidly changing field due to the ever changing developments in technology. There are a number of platforms that PA is currently based on, and each platform is related to the underlying technology. Crop, soil, and positioning sensors include both remote and vehicle-mounted, "on-the-go" sensors that detect soil texture, soil moisture levels, crop stress, disease and weed infestations (Grisso et al. 2005b).

Global positioning systems (GPS) were initially introduced for military purposes (Auernhammer 2001). The combination of increased precision and reduced cost of GPS receivers, which use satellite signals to locate an observer, have made them more affordable for commercial farming enterprises (Sylvester-Bradley et al. 1999). GPS is one key factor that has allowed precision technology to progress to current levels (Kitchen et al. 2002). GPS has enhanced the ease and versatility of spatial data acquisition and also diversified the approaches by which it is integrated with remote sensing and geographic information systems (Sood et al. 2015). Through GPS technology, satellites broadcast signals allow GPS receivers to calculate geographic coordinates (latitude and longitude) and record the position of interest. This allows crop and soil measurements to be mapped in real time (Adamchuk et al. 2004), and accuracies of 1–5 m are now possible. The information provided through GPS ties together all relevant layers of information obtained for a field, such as soil type, nutrient levels, etc. (Kitchen et al. 2002). GPS systems are widely available and their potential is growing.

There are many types of sensors now in use in PA. Remote sensing involves collecting data from a distance using specialised equipment that can be hand-held devices, mounted on an aircraft or satellite-based (Davis et al. 1998). Robust, low-cost, and, preferably, real-time sensing systems are needed for implementing various PA technologies. Remote sensing has a wide range of potential applications including detection of crop stress; monitoring variability in crops, soils, weeds, insects, and plant disease; detection of unusual conditions, such as broken drainage tiles or crop injury during cultivation; and yield estimation, which is highly dependent on the type and variety of crop and GIS applications (Strickland et al. 1998). Remote sensing provides the solution of monitoring the spectral and spatial changes over time at high resolution by showing spatiotemporal changes that provide a benchmark to understand the variability that has occurred over a period of time (Sood et al. 2015). Remote sensing has therefore been used operationally for pre-harvest forecasting of yield. The technology has been used with other models to calculate crop growth based on meteorological data, which is then used to predict yield as a function of biomass growth rate. Remote sensing has also been used to predict parameters such as moisture content, crop phenology, crop growth and evapotranspiration (Moran et al. 1997).

Yield monitors are mounted on equipment such as combine harvesters. Yield monitors are made of several components; these include a data storage device, user interface (display and key pad), and a console located in the combine cab, which controls the integration and interaction of these components (Grisso et al. 2005a). Grain yields are measured using yield sensors-impact or mass flow sensors, weight-based sensors, optical yield sensors, and γ-ray sensors.

Field sensor systems also allow users to record visual observations on crop growth, weeds, diseases, or other anomalies while walking in the fields (Zhang et al. 2002). Plant chlorophyll reflectance is the measurement used because it correlates closely with the nitrogen content of the plant and with the resulting plant mass. Reflectance in the near infrared area is easily distinguishable from surrounding plant matter and soil. In combination with type-specific growth functions for individual plants, growth deficits may be detected and remedied by real-time application of nitrogen-fertilisers (Auernhammer 2001). Soil, crop and anomaly sensors have also been developed (Zhang et al. 2002).

### 9.2.2 Machine Controls

Machine controls are used to guide field equipment and can vary the rate, mix, and location of water, seeds, nutrients, or chemical applications (Grisso et al. 2005b). Variable-rate application (VRA) or variable rate technology (VRT) is a process where crop production input rate is changed within a field in response to variable factors that affect the optimum rate of application (Sawyer 1994; Bolotova 2006). The approach allows farmers to identify the characteristics of a field in detail and then correlate specific features to specific geographic locations, thereby making it possible to focus on areas that need attention and treat them according to yield potential (Kincheloe 1994).

According to Grisso et al. (2011), VRA technologies can be used with or without a GPS system and there are two basic technologies for VRA: map-based and sensor-based. To develop a prescription map for nutrient VRA in a particular field, the map-based method could include these steps:

- Perform systematic soil sampling (and lab analysis) for the field.
- Generate site-specific maps of the soil nutrient properties of interest.
- Use an algorithm to develop a site-specific nutrient prescription map.
- Use the prescription map to control a fertilizer variable-rate applicator.

A positioning system is used during the sampling and application steps to record the location of the sampling points in the field and to apply the prescribed nutrient rates in the appropriate areas of the field (Grisso et al. 2011).

The sensor-based method provides the capability to vary the application rate of inputs with no prior mapping or data collection involved. Sensors on the applicator measure soil properties or crop characteristics "on the go". Based on this continuous stream of information, a control system calculates the input needs of the soil or plants and transfers the information to a controller, which delivers the input to the location measured by the sensor (Grisso et al. 2011; Bakhtiari et al. 2013).

### 9.2.3 Computer-based Systems

Computer-based systems include GIS maps and databases that use sensor information to "prescribe" specific machine controls (Grisso et al. 2005b). Computer software including spreadsheets, databases, GIS, and other types of application software are readily available and most are easy to use (Grisso et al. 2005).

GIS can be used to produce images, drawings, animations, and other cartographic products. Natural inputs in farming can be better understood and managed with GIS applications such as crop yield estimates, soil amendment analyses, and erosion identification and remediation (Sood et al. 2015). A GIS for crop production might include information from various sources pertaining to field history; input operations, GPS-based yield maps and soil surveys, aerial photography, satellite imagery, and pest or pathogen scouting data (Strickland et al. 1998).

## 9.3 Operations Benefitting from Precision at the Farm Level

Precision in field operations in crop production is a cornerstone of profitable and sustainable farming. With high precision, all production resources are used more efficiently and the farmer realizes increased profits. Precision farming allows better management of spatial and temporal variability of production factors on the farm through scheduled and spot application of agricultural inputs. Studies have shown evidence of differences between owners and soil types and the varied response of soil types to weather and the potential for improvement in irrigation practices using temporal and spatial patterns (Kerry et al. 2017). A number of field operations require high precision in both commercial and small-scale farming sectors.

Preparation of the soil for planting is critical for establishing the appropriate tilth that allows good soil-seed contact at planting. A well-prepared soil is the beginning of good crop establishment, which directly influences the final yield. With good soil-seed contact, faster seed germination is facilitated whenever adequate soil moisture is available. A well-prepared soil allows the correct planting depth to be achieved for a selected crop. Planting depth depends on seed size of a selected crop and is a critical factor that influences crop establishment and the final yield (Lafitte 1994).

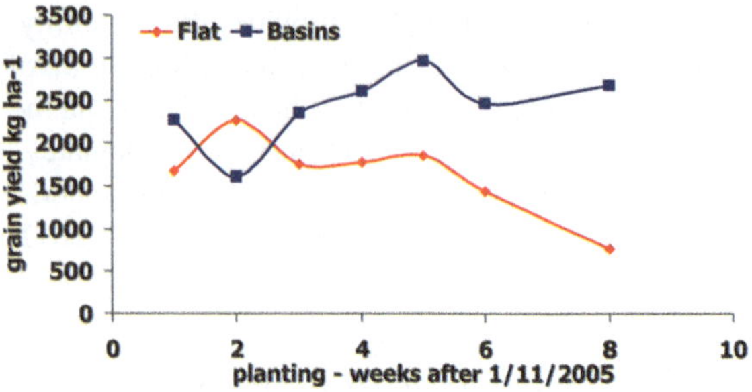


**Fig. 9.1** Observed variation in smallholder cereal grain yield response to planting dates under conventional (flat) and conservation agriculture (basins) cropping systems in southern Zimbabwe. *Source* Twomlow et al. (2008)

Crop density is often related to the final yield, and correct seed spacing is therefore critical on the farm and dependent on the available rainfall, soil type and selected crop varieties (Twomlow et al. 2006). Crop density can be reduced if planting depth and the prevailing soil moisture conditions are not ideal for germination. Crop varieties are developed for different agro-ecological regions, and farmers should select varieties suitable for their area. High-quality seed also supports good crop establishment and hence correct density in the field.

The final crop yield is directly influenced by planting time of a selected crop in a growing season. Delaying planting results in significant crop yield losses especially where small-scale farmers rely on conventional practices (Fig. 9.1).

Under sub-humid conditions of Southern Africa, every week's delay in maize planting reduces yield by 5% (Shumba et al. 1989). In the smallholder sector, PA practices such as basin planting increase the sowing window, which in turn benefits the crop yield (Fig. 9.1). Planting basins are hand dug holes spaced at 75–90 cm × 50–75 cm for seed and basal fertilizer placement during sowing. Where irrigation is available, precision in the timing of planting can be increased. Under rainfed systems, farmers need to rely on seasonal rainfall forecasts in order to increase precision on planting timing.

Precision in fertilization starts with the selection of the correct plant nutrients required to support the growth of a selected crop (Bationo et al. 2012). Through soil analysis, limiting nutrients can be identified and their required quantities determined. The right fertilizer can then be purchased to supply the limiting plant nutrients. The right fertilizer quantities should be applied at the appropriate crop growth stage. Applying fertilizers at the wrong growth stage promotes inefficient utilization of nutrients by crops. Fertilizers can be split applied at different crop stages in order to improve synchrony between crop demand and soil supply during the growing season. The split application also reduces nutrient losses, e.g. nitrogen (N) and phosphorus (P), through leaching and erosion. Fertilizers such as urea (46% N) should be covered soon after application while others can be applied on the soil surface without covering. Basal fertilizers should be applied at planting in order to maximize utilization of nutrients such as P and N, which are critical for early crop development (Lafitte 1994).

Weed competition with crops for plant nutrients, soil moisture and sunlight significantly reduces the final yield. Timely weed control, either manually or by herbicides or a combination of the two approaches, is paramount to avoiding yield losses in the farming system (Table 9.1).

Delayed and inefficient weed control can result in 30% loss in the potential yield of cereals grown in sub-Saharan Africa (Twomlow et al. 2006). Where herbicides are used, the correct herbicide should be applied depending on the existing weed spectrum and crop combinations on the farm. The application rate of a herbicide will depend on the weed spectrum, growth stage of weeds, as well as the soil type on the farm. When weeds have grown bigger, a higher application rate of some herbicides can be used. Calibration of spraying equipment is important to avoid under or over application of a herbicide. Prevailing weather conditions should be considered before a herbicide is applied to ensure the effectiveness of the chemical on weed plants. Herbicides should not be applied just before rain or on a windy day because they will not be effective in controlling weeds.

The commercial farming sector requires water resources for full or supplementary irrigation while the smallholder sector is largely a rainfed system. Maximum crop yield should be produced from every drop of water applied through irrigation in order to realize profits from commercial crop production (Dennis and Nell 2002). This is particularly important in the face of climate change and dwindling water resources worldwide. Precision in irrigation should be site specific, depending on topography, soil type, crop type and the potential for environmental damage from the farming operations. Factors to consider include how much water to apply, how to apply it and when irrigation should occur on the farm.

A precise assessment of crop water requirement is the first step in ensuring that adequate water is applied to the crop and that yield penalties from soil moisture stress are minimized. Over irrigation and under irrigation are equally damaging to the crop and its environment, and should be avoided. Similarly, the selected irrigation method should minimize water losses from the cropping area while supplying water uniformly across the field. Irrigation water should be applied when required by the crop, and timely application is critical in order to synchronize water supply with fertilization because nutrients can be lost through erosion or leaching.

**Table 9.1** Comparative performance of maize in high and low rainfall seasons as influenced by weeding time under smallholder farming conditions

| Time of weeding in weeks after emergence (WACE) | High rainfall season (881 mm) | | Low rainfall season (507 mm) | |
|---|---|---|---|---|
| | Grain yield (50 kg bags/ha) | % Yield reduction due to delayed weeding beyond 2 WACE | Grain yield (50 kg bags/ha) | % Yield reduction due to delayed weeding beyond 2 WACE |
| 2 | 108 | – | 44 | – |
| 4 | 50 | 54 | 34 | 23 |
| 6 | 22 | 80 | 6 | 86 |
| 8 | 8 | 93 | 4 | 91 |
| No weeding at all | 4 | 96 | 2 | 95 |

*Source* Mabasa and Nyahunzvi (1994)

Irrigation scheduling is, therefore, a fundamental activity in the commercial farming sector and an important element in precision agriculture.

## 9.4 Benefits of Precision in Commercial and Smallholder Farming Sectors

In Africa, significant benefits can be accrued by commercial and small-scale farmers through the implementation of precision agriculture practices. The benefits on a farm can be crop, economic and environment related and all are important for the farmer and beyond the farm.

### 9.4.1 *Crop Related Benefits*

Precision agriculture practices increase nutrient and water use efficiencies resulting in high crop yields in the farming system (Dobermann and Nelson 2013). Good early crop establishment is achieved when appropriate planting time and depth, as well as the plant nutrient and soil moisture conditions, are ideal. During fertilization, the correct fertilizer quantity is placed in the right place relative to the plant and at the correct crop growth stage (Ibrahim et al. 2015). Water is precisely applied through irrigation and where rainfed cropping is practised, *in situ* rainwater capturing practices ensure that soil moisture is available close to the crop. PA practices such as no-till and conservation agriculture improve soil quality over time. Improvements in soil structure through reducing tillage allows higher rain and irrigation water infiltration into the soil, organic matter accumulation and increased soil biodiversity (Leech and Newbold 2012; Thierfelder et al. 2014). These improved soil conditions are critical in buffering agroecosystems against high variability in rainfall and other climate elements, especially in rainfed systems. Timely and effective weed control reduces competition for plant nutrients and soil moisture, allowing all resources to be channelled towards the growing crop. This results in increased water and nutrient-use efficiency in the farming system. When these practices are applied together crop yields are increased substantially on the farm. With PA, production resources such as labour and mineral fertilizers are efficiently utilized thereby allowing more land to be put into production using a given set of agricultural inputs purchased by the farmer (Knight and Malcom 2009).

### 9.4.2 *Economic Benefits*

PA can be a pillar of efficient production in both small-scale and commercial agroecosystems leading to substantial savings in agricultural inputs (Jensen et al.

2012; Takacs-Gyorgy et al. 2013). Economically, precision agriculture saves costs on farming inputs because they are applied as needed in the cropping system selected on the farm (Tweteen 1996; Silva et al. 2007), although in some studies farmers felt PA machinery was too expensive (Reichardt and Jürgens 2009). PA can result in up to 60% savings in agro-chemicals such as pesticides and close to 30% savings in mineral fertilizers (Batte and Van Buren 1999; Rider et al. 2006). PA practices also reduce the risk of over or under applying production inputs such as fertilizers and agro-chemicals and in some studies PA has shown a good potential for in-season site-specific N management in small scale farming systems (Cao et al. 2012). Therefore, for smallholder farmers, precision in the use of fertilizers and herbicides reduces losses and therefore lowers input costs (Twomlow et al. 2010). Use of well-calibrated farm equipment reduces application costs of fertilizers, herbicides and other agro-chemicals used for crop production. In commercial and smallholder farming sectors, precision agriculture practices increase the productivity of labour available on the farm because of the increased output from the labour invested in farming operations. Ultimately farm profits are increased because of the increased output per unit input invested into production by the farmer.

### *9.4.3 Environmental Benefits*

Precision is more critical in the commercial farming sector where crop production involves high input use every year, e.g. fertilizer, herbicides and pesticides. Pollution of ground and surface water resources by agrochemicals is of great concern worldwide. Precision agriculture practices play a critical role in reducing over application of chemicals that are harmful to the environment (Oliver et al. 2013). With precision agriculture, production can be intensified on unit area and this slows down the opening up of new lands for crop production (Dobermann and Nelson 2013). Soil degradation through erosion is also minimized on and around farms where precision agriculture practices are implemented. Precision agriculture is therefore fundamental for conserving natural resources on and around the farming areas.

## 9.5 Precision Agriculture and Food Security

The world population is projected to reach 9 billion by 2050 (Godfray et al. 2010), while sub-Saharan Africa will have 2 billion people by the same year (FAO 2015). The current population growth should be accompanied by higher growth rates in food production and accessibility. Precision agriculture tools and practices have a significant role to play in reducing world food deficits in both developed and developing countries. Precision agriculture enables both smallholder and commercial farmers to increase crop yields by reducing the yield gap of the major food crops such as maize, rice, wheat and pulses (Fig. 9.2).

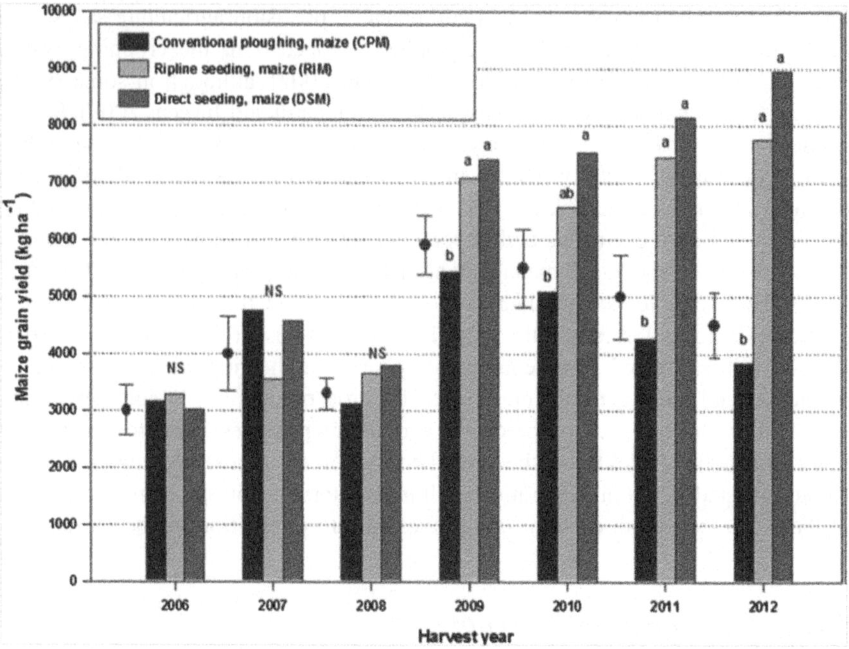


**Fig. 9.2** Contribution of precision agriculture practices (animal traction ripline and direct seeding) towards increased maize yields over time under smallholder farming conditions in southern Africa. *Source* Thierfelder et al. (2014)

Animal traction ripping and direct seeding involves the use of ripper tines and direct planters for opening planting furrows and placing seed and basal fertilizer in the furrows during sowing. The use of PA-based practices provides better and more accurate management of farm operations resulting in an overall increase in productivity and profitability of the farming systems. This, in turn, provides incentives for farmers to invest in farming, thereby increasing crop outputs which ultimately reduce food deficits. When farming becomes profitable, farmers can invest in high-value nutritious crops, and this increases both nutrition and food security, particularly in the developing countries. High-quality crops are also maintained because practices such as fertilization and use of agrochemicals are applied as required in the cropping system. Precision in fertilization of major cereals has shown an increase in both fertilizer use efficiency and final yield (Talebpour et al. 2015). Similarly, PA tools using indices such as Normalized Difference Vegetation Index (NDVI) have indicated increased grain protein content in wheat with precise fertilization practices.

Rainfall and temperature are major climatic factors that cause stress on food production systems. Precision in planting methods and timing, fertilization and weed control reduces production risks associated with these climatic factors (Lafitte 1994). This increases the opportunity to achieve high crop yields even in growing seasons with unfavourable climatic conditions.

Food production depends on the available water resources for the commercial sector and rainfall for the small-scale farming system. Worldwide, efficient use of available water in agriculture is now paramount given the increasing demands by other non-agricultural sectors. Irrigation methods such as drip and pivot systems efficiently deliver water to crops and hence increase yield and consequently food security.

## 9.6 Precision Agriculture in Africa: Case Studies

There is evidence of application of PA in African commercial farming systems, but the use of PA in smallholder agriculture has been negligible (Lowenberg-DeBoer and Erickson 2010). Searching for the "appropriate PA technology" for small farms is a real challenge faced by scientists and engineers (Mondal and Basu 2009). Yet despite these challenges, there are promising success stories for PA application in both the commercial and smallholder farming sectors in Africa. Two successful case studies are demonstrated.

### *9.6.1 Case Study 1: FruitLook, South Africa*

FruitLook is a state-of-the-art information technology that helps deciduous fruit and grape farmers to be water efficient and climate-smart.

Water is critical for agricultural productivity. In South Africa, water is a scarce resource that is increasingly threatened by pollution, rising demands and wastage (e.g. through inadequate maintenance of water service infrastructure). Farmers in the Western Cape Province experience increased competition for water from other sectors while also having to cope with changing rainfall patterns. Climate change projections for the Western Cape suggest a warming of 1.5–3 °C by around 2050. Many parts of the province will experience more hot days, fewer cold days and increased evaporation. In addition, climate models also project a reduction in winter rainfall across the province. Hence, while the need for irrigation increases, the replenishment of existing water sources (e.g. rivers, ground water and dams) becomes less certain.

To maintain and increase their agricultural productivity now and in the future, farmers must increase their water use efficiency. Simply put, farmers need to achieve "more crop per drop". The Western Cape Department of Agriculture (WCDoA) is offering a state-of-the-art, PA-based tool, called FruitLook, to deciduous fruit and grape farmers in the Western Cape. FruitLook allows these farmers to improve their water use efficiency using information from spatial data derived from remote sensing.

Using satellite technology, FruitLook provides weekly, semi-real time information on crop growth, evapotranspiration deficits and crop nitrogen status for irrigation blocks in orchards and vineyards in key growing areas of the Western

Cape. This quantitative, spatial information on water, vegetation and climate assists farmers to better understand the effects of their water use and crop management decisions and to reduce costs by saving on inputs (such as water, fertilisers and electricity). Through the spatial identification of problems or unusual events, the technology can also improve the quantity and quality of yields and therefore increase profit. To have access to the technology, farmers and other interested parties have to register on the FruitLook web portal (www.fruitlook.co.za). On the portal, farmers can demarcate their irrigation blocks, analyse crop growth and water status over time during the growth season (October to April), as well as compare crop development for different growing seasons.

The current total benefits (cost savings and increased revenue) for the Western Cape fruit farmers using PA is estimated at R64 million to R390 million per annum. (Currently, the WCDoA offers the FruitLook services free of charge to the farming community.) In the long term, FruitLook is expected to become a commercial product, where farmers will need to pay to obtain spatial and temporal information on growth, moisture and minerals.

For more information on FruitLook contact André Roux, Director: Sustainable Resource Management, Department of Agriculture Phone: +27 21 808 5010, E-mail: andrer@elsenburg.com or Caren Jarmain, Independent Researcher Email: cjarmain@gmail.com Phone: +27 21 880 1008. FruitLook Websites: www.fruitlook.co.za; www.elsenburg.com. The full version of the report on this case study is available at http://www.greenagri.org.za/assets/documents-/SmartAgri/Case-Studies/1.-Case-Study-FruitLook-FINAL.pdf.

### 9.6.2 *Case Study 2: Chameleon and Wetting Front Detector Sensors (Mozambique, Tanzania and Zimbabwe)*

Chameleon and Wetting Front Detector Sensors enable small scale farmers in Mozambique, Tanzania and Zimbabwe to cut down irrigation frequency fifty times and double productivity.

A Chameleon and the Wetting Front Detector (WFD) are relatively low-cost devices developed at the Commonwealth Scientific and Industrial Research Organisation (CSIRO) in Australia for aiding and supporting good irrigation scheduling and soil nutrient management. The Chameleon consists of sensors which are buried at different soil depths relative to rooting depths of crop plants during development. To take readings the chameleon sensors are connected to a solar-powered mobile Chameleon reader (Plate 9.1). The soil moisture content readings are interpreted by the reader in terms of colours (blue = wet; green = drying; red = dry). Generally, blue corresponds to the matric potential of less than 25 kPa, green corresponds to 25–50 kPa, while red corresponds to the matric potential of greater than 50 kPa.

**Plate 9.1** Chameleon sensors (blue wires) buried below the soil surface in a tomato crop field and connected to the solar-powered chameleon reader with different colours (blue, green, red). Picture by Ikenna Mbakwe (2014)

The WFD is a funnel-shaped container that is buried open end up in the soil and works on the principle of convergence (Stirzaker 2003). The WFD comprises a specially shaped funnel, a filter and a mechanical float mechanism (Fig. 9.3).

The free water produced at the base of the funnel flows through a filter into a chamber where it activates an electrical float switch (the detector is "tripped"). The float switch may be connected to an alarm, irrigation controller or connected in series with a solenoid valve so that the irrigation is switched off automatically (ibid). WFDs are usually used in pairs—one that makes immediate contact with the

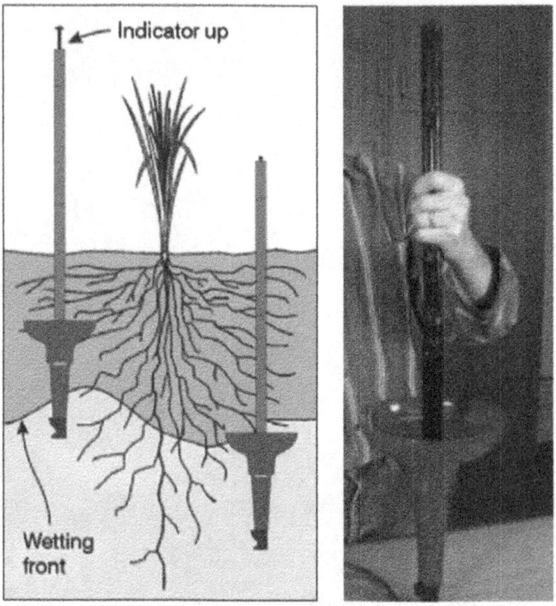


**Fig. 9.3** A FullStop Wetting Front Detector (WFD). The funnel part is buried in the soil with the black tube protruding above the soil surface. When a wetting front reaches the detector a red indicator pops up (Stirzaker 2005)

plant or buried about one third and the other one about two thirds of the depth in the active root soil environment.

Both, the Chameleon and WFD were adopted by the Australian Centre for International Agricultural Research (ACIAR) funded project: "Small Irrigation Project in Africa" to facilitate farmer adaptive learning on soil moisture and nutrient management. The project (FSC-2013-006) "Increasing Irrigation Water Productivity in Mozambique, Tanzania and Zimbabwe through On-Farm Monitoring, Adaptive Management and Agriculture Innovation Platforms (Small Irrigation Project in Africa)" has been implemented in two irrigation schemes in each of the three project countries. About sixty farmer plots in the Kiwere (Tanzania), 25 de Setembro (Mozambique), Mkoba and Silalatshani (Zimbabwe) irrigation schemes have been equipped with the Chameleon and WFD to demonstrate the technology and to train farmers on how to use the sensors to schedule irrigation and monitor soil nutrients.

The monitoring of the Chameleon and WFD in farmers' plots in each research country is conducted by trained field extension officers in collaboration with the farmers. The Chameleon records the colours of the sensors at different depths, e.g., 20, 30, 40 and 50 cm below the soil surface, while WFDs measure nitrates and electrical conductivity (EC) using nitrate and EC meters. The measurement of nitrates and EC is taken from the water that is extracted from the bottom of the WFD's funnel. Nitrates are used as an indicator of the level of fertilisers in the soil while EC indicates the level of salinity. The measured values are compared with existing established ranges in order to be able to provide proper advice to the farmer (Table 9.2). The measurements are taken twice a week and the research project

**Table 9.2** Levels of nitrate and salt in the soils

| Parameter | Range | Description |
|---|---|---|
| Nitrate | <25 mg/L | Nitrate level is low |
| | 25–100 mg/L | Nitrate level is good |
| | >100 mg/L | Nitrate level is high |
| Salt | <2 dS/m | Salinity acceptable for most plants |
| | 2–4 dS/m | Salinity a potential problem |
| | >4 dS/m | Salinity too high for most plants |

*Source* Pittock and Ramshaw (2016)

encourages the farmers' presence in their plots during the exercise whenever possible. Once the measurements have been taken, the results are immediately communicated to the farmer who owns the plots, and the extension officer advises about the next irrigation event and other agronomic practices based on the measurements and field observations. For farmers who are not present on the plots, communication is achieved through a mobile phone. Other information recorded as part of the monitoring includes the date of last irrigation events, dates of agrochemical application (fertiliser, pesticide) and a picture illustrating the condition of crops on the plot.

The impacts from the use of the tools have been very remarkable in all three project research countries. Before the project, in the Kiwere irrigation scheme in Tanzania, farmers' decisions regarding irrigation relied on observing soil appearance ("perceived dry or wet"), but now the Chameleon—popularly known by farmers as *Kinyonga* (Swahili for Chameleon)—indicates whether or not to irrigate. Previously, farmers irrigated on average every two days but now they have been irrigating only once or at most twice a week. These changes in irrigation practices occurred just one year after starting to use the tools and learning from the records of the Chameleon and WFDs. Overall, these farmers have reduced irrigation frequency by almost fifty percent, and by achieving this they have been able to save labour, freeing it for other household economic activities such as micro businesses. Further, water use conflicts between farmers in the schemes have been eliminated.

In addition to the reduction of water used for irrigation, the productivity of water and land have doubled or more than doubled (Fig. 9.4). Our understanding of this impact is that the farmers are now able to optimally manage fertilizer and soil nutrients using these tools, and so avoid leaching of the nutrients beyond the root zone through excessive irrigation. The increase of crop production has provided a considerable motivation to nearby farmers to adopt changed irrigation practices.

The benefits of using these tools are spreading amongst many non-trial scheme farmers to the extent that they are now asking the project whether they can also have them on their farms. The use of the Chameleon and WFDs in the project area has proven within a short time to have positive impacts; e.g. water savings, labour savings and increased crop production. However, from survey results, it was found that other non-water related factors were also important, e.g. access to inputs, farm equipment, transportation, value-adding opportunities and functional markets,

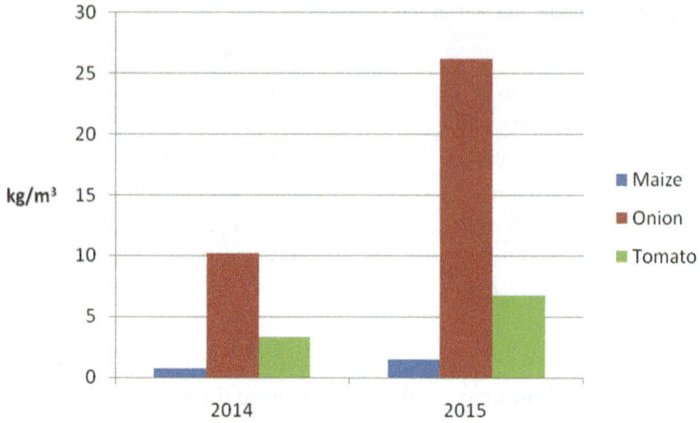


**Fig. 9.4** Water productivity of maize, onion and tomato (kg $m^{-3}$) in Kiwere Scheme before (2014) and after (2015) implementation of the Chameleon and WFDs. *Source* Pittock and Ramshaw (2016)

which impeded the ability to increase production and farm profitability (Mdemu et al. 2017).

A major challenge to the implementation of the Chameleon innovations, however, is the lack of an assured supply of the devices to the farmers in the schemes or nearby when they need them. At the moment the sensors are fabricated in Australia and ordering logistics are cumbersome for farmers. Further, the cost of shipping and clearance in African countries may exceed the actual cost of the tools.

The project is funded by Chameleon and the Australian Centre for International Agricultural Research (ACIAR). For more information contact Makarius Mdemu (Ph.D.): Project Team Leader for Tanzania. Institute of Human Settlement Studies (IHSS), Ardhi University, P.O. Box 35176, Dar es Salaam, Tanzania. E-mail: mak_mdemu@yahoo.co.uk.

## 9.7 Conclusions

Precision farming has seen strides in terms of technology development and application in farming, particularly in the developed world. State of the art technologies in GPS, remote sensing, crop, soil, and positioning sensors, machines controls and computer based systems have been developed. High costs and knowledge requirements, unavailability of many services, and uncertain benefits previously seemed to preclude any possibility of precision farming in developing countries. However, there is now substantial evidence of PA technology adoption and potential in developing countries, as shown by the two case studies presented in this chapter; as well as recent pilot tests conducted by the Food and Agriculture Organisation on rice production systems in developing countries.

The application of the balanced soft and hard PA technologies based on the needs and specific socioeconomic conditions of a country makes PA suitable not only for developed countries but also for developing countries and can work as a tool to reduce the gap in food production and technology between the developed world and the rest. However, it should also be noted that not all PA techniques will align with sustainable pathways, such as conservation agriculture. That makes PA application site and need specific in many instances. There is also a wide variability of techniques, some suitable for commercial farming while others are more inclined to smallholder farming. The choice of techniques will ultimately depend on affordability.

In Africa, the implementation of PA has so far been mainly focused on technologies that reduce water use and input costs. Efficient use of available water in agriculture is crucial, given the increasing demands by other non-agricultural sectors and drought threats. Farmers have invested in precision irrigation methods such as drip and pivot systems to efficiently deliver water to crops, increasing yield and consequently improving food security. There are examples of successful PA applications in both the commercial and smallholder sector in Africa. While for commercial farmers, adoption of PA has been largely quick and welcome, there may be some challenges for smallholder farmers such as the limited local availability of components as shown by the case of the Chameleon technology in Tanzania.

## References


Adamchuk, V. I., Hummel, J. W., Morgan, M. T., & Upadhyaya, S. K. (2004). On-the-go soil sensors for precision agriculture. *Computers and Electronics in Agriculture, 44*(1), 71–91.

Auernhammer, H. (2001). Precision farming-the environmental challenge. *Computers and Electronics in Agriculture, 30*(1), 31–43.

Bakhtiari, A. A., & Hematian, A. (2013). Precision farming technology, opportunities and difficulty. *International Journal for Science and Emerging Technologies with Latest Trends, 5* (1), 1–14.

Bationo, A., Fairhurst, T., Giller, K., Kelly, V., Lunduka, R., & Mando, A. (2012). Handbook for integrated soil fertility management. *Africa Soil Health Consortium* (CAB International 156).

Batte, M. T., & VanBuren, F. N. (1999). Precision farming-factors influencing profitability. In *Northern Ohio Crops Day meeting*, Wood County, Ohio (Vol. 21).

Bolotova, Y. (2006). Crop production using variable rate technology for P&K in the United States Midwest: Evaluation of profitability. American Agricultural Economics Association (New Name 2008: Agricultural and Applied Economics Association). Annual meeting. Long Beach, CA No. 21098:23–26 July.

Bongiovanni, R., & Lowenberg-DeBoer, J. (2004). Precision agriculture and sustainability. *Precision Agriculture, 5*(4), 359–387.

Cao, Q., Cui, Z., Chen, X., Khosla, R., Dao, T. H., & Miao, Y. (2012). Quantifying spatial variability of indigenous nitrogen supply for precision nitrogen management in small scale farming. *Precision Agriculture, 13*(1), 45–61.

Davis, G., Casady, W. W., & Massey, R. E. (1998). *Precision agriculture: An introduction*. Columbia: Extension publications, (MU).

Dennis, H. J., & Nell, W. T. (2002). Precision irrigation in South Africa. In *A Paper Presented at the 13th International Farm Management Congress*, 7–12 July 2002. Wageningen, The Netherlands.

Dobermann, A., & Nelson, R. (2013). *Opportunities and solutions for sustainable food production*. United Nations: Sustainable Development Solutions Network.

FAO. (2015). *The state of food and agriculture. Social protection and agriculture: Breaking the cycle of rural poverty* (p. 151). Food and Agriculture Organization of the United Nations. Rome, Italy.

Godfray, H. C. J., Beddington, J. R., Crute, I. R., Haddad, L., & Lawrence, D. (2010). Food security: The challenge of feeding 9 billion people. *Science, 327,* 812–818.

Grisso, R. D., Alley, M. M., & Heatwole, C. D. (2005). Precision Farming Tools. Global Positioning System (GPS).

Grisso, R. D., Alley, M. M., & McClellan, P. (2005a). Precision Farming Tools. Yield Monitor.

Grisso, R. D., Alley, M. M., McClellan, P., Brann, D. E., & Donohue, S. J. (2005b). Precision farming a comprehensive approach.

Grisso, R. D., Alley, M. M., Thomason, W. E., Holshouser, D. L., & Roberson, G. T. (2011). Precision farming tools: Variable-rate application.

Ibrahim, A., Pasternak, D., & Fatonji, D. (2015). Impact of depth of placement of mineral fertilizer micro-dosing on growth, yield and partial nutrient balance in pearl millet cropping system in the Sahel. *Journal of Agricultural Science, 153,* 1412–1421.

Jensen, H. G., Jacobsen, L. B., Pedersen, S. M., & Tavella, E. (2012). Socioeconomic impact of widespread adoption of precision farming and controlled traffic systems in Denmark. *Precision Agriculture, 13*(6), 661–677.

Kerry, R., Goovaerts, P., Giménez, D., & Oudemans, P. V. (2017). Investigating temporal and spatial patterns of cranberry yield in New Jersey fields. *Precision Agriculture, 18*(4), 507–524.

Kincheloe, S. (1994). Tools to aid management: The use of site specific management. *Journal of Soil and Water Conservation, 49*(2), 43.

Kitchen, N. R., Snyder, C. J., Franzen, D. W., & Wiebold, W. J. (2002). Educational needs of precision agriculture. *Precision Agriculture, 3*(4), 341–351.

Knight, B., & Malcolm, B. (2009). A whole-farm investment analysis of some precision agriculture technologies. *Australian Farm Business Management Journal, 6*(1), 41.

Kushke, I., & Jordaan, J. (2017). Agriculture: 2017 Market Intelligence Report. [Online]. http://www.greencape.co.za/assets/Uploads/GreenCape-Agri-MIR-2017-electronic-FINAL-v1.pdf. Accessed 8 Sep 2017.

Lafitte, H. R. (1994). *Identifying production problems in tropical maize: A field guide*. DF, CIMMYT: Mexico.

Leech, P., & Newbold, A. (2012). Agricultural Engineering: a key discipline for agriculture to deliver global food security. A status report developed by IAgrE in response to the UK Government's Foresight Project: Global Food and Farming Futures. Cranfield, UK. https://iagre.org/kcfinder/upload/files/documents/3315_IAG_Global_Food_Security_v11.pdf. 58p.

Lowenberg-DeBoer, J., & Erickson, B. (2010). The search for the killer app: Precision farming in Africa. *Georgetown Journal of International Affairs, 11*(2), 107–116. http://www.jstor.org/stable/43133849.

Mabasa, S., & Nyahunzvi, S. (1994). Maize competition in communal areas in three agro-ecological zones of Zimbabwe. In Jewell, D. C., Waddington, S. R., Ransom, J. K., & Pixley, K. V. (Eds.), *Maize for stress environments. Proceeding of the fourth Eastern and Southern Africa regional maize conference*, (Vol. 28) March-1 April 1994. Harare, Zimbabwe.

McBratney, A., Whelan, B., Ancev, T., & Bouma, J. (2005). Future directions of precision agriculture. *Precision Agriculture, 6*(1), 7–23.

McLoud, P. R., Gronwald, R., & Kuykendall, H. (2007). *Precision agriculture: NRCS support for emerging technologies* No. 1. Agronomy Technical Note.

Mdemu, M. V., Mziray, N., Bjornlund, H., & Kashaigili, J. J. (2017). Barriers to and opportunities for improving productivity and profitability of the Kiwere and Magozi irrigation schemes in Tanzania. *International Journal of Water Resources Development, 33*(5), 725–739.

Mondal, P., & Basu, M. (2009). Adoption of precision agriculture technologies in India and in some developing countries: Scope, present status and strategies. *Progress in Natural Science, 19*(6), 659–666.

Moran, M. S., Inoue, Y., & Barnes, E. M. (1997). Opportunities and limitations for image-based remote sensing in precision crop management. *Remote Sensing of Environment, 61*(3), 319–346.

Oliver, M. A., Bishop, T. F. A., & Marchant, B. P. (2013). *Precision agriculture for sustainability and environmental protection. Earth Scan Food and Agriculture*. Abingdon: Routledge.

Pittock, J., & Ramshaw, P. (2016). Annual report: Increasing irrigation water productivity in Mozambique, Tanzania and Zimbabwe through on-farm monitoring, adaptive management and agricultural innovation platforms. Project number, FSC-2013–006.

Reichardt, M., & Jürgens, C. (2009). Adoption and future perspective of precision farming in Germany: Results of several surveys among different agricultural target groups. *Precision Agriculture, 10*(1), 73–94.

Rider, T. W., Vogel, J. W., Dille, J. A., Dhuyvetter, K. C., & Kastens, T. L. (2006). An economic evaluation of site-specific herbicide application. *Precision Agriculture, 7*(6), 379–392.

Rilwani, M. L., & Ikhuoria, I. A. (2006). Precision farming with geoinformatics: A new paradigm for agricultural production in a developing country. *Transactions in GIS, 10*(2), 177–197.

Sawyer, J. E. (1994). Concepts of variable rate technology with considerations for fertilizer application. *Journal of Production Agriculture, 7*(2), 195–201.

Shumba, E. M., Waddington, S. R., & Rukuni, M. (1989). Delayed maize plantings in a smallholder farming area of Zimbabwe: Problem diagnosis. *Zimbabwe Journal of Agricultural Research, 27,* 103–112.

Silva, C. B., do Vale, S. M. L. R., Pinto, F. A., Müller, C. A., & Moura, A. D. (2007). The economic feasibility of precision agriculture in Mato Grosso do Sul State, Brazil: A case study. *Precision Agriculture, 8*(6), 255–265.

Smith, R., & Baillie, J. (2009). Defining precision irrigation: A new approach to irrigation management. In *Irrigation Australia 2009: Irrigation Australia Irrigation and Drainage Conference: Proceedings* (pp. 1–6). Irrigation Australia Ltd.

Sood, K., Singh, S., Rana, R. S., Rana, A., Kalia, V., & Kaushal, A. (2015). Application of GIS in precision agriculture. In *Paper presented as lead lecture in national seminar on "Precision farming technologies for high Himalayas*" 04–05 October, 2015, organized by Precision farming development centre and High Mountain Arid Agriculture Research Institute, Leh, Ladakh, Jammu and Kashmir. India 8–16.

Stirzaker, R. (2005). Managing irrigation with a wetting front detector. *UK Irrigation, 33,* 22–24.

Stirzaker, R. J. (2003). When to turn the water off: Scheduling micro-irrigation with a wetting front detector. *Irrigation Science, 22,* 177–185. https://doi.org/10.1007/s00271-003-0083-5.

Strickland, R. M., Ess, D. R., & Parsons, S. D. (1998). Precision farming and precision pest management: The power of new crop production technologies. *Journal of Nematology, 30*(4), 431.

Sylvester-Bradley, R., Lord, E., Sparkes, D. L., Scott, R. K., Wiltshire, J. J. J., & Orson, J. (1999). An analysis of the potential of precision farming in Northern Europe. *Soil Use and Management, 15*(1), 1–8.

Takács-György, K., Lencses, E., & Takács, I. (2013). Economic benefits of precision weed control and why its uptake is so slow. *Studies in Agricultural Economics, 115*(1).

Talebpour, B., Turker, U., & Yegul, U. (2015). The role of precision agriculture in the promotion of food security. *International Journal of Agricultural and Food Research, 4*(1), 1–23.

Thierfelder, C., Rusinamhodzi, L., Ngwira, A. R., Mupangwa, W., Nyagumbo, I., Kassie, G. T., et al. (2014). Conservation agriculture in Southern Africa: Advances in knowledge. *Renewable Agriculture and Food Systems, 30,* 328–348.

Tweteen, L. (1996). Is precision farming good for society? An economist's view. *Better Crops, 80* (3), 3–5.

Twomlow, S., Hove, L., Mupangwa, W., Masikati, P., & Mashingaidze, N. (2008). Precision conservation agriculture for vulnerable farmers in low potential zones. In *Proceedings of the workshop on increasing the productivity and sustainability of rain-fed cropping systems of poor, smallholder farmers*. Tamale, Ghana, 22–25, September 2008.

Twomlow, S., Rohrbach, D., Dimes, J., Rusike, J., Mupangwa, W., Ncube, B., et al. (2010). Micro-dosing as a pathway to Africa's green revolution: Evidence from broad-scale on-farm trials. *Nutrient Cycling in Agroecosystems, 88,* 3–15.

Twomlow, S. J., Steyn, J. T., & du Preez, C. C. (2006). Chapter 19: Dryland farming in Southern Africa. In Petersen, G. A., Unger, W. P., & Payne, W. A. (Eds.), *Dryland agriculture (Agronomy monograph)* (2nd Ed. No. 23, pp. 769–836), Madison, Wisconsin: American Society of Agronomy.

Whelan, B. M., & McBratney, A. B. (2000). The "null hypothesis" of precision agriculture management. *Precision Agriculture, 2*(3), 265–279.

Zhang, N., Wang, M., & Wang, N. (2002). Precision agriculture—A worldwide overview. *Computers and Electronics in Agriculture, 36*(2), 113–132.

# Part III
# Ecosystems

# Chapter 10
# Introduction to Part III: Functions and Services of Ecosystems

**Mao Amis**

**Abstract** Global climate change is accelerating at an unprecedented rate, threatening ecosystem functions and the vital services they provide.

Global climate change is accelerating at an unprecedented rate, threatening ecosystem functions and the vital services they provide. There are many processes within ecosystems that influence important factors such as soil fertility and water quality that are important for human wellbeing. Understanding ecosystem response to environmental change is important for developing effective adaptation strategies. However, ecosystems are complex systems, which are constantly interacting with the climatic, social and economic system. Hence studying natural ecosystems response in isolation of the complexity brought about by these interactions is fruitless.

The concept of complex socio-ecological systems recognises that people and nature are interlinked and that humans must be regarded as part of nature instead of an externality. It therefore means that effective response to environmental change requires interdisciplinary approaches at multiple scales and levels. Such socio-ecological systems can be defined at multiple spatial, temporal and organisational scales. For example, from a social dimension, human behaviour has significant impact on land use trends, migration and value systems. This human behaviour can inturn result in negative environmental impact or stressors, such as biodiversity loss, land cover change, logging and illegal hunting. From an ecological perspectives, these stressors may lead to species extinction and loss of vital ecosystem functions like primary productivity, soil fertility and water quality.

M. Amis (✉)
African Centre for a Green Economy,
Cape Town, South Africa
e-mail: mao@africancentre.org

M. Amis
Graduate School of Business, University of Cape Town,
Cape Town, South Africa

 
P. Mensah et al. (eds.), *Systems Analysis Approach for Complex Global Challenges*,
https://doi.org/10.1007/978-3-319-71486-8_10

In order to sustain the impact of environmental change, socio-ecological systems must be able to effectively respond to perturbations so that they bounce back if they experience unexpected shocks to the system, which constitute resilience. Resilient ecosystems enable effective adaptation to global environmental change, recognising that it's virtually impossible to halt the impact of global change, as it is already well under way, however the ability to respond effectively remains an important factor.

Viewing natural ecosystems as complex socio-ecological systems has several advantages, including; providing a structure for complex processes, considers multiple drivers such as anthropogenic and biophysical factors it involves inter-disciplinary collaboration for effective ecosystem analysis and helps to assess the specific context of public goods and services.

Over the years our understanding of socio-ecological systems has improved significantly, enabling the development of effective climate change adaptation strategies. However, major research gaps still exist, and translating the learning into effective policies is still a challenge. The three chapters presented in this section attempt to respond to some key questions on socio-ecological systems, which is important for developing management options in practice. The contribution by (Scharler et al. 2018) unpacks resilience measures in ecosystems and socioeconomic networks. The issue of complexity and stability in adaptive ecological networks is explored in the contribution by (Landi et al. 2018) Finally, Banasiak et al. 2018, present a theory for aggregating methods for analysis of complex multiple scale systems.

In conclusion, ecosystems provide vital goods and services to humanity, it is our endeavour to ensure that these ecosystem services are secured in the face of adverse global change. However, progress can only be made when a systems approach is used to develop understanding of the inherent complexity of socio-ecological systems. Policy responses also need to be designed in a manner that recognise these complexity, for example by ensuring that in policy development both top down and bottom up strategies should be mainstream.

## References

Banasiak, J., Falkiewicz, A., Tchamga, M. S. S. (2018). Aggregation methods in analysis of complex multiple scale systems. In P. Mensah, S. Hachigonta, D. Katerere & A. Roodt (Eds.), *Systems Analysis Approach for Complex Global Challenges*. https://doi.org/10.1007/978-3-319-71486-8_5.

Landi, P., Henintsoa, M. O., Brännström, Å., Hui, C., Dieckmann, U. (2018). Complexity and stability of adaptive ecological networks. In P. Mensah, S. Hachigonta, D. Katerere & A. Roodt (Eds.), *Systems Analysis Approach for Complex Global Challenges*. https://doi.org/10.1007/978-3-319-71486-8_5.

Scharler, U. M., Fath, B. D., Banerjee, A., Fang, D., Mukherjee, J., Xia, L. (2018). Resilience measures in ecosystems and socio-economic networks. In P. Mensah, S. Hachigonta, D. Katerere & A. Roodt (Eds.), *Systems Analysis Approach for Complex Global Challenges*. https://doi.org/10.1007/978-3-319-71486-8_5.

# Chapter 11
# Resilience Measures in Ecosystems and Socioeconomic Networks

**Ursula M. Scharler, Brian D. Fath, Arnab Banerjee, Delin Fang, Le Feng, Joyita Mukherjee and Linlin Xia**

**Abstract** *Background and Significance of the topic*: This chapter contributes to the documentation of novel network-based resilience concepts to socio-ecological systems. Although the resilience concept has been studied in depth in ecological systems, it surely has relevance outside this area and in recent years has been a main domain of study for socioeconomic systems. This chapter provides an overview of the application of resilience concepts in ecology, with a particular focus on the application of two methods developed using ecological network analysis. *Methodology*: The first method uses information-theory based network analysis to ascertain the trade-off between efficiency and redundancy in networks (in terms of the structure and flows). The second method uses an energy-flow based method to assess keystoneness and the direct and indirect relations in the networks. *Application/Relevance to systems analysis*: Earlier work using information-theory based network analysis has shown that ecological systems display a robust balance between efficiency and redundancy in networks (in terms of the structure and flows) thereby bestowing them with robust and resilient features. Results indicate that a dam ecosystem in southwest China falls just short of the optimum but suffers

U. M. Scharler (✉)
School of Life Sciences, University of KwaZulu-Natal,
Durban 4001, South Africa
e-mail: scharler@ukzn.ac.za

B. D. Fath
Department of Biological Sciences,
Towson University, Towson, MD 21252, USA

B. D. Fath
Advanced Systems Analysis Program,
International Institute for Applied Systems Analysis, Laxenburg, Austria

A. Banerjee · J. Mukherjee
Ecological Modelling Laboratory, Department of Zoology,
Visva-Bharati University, Bolpur, India

D. Fang · L. Feng · L. Xia
State Key Laboratory of Water Environment Simulation,
School of Environment, Beijing Normal University, 100875 Beijing, China

 
P. Mensah et al. (eds.), *Systems Analysis Approach for Complex Global Challenges*,
https://doi.org/10.1007/978-3-319-71486-8_11

substantial loss of robustness when the phytoplankton community is perturbed. Application to a virtual water network shows the system is not near the robustness peak. Using the energy-flow based method, a South African estuary showed alteration of the keystone species depending on the seasonality; a land use change model of Beijing showed a decrease in mutualism due to urban expansion. *Policy and/or practice implications*: The case studies presented illustrate the application of ecological network analysis. Positive and negative relations between sectors of ecosystems or economic systems highlight the influence of various species and economies on one another, resulting in a comprehensive picture of relations, impacts and therefore management options to achieve balance between sectors. *Discussion and conclusion*: Overall, networks provided a useful model to illustrate system resilience measures, and other system analysis methods of direct and indirect impacts of system components on each other.

## 11.1 Introduction

In recent decades, resilience has become an important concept as ecosystems and socioeconomic systems are adversely impacted following chronic and acute resource exploitation stemming from economic policy based on growth rather than on environmental sustainability. Early warnings on the consequences of unlimited economic growth were already issued in the early 1970s by the Club of Rome (Meadows et al. 1972) on the limits that the Earth sets in the sustenance of an exponentially growing human population and its resource demands. These limits are tracked through measures such as ecological footprint in terms of human resource use, the rate of which has surpassed that of their renewal in the 1980s (Wackernagel et al. 2002). Recent investigations have looked at how human society might flourish within these biophysical limits by following general patterns and principles of ecological systems (Jørgensen et al. 2015).

The continued growth of the human population depends on the provision of many ecosystem services, mostly at no cost, but nevertheless vital for the survival of humankind. These include, for instance, pollination, availability of water and its purification, climate regulation, nutrient cycling, as well as direct provision of water, air, food, and building materials (MEA 2005). As such, continuation of ecosystem services is dependent on the sustained functioning of the ecosystems and the continued existence of its habitats and species. In this context, it is highly important to have a means to identify the space within which ecosystems function. How successful ecosystems are in remaining in this space depends on their degree of resilience, which is therefore an important trait to define.

Resilience of systems relies on the interrelationship of processes and interactions between groups, and is therefore, by definition, a system-level concept. The focus of this chapter is to review definitions of resilience on the systems-level and its related concepts, including a brief historical background over the past decades. Within this framework, we discuss various methodologies to analyse resilience, with a more detailed account of system analyses methodologies of networks for different types of systems (e.g., ecological, economic, social). We furthermore present several examples of their application to ecological and Socioeconomic systems, and discuss gaps in methodology and drawbacks of current methods, as well as future research and applications.

## 11.2 Resilience Concepts (Ecological and Socioeconomic)

Folke (2006) illustrates several concepts of resilience and portrays each as a stage in the development towards a more comprehensive understanding of the term within the context of large systems. Starting as a concept in ecology during the 1970s, resilience encompassed mainly the stability of smaller systems with limited players. This early engineering resilience (Folke 2006) mainly focussed on the return to a certain stable and constant state following a disturbance. Early work in ecology has intensely studied the latter, and used the term resilience and stability interchangeable. Beginning with population studies, often of simple predator-prey systems oscillating through cycles of high and low prey and predator abundances in a typical Lotka-Volterra (LV) fashion, stability was investigated in terms of maintaining the interplay between predators and preys by avoiding extinctions of either. Such studies gained wide recognition especially among mathematical ecologists, developing further on including important feedbacks with the environment, multiple food sources and predators (e.g., Chen and Cohen 2001). Later on, such studies were expanded towards communities and food webs, where stability has often been interpreted as the ability to return to an equilibrium starting point after impacts on the number of species, function and population sizes in the systems investigated (Moore and de Ruiter 2012). When such systems are viewed as interaction matrices between the components of the system, the interaction values at equilibrium can be disturbed and their return to equilibrium calculated. Within the LV framework, systems of equations representing predators and prey were established to investigate this return to equilibrium; namely, if the eigenvalues of its Jacobian matrix are negative for its real parts, then they describe the weakening of the disturbance over time which then brings the system back to equilibrium. Such return times to equilibrium were for instance investigated for three and four species systems, and it was apparent that return to equilibrium was faster in highly productive systems compared to those featuring lower productivity (Moore et al. 1993).

Later research was seeking to connect an ecosystem's resilience and stability to its biodiversity, thereby also connecting human impacts on species loss and invasive species to resilience (e.g., Chapin III et al. 2000; Dudgeon et al. 2006; May 1972). Work on the influence of species diversity on the resilience and stability of ecosystems has a rich history. In the 20th century, species diversity was believed to have a positive impact on stability (e.g. Elton 1958; MacArthur 1955), and a diversity-stability debate over the past decades has hardly changed this view, although further details on various forms of diversity, and natural variability in ecosystems has become available (e.g., McCann 2000; Naeem et al. 1999; Tilman 1999). Growing concerns about species and habitat loss, and the introduction of invasive species continuously energise the debate by giving rise not only to a myriad of theoretical studies, but also to experimental studies explicitly connecting species diversity and functional diversity to resilience, in short and long term laboratory and field experiments (e.g., Gamfeldt and Hillebrand 2008; Hughes et al. 2003; Loreau 2000; Müller et al. 2016; Tilman 1999).

Changes on the ecosystem level following species loss may be averted by a rich taxonomic diversity, if taxonomic diversity is reflected in a system's functional diversity (McNaughton 1977). On the other hand, if only rare species fulfil certain functions, their loss may be detrimental to system function (Chapin III et al. 2000). Hampering the application of such deliberations is of course the lack of comprehensive a priori knowledge on the consequences of the removal of a taxonomic species. Assessing species' functions and their contribution to ecosystem resilience and human livelihoods may therefore be a more effective application of conservation efforts (Dudgeon et al. 2006). Or, as Mori et al. (2013) point out, assessing responses to anthropogenic impacts by evaluating the response diversity of an ecosystem may be a better gauge of its resilience. Species that may respond differently to disturbances, besides having the same functional diversity in the ecosystem, may thus be important factors of determining an ecosystem's resilience and its adaptive capacity (Elmqvist et al. 2003). Lower response diversity may diminish a system's resilience, whereby the relationship between the response diversity and increasing impact is of higher importance than a certain response diversity calculated for instance at the start of an impact (Mori et al. 2013).

The resilience of ecosystems has been depicted as a ball residing in a basin (Scheffer et al. 2001), where the shape of the basin discerns the extent of resilience. A flat basin indicates low resilience, whereas a cup-shaped basin indicates higher resilience. A system with high stability would reside at the lowest point of the basin, whereas a system with high resilience may occupy various points of the basin. Using a ball and basin imagery, it is easy to envision a landscape that has multiple basins each one representing different stability regimes.

The seminal work of Holling (1973) recognised already that the concept of resilience had more to offer in addition to the concept of stability, in that it provided a way to describe persistence under conditions of disturbances and variability. The focus shifted from perceiving a certain state with little fluctuation as desirable to the ability of withstanding perturbations and persisting over time (Holling 1973). Expanding this concept to multiple equilibria allowed the exploration of a wider range of functioning and stable states. This is termed ecological or social resilience, depending on the type of system explored and opens the possibility for regime shift from one stable state to another. Disturbances such as reductions in functional and response diversities may move the ecosystem towards the edge, or outside of its basin, inducing the so-called regime shift (e.g., Folke et al. 2004). Also, the alteration of the duration and frequency of existing natural variability through human activities may reduce the resilience and induce regime shifts. Following a regime shift, ecosystems have been shown to operate in spaces of stable states, and to switch between certain alternative states following disruption of the current state (Scheffer et al. 2001). The new states are deemed stable when emergent negative feedbacks operate to maintain the system in this new state. In this manner, a hysteresis may develop such that the same shock that brought the system to the new state might not be sufficient to dislodge it back again.

Following a reduction in resilience leading to a regime shift, a reconfiguration of the original energy flow pathways to a different configuration (e.g. species composition) may occur. Classic examples are of coral reefs, switching to algae dominated reefs, following overfishing of herbivorous fish and nutrient input. In shallow lakes, a switch from a clear water and macrophyte dominated state to a turbid, phytoplankton dominated state occurs after increased anthropogenic nutrient input (Scheffer et al. 2001). Once an alternative state is reached, considerable effort is needed to create conditions that occurred long before the switch occurred, or may not be possible at all, due to hysteresis effects. In such alternative states the ecosystem may not be able to provide the same ecosystem services any longer to the same degree, on which humans depend upon.

Not only did these studies elucidate the concepts of stability and resilience, they also fostered an in depth understanding of different kinds of diversities—species, functional, trophic—and their role in the resilience and resistance of ecosystems. Nowadays information is available on specific invasive species, or on the consequences of a species lost, and this knowledge on the complex interactions and on the consequences if they disappear has dramatically increased the understanding of the functioning of real world systems, and additionally of theoretical and applied models. In addition to species, community, and ecosystem considerations, the impact of humans, directly, and on their economy and society (Chapin III et al. 2000) has moved to the center of this debate.

In this wider context, social-ecological resilience includes additional concepts such as feedbacks within a system, and across various spatial and temporal scales (Folke 2006). Such scales are interlinked in systems with a panarchic setup, cycling repeatedly through exploitation, growth, collapse, and reorganisation phases across scales (Holling 1986; Gunderson and Holling 2002). The changes a system undergoes in this arrangement are part of its resilience as it can self-organise, can stay in one state depending on the amount of disturbance, and most importantly, provides opportunity for adaptation to different influences (Carpenter et al. 2001). Resilience has been proposed as the capacity of a system to navigate all stages of the adaptive cycle (Fath et al. 2015). This model was applied to the survival of firms in a socioeconomic system, with indication of how preparedness for each stage must be cultivated in each of the other preceding stages continuously, not simply the stage immediately prior. For example, to manage the collapse phase, one should already reduce possible fault cascades during the growth phase to prevent crises from spreading throughout the system; enhance cohesive leadership during the conservation stage; identify and maintain vital functions during the collapse itself; and, learn improvisation during the reorganization stage (Fath et al. 2015). Resilience has thus developed into a concept that encompasses systems in their entirety throughout time and space, rather than being restricted to understanding the stable states which a system should return to after perturbation (Kharrazi et al. 2016).

Judging the state of resilience of a system, characterising the space within which it operates, and defining the borders of this state (ranges of variability) is thus of importance when certain ecosystem services produced by certain states are desired not to be lost. It is equally important for systems heavily influenced by anthropogenic designs, such as production ecosystems, to maintain resilience in order to ensure food security. However the continued functioning of such systems usually depends on considerable external influences that uphold its production value, including the use of fertilisers, pesticides, water, or fossil fuels. Due to this, the resilience of such a system is upheld by anthropogenic influence, and is termed 'coerced' resilience (Rist et al. 2014). The production systems are held in a certain basin of attraction, and therefore the state of resilience, by the anthropogenic activities, often in otherwise unstable states, and bringing forth questions on their sustainability (Rist et al. 2014). The value of resilience of production and other ecosystems has found its way into the economic valuation of ecosystems, planning land-use according to bundles of ecosystem services, rather than too few, in order to maintain resilience of the ecosystem (Admiraal et al. 2013). The optimisation of an ecosystem's services, and thus its value, may, according to Admiraal et al. (2013), be informed by resilience theory and incorporate functional diversity.

## 11.3 Resilience and Networks

Resilience on the systems-level can be studied by depicting the system as a network in which all actors in the system are nodes, and their direct interactions are links. When links have quantifiable attributes (e.g., biomass transfer, goods or money exchanges, or interactions between people), so called weighted networks describe the system (Fath et al. 2007). Resilience in networks has been studied at the systems-level in the form of system indicators, or at the level of how indirect effects are propagated through the system. The importance of nodes (e.g. keystone species) and links (e.g. indirect propagation, redundant pathways) within the system are then quantified (see methodology and examples below and in Sect. 11.4). A related series of talks on resilience and networks, which is found here: www.fas-research.com/resilience/, may be of interest.

In the field of ecology, certain network configurations have been put forward as providing stability in networks (and therefore increasing its resilience). These include for instance the relation of weak links (low weight) to strong links (large weight) in a network, and especially when weak links are configured into large cycles a stabilising effect can be apparent (Neutel et al. 2002). Stability has also been investigated in densely connected sub-sections, or small modules of networks, and found that small predator-prey modules stabilise networks (Allesina and Pascual 2008). With this knowledge, the necessity of long cycles for stability (Neutel et al. 2002) is disputed, as well as the necessity for dominance of weak cycles since the predator-prey loops have a relatively large weight in the system (Allesina and Pascual 2008). In a more in-depth study on the role of modularity in networks, a less clear result emerged, with stability of networks depending on its modularity only under certain conditions, namely the size of the subsystems and the mean interaction strength (Grilli et al. 2016).

In addition to studies on the dynamic stability of networks calculated from link weight, loop length, and modularity, resilience measures can also be calculated for networks from its weighted-link distribution between all pairs of nodes. Thus, it is not necessarily restricted to certain types of interactions (predator-prey), or certain loop configurations. Therefore, the approach here does not follow directly from the use of ecological networks to assess subsystem stability, but rather builds on whole-system energy-flow and information-theory based ecological network analysis.

Ecological Network Analysis (ENA) is one way to analyse both the direct and indirect interactions in such networks. The network configuration that comprises a certain number of nodes and links, and often a weighted link distribution, are at the centre of the ENA to calculate descriptors (including resilience) for any type of network. Such resilience measures are based on constraints on transfers of material flow along links between a source node's output and a recipient node's input (ascendency based, Ulanowicz 1986, 2009), and have been applied to ecological and socioeconomic networks (e.g. Chrystal and Scharler 2014; Goerner et al. 2009; Kharrazi et al. 2013).

Resilience measures that describe the constraint of flows along links or pathways in networks are dependent on the network's connectivity, and the respective interaction between pairs of nodes (Ulanowicz 1986). Of prime interest regarding the interactions is the degree of uniformity of the flow distribution. When the total output of a source node and the total input into a recipient node are the same, the constraint on the flow is maximal. In contrast, when a source node donates only part of its output to a recipient node, and the latter receives more input from other nodes, the interaction is much less constrained. Such flow and interaction distributions can be quantified by defining the probability of input into a node and the output from the source node. These two probabilities are the same in the example of maximal constrained flows, but different in the example of the less constrained interaction between two nodes. Therefore, whenever flows are maximally constrained, each node only has one output to a receiving node, and only one input from a source node. Such systems are constrained to such an extent that there is no resiliency left in the system in case of perturbations—there is no possibility to channel energy along any other link in the system should a particular link be lost. The receiver node thus loses its entire input. In the less constrained network, nodes are more diversely interlinked, which leads to parallel pathways in that an output from a source node can reach a recipient node by different pathways, also via other nodes, or 'detours'. Such pathways build redundancy into a network and its resilience increase as it is has a higher chance to buffer against a link loss. Such a state is therefore more desirable compared to a maximally constrained state. On the other extreme, however, a minimally constrained state prevents efficient transfers along links, a high dependency of a particular node on many other nodes, and consequently reduced functioning and a lower resilience (Goerner et al. 2009; Ulanowicz 2009). For instance, higher trophic levels in food webs, or receivers at the end of product chains, are not well served with minimally constrained networks.

Mathematically, this constraint-induced trade-off between redundancy and efficiency can be calculated using information theory. Specifically, according to Rutledge et al. (1976) the information can be determined as the reduction of uncertainty. Using the conditional probabilities of flows based on a particular network configuration, we arrive at the following equations (Ulanowicz 2004):

$$H = -\sum_{i,j}\left(\frac{T_{ij}}{T..}\right)log\left(\frac{T_{ij}}{T..}\right). \tag{11.1}$$

$$\Phi = -\sum_{i,j}\left(\frac{T_{ij}}{T..}\right)log\left(\frac{T_{ij}^2}{T_{i.}T_{.j}}\right). \tag{11.2}$$

$$AMI = \sum_{i,j}\left(\frac{T_{ij}}{T..}\right)log\left(\frac{T_{ij}T..}{T_{i.}T_{.j}}\right). \tag{11.3}$$

where $H$ is the total system flow diversity, $\Phi$ is the system redundancy, $AMI$ is the system average mutual information, $T_{ij}$ the flow from node $i$ to $j$ and $T..$ the sum of all internal and boundary flows (referred to as Total System Throughput—TST). It has been shown (Ulanowicz 1986) that H = Φ + AMI. These quantities were used to create a metric that can assess the trade-off between the system redundancy (high number of pathways with more uniform flow) versus systems with high mutual information (articulated pathways with more asymmetric flow). This new metric was first termed "fitness for change" (Ulanowicz 2009; Ulanowicz et al. 2009), subsequently sustainability (Goerner et al. 2009), and lastly system robustness (e.g. Mukherjee et al. 2015; Kharrazi et al. 2016). Quantitatively, it was derived by multiplying the ratio of AMI/H by the Boltzmann measure of disorder (−k log(a), Ulanowicz 2009) and is here termed system robustness, designated $R$:

$$R = -\alpha log(\alpha)$$

where α = AMI/H.

The robustness curve shows increasing degree of order (higher value of AMI relative to H) on the x-axis as determined by AMI/H (Fig. 11.1). Work on empirical ecosystem networks revealed that their sustainability values clustered around a narrow region near the maximum of the curve, and this region has been referred to as "The Window of Vitality". This window describes an optimum balance between redundancy and efficiency in a network that results in the highest values of sustainability or resilience. To the contrary, sustainability is lowest for both highly and minimally constrained networks compared to intermediate constrained networks (Ulanowicz 2009; Ulanowicz et al. 2009; Goerner et al. 2009; Fig. 11.1).

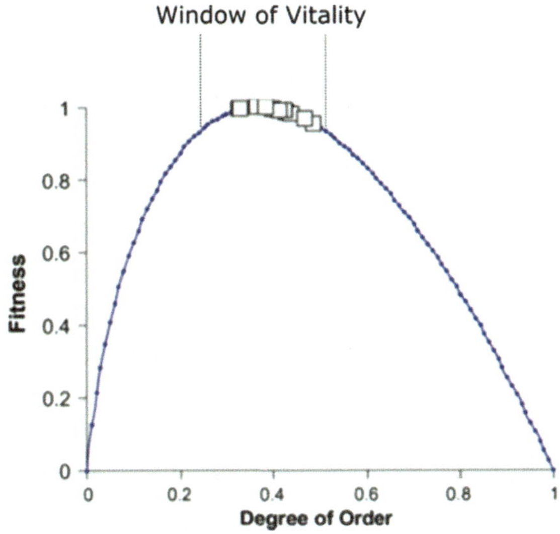


**Fig. 11.1** Robustness (Fitness in Ulanowicz 2009) as a function of efficiency and resilience. X-axis: AMI/H. Y-axis: $-\alpha\log(\alpha)$. Reprinted from *Ecological Modelling*, with permission from Elsevier

Ecological relations are direct and indirect processes that have effects on the resilience of ecosystems, as they exert some control on the interaction between components of the system. Indirect effects in ecosystems play an important part in their functioning, as they depict the wider impact in the system of changes to specific nodes or links. Examples of indirect effects include node-node interactions, trophic cascades, apparent competition, indirect mutualism and others (Wootton 1994; Higashi and Nakajima 1995; Szyrmer and Ulanowicz 1987; Shetsov and Rael 2015; Salas and Borrett 2011; Higashi and Patten 1989). Many of the studies attribute a dominance of indirect effects in ecosystems over direct effects.

In this chapter, we explore two different categories of indirect effects, firstly the influence of keystone species (Libralato et al. 2006) on ecosystems and their resilience, specifically changes in biomass and consequently the related flows. Secondly, we explore indirect effects including direct and indirect competition and mutualism, and other interrelations effects (combinations of +, −, 0 relations) (Ulanowicz and Puccia 1990; Fath 1998).

For networks, indirect effects can be calculated as follows. First, the direct negative impacts of the consumer on the resource, and the positive impacts of the resource on the consumer are calculated from the flow matric with elements $T_{ij}$ representing the flow from node $i$ to node $j$.

$g_{ij} = \frac{T_{ij}}{\sum_k T_{kj}}$ represents the amount of all prey ($i$) consumed by a predator ($j$), whereas $g_{ij} = \frac{T_{ij}}{\sum_m T_{im}}$ represents the portion of the prey's production that is consumed by all predators.

The net impact is achieved by subtraction: $q_{ij} = g_{ij} - f_{ij}$, and the resulting $Q$ matrix represents all direct impacts between two nodes. Second, the matrix of direct and indirect impacts, or total impact between two nodes, is calculated by $M = \sum_{h-1}^{\infty} Q^h$, representing all powers of $Q$, and therefore all pathway lengths between any two nodes (Ulanowicz and Puccia 1990).

The elements of the $M$ matrix can then be used to calculate the keystoneness of a node ($KS_i$) in the network (Libralato et al. 2006) as:

$$KS_i = \log(\varepsilon_i(1 - p_i),$$

where$\varepsilon_i = \sqrt{\sum_{j\neq1}^{n} m_{ij}^2}$, not including the effect of a node on itself (i.e. $m_{ij}$) Here, $m_{ij}$ represents the entries of the $M$ matrix of total impacts, and $p_i$ the contribution of a node's biomass to the total biomass of the network:

$$p_i = \frac{B_i}{\sum_k B_k} \quad \text{(Power et al. 1996)}.$$

These approaches, employing the information-theory based trade-off of redundancy and efficiency and the energy-flow based indirect relations assessment (i.e., keystoneness), are used in a number of case studies involving ecological and socio-ecological systems.

## 11.4 Case Studies

Here we present several case studies that illustrate various applications of systems analysis and the exploration of the resilience concepts. These include two ecological examples, that of Manwan Dam, China and Mdloti Estuary, South Africa, which were both investigated in terms of their resilience over a time series of data and under perturbation. Furthermore, a third example included a socioeconomic system, the Heihe River Basin, China, as a Virtual Water Network. Indirect effects, on the other hand, were investigated for an ecological system (keystone species, Mdloti Estuary) and a socioeconomic system (mutualisms and competition on the systems-level) in Beijing, China.

### *11.4.1 Ecosystem Perturbation Examples: China and South Africa*

We engaged in analysing ecological and economic networks for their robustness, manipulated networks to analyse their changing robustness, and described the related values of Average Mutual Information (AMI), Total System Throughput (TST), Flow Diversity (H), Ascendency (A = AMI × TST) and Development Capacity (DC = H × TST) (Ulanowicz 1986). Two types of ecological systems included the South African Mdloti estuary and Manwan Dam, China, and the socioeconomic system representing the Virtual Water Network of the Heihe River Basin, China.

#### 11.4.1.1 Manwan Dam, China

In Manwan Dam, a reservoir on the Lancang River in southwest China, we investigated the effects of perturbation on robustness indicators by comparing the original network to perturbed networks in which (1) the flow structure of the food web was changed and (2) species were removed from the system. The network representing Manwan Dam (Chen et al. 2011) featured ten compartments, 41 within-system energy flows (kJ $m^{-2}$ $y^{-1}$), three boundary inputs, and ten boundary losses. Flows to and from the basis of the food web, i.e., detritus and phytoplankton, were removed one by one to examine the effect on overall system robustness. As a first scenario, we removed flows to and from the primary producer and detritus compartments (Fig. 11.2). Removal of the largest flow in the network —input into phytoplankton (Flow #5)—resulted in the largest drop in the system robustness. This flow into phytoplankton is the main energy source of the system and its loss therefore substantially decreased robustness of the system. Removal of the output flow from detritus (Flow #3), flow from phytoplankton to detritus (Flow #2), input flow into detritus (Flow #1), and output flow from phytoplankton (Flow #4), resulted in a sequential increase in the system robustness compared to the unperturbed ecosystem. This may be attributed to the fact that the four flows have medium flow values, and their removal led the flow distribution in the food web to become more uneven compared to the original unperturbed network.

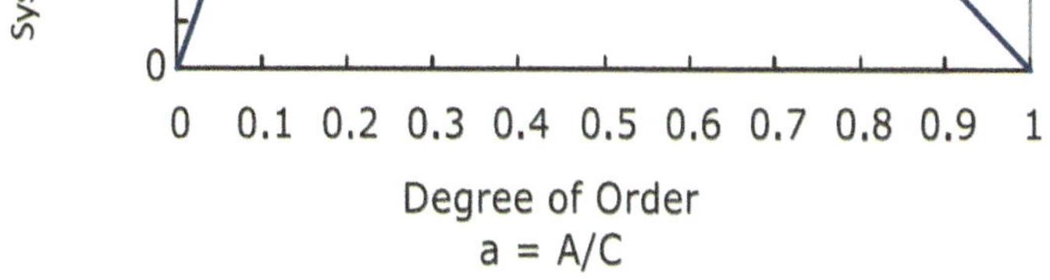


**Fig. 11.2** Robustness values for the original unperturbed Manwan Dam network (green), after the removal of the input flow into detritus (1), of the flow phytoplankton → detritus (2), of the output from detritus (3), of the output from phytoplankton (4) and of the input into phytoplankton (5)

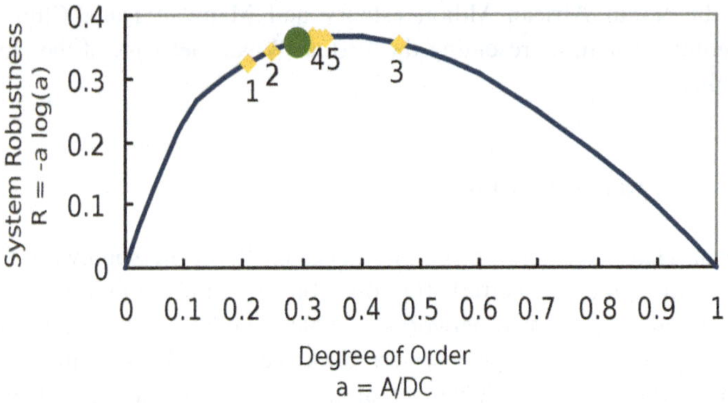


**Fig. 11.3** Robustness values for the original unperturbed Manwan Dam network (green), after the removal of all boundary flows (1), removal of the phytoplankton node (2), detritus node (3), zooplankton node (4) and other microalgae node (5)

In the next scenario, the system was isolated from its environment by removing all boundary flows (Fig. 11.3). The most obvious impact of flow removal on robustness was apparent not from a compartment removal but the boundary flow removal (1), which represents isolating the system from its environment (Fig. 11.3). Thereafter, we removed nodes from the system, as an extreme scenario of perturbation. Removal of the phytoplankton node from the system led to the loss of an important energy source to the rest of the system, which also decreased its robustness. Compared with the scenario of losing total input and output, the robustness variability is relatively low, indicating that primary production is not the only source of energy to the food web. Also, the flow distribution became less uneven by removing the phytoplankton node compared to removing all boundary flows. When we removed detritus (3) in the system, as the most connected compartment in the system, the flow distribution became more efficient and the robustness of the system decreased slightly (Fig. 11.3).

#### 11.4.1.2 Mdloti Estuary, South Africa

Even naturally stressed ecosystems such as estuaries show various degrees of resilience. As they undergo a constant change in the physical environment, certain species have adapted to cope with changing salinity and conductivity, water levels, turbidity and other physical factors. It is often difficult to distinguish impacts in such systems as natural and anthropogenic impacts are difficult to discern (Elliott and Quintino 2007). Scenarios of real-world impacts on estuaries concerning the base of the food web (primary producers, detritus) and fishing impacts revealed interesting results on the robustness in the South African subtropical Mdloti estuary (Mukherjee et al. 2015). For this purpose, networks representing the carbon exchanges among compartments for two different seasons [dry, wet; described in Scharler (2012)] were transferred to the Ecopath software (Christensen and Walters 2004), in which the changes to the food web were implemented. The impact of perturbations on the food web network was investigated and assessed using the information theory based indices AMI, H, and Robustness. We considered the sensitivity of the network under three scenarios: (a) autotrophic biomass increase and decrease from 10 to 99 in 10% intervals; (b) an increase in fish yield from 10 up to 99% in 10% intervals; (c) an increase and decrease of the detritus import to the system from 10 to 99% in 10% intervals. The changes in network indices AMI and H (Ulanowicz 1986) and robustness (Ulanowicz et al. 2009) were calculated.

Robustness is a state of system's health which indicates the balance between system efficiency and redundancy (see also above). All indices indicated that robustness of the system increased with a rise in autotroph biomass (Scenario a). In the fish harvest scenario (Scenario b), robustness of the system decreased as the top predators (fish) are reduced in biomass, and consequently their throughput. In the detritus import scenario (Scenario c), the relationship of robustness and A/C show a similar pattern in the two seasonal networks (Fig. 11.4). In both phases, robustness decreases with the perturbations, and the ratio of the network indices showed a downward trend in each scenario (Fig. 11.4). System robustness increased with an increase in autotrophic biomass, decreased with larger fish harvest, and also decreased with change in detritus import. Of the three different scenarios, the robustness values did not differ much from the original values in the autotrophic biomass scenario but changed comparatively more in the other two scenarios—fish harvest and detritus import. These results clearly indicate that the robustness measure is able to reflect the information about the system even when it faces both smaller and larger scale perturbations. The system maintained a balance in the face of stress, and only beyond a threshold are significant changes in robustness apparent.

### 11.4.2 Socioeconomic Example: Water Network in China

The same information theory-based network resilience measures were applied to a socioeconomic system, a Virtual Water Network (VWN) from the Heihe River basin in China (Fang et al. 2014). Unlike fresh water management strategies, which mainly focus on the efficiency of the target production processes of water utilisation, the VWN provides an integral view via the link of virtual water flows with different socioeconomic activities. The notions of virtual water flows provide important indicators to manifest the water consumption and allocation between different sectors via product transactions, as all production processes utilise water to varying degrees. During this study of Ganzhou District in the Heihe River Basin, we investigated configurations of a virtual water network (VWN) to identify the water network efficiency and stability in the socioeconomic system.

The system was divided into six sectors, representing the Ganzhou District economy. These include farming (1), livestock (2), other agricultural (3), industry

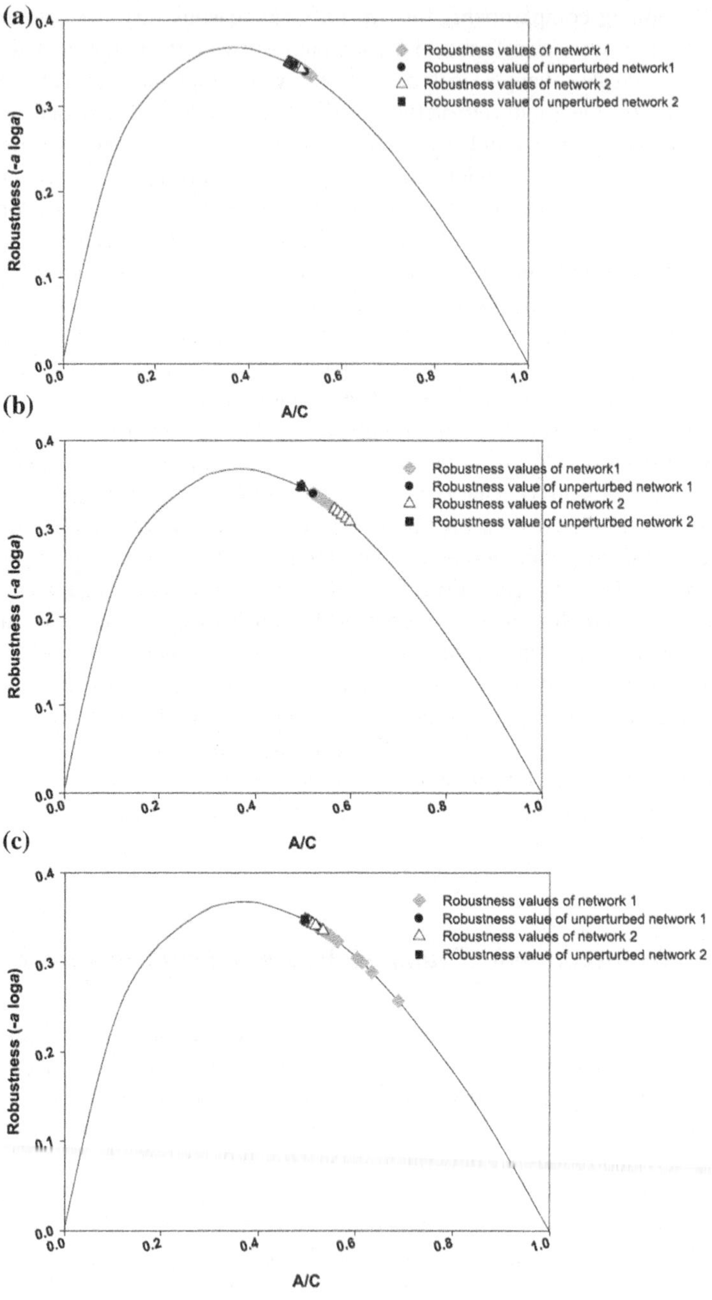


**Fig. 11.4** Degrees of order (AMI/H, or A/C) and corresponding magnitudes of robustness for the Mdloti estuarine system for networks of the open (network 1) and closed (network 2) phases for **a** autotrophic biomass change scenario, **b** fish harvest scenario, and **c** detritus import scenario. Reprinted from Mukherjee et al. (2015), with permission from Elsevier

(4), construction (5), and services (6), the three sectors related to agriculture emphasize the important role of this sector in the district (Fang et al. 2014), and also its comparatively high water consumption (Cheng et al. 2006). The system was then analysed over four phases representing the period 2002–2010. Results based on ENA show that the agricultural sector is the major local economic sector with high-intensity water consumption. Changes in the network structure occurred in the years 2003, 2005, and 2007. The number of links declined from 31 in 2002, to 30 in 2003/4 with the disappearance of the link from livestock (2) to other agriculture (3). The lowest number of links (26) occurred from 2005 to 2007 as the flows from livestock (2) to farming (1), other agricultural (3) to farming (1), other agricultural (3) to livestock (2), and the self-loop of construction sector (5) disappeared. In the last period (2008–2010), the flow from livestock (2) to other agricultural (3) is again present. Overall, the main changes in links occurred among the three agricultural sectors, i.e., farming (1), livestock (2), and other agricultural (3).

Although the total fresh water consumption declined over the time period of the studies, the efficiency of VWN was still low, and overall robustness declined over the study period (Fig. 11.5). Due to the low efficiency, the robustness of VWN remains on the left-side of the robustness curve (Fig. 11.6), implying that the VWN has a higher redundancy level due to stable circulation among various sectors but lower system efficiency, which is contrary to the intuition that human-designed networks should be more efficient. This occurs because, except for the freshwater flowing through specific supply chains, the water hidden in the products or services is circulating within diverse sectors with more exchanges along various pathways, which results in higher redundancy. The current policies for water management emphasize the control of the total amount of water consumption (with the aim of reduction), ignoring the balance between efficiency and redundancy for virtual water circulation within the socioeconomic system.

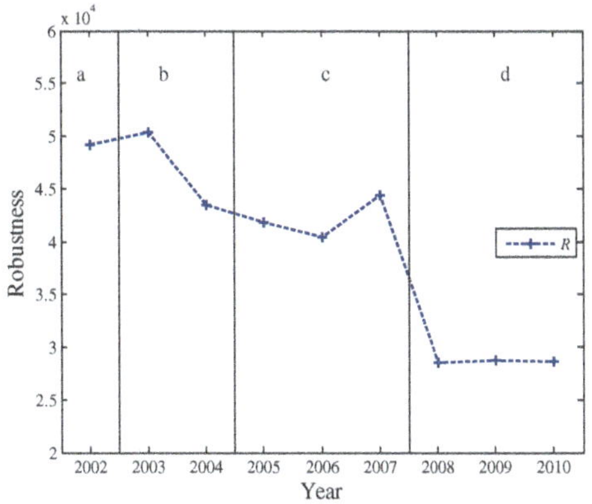


**Fig. 11.5** Trend in robustness values for the socioeconomic water system for the four periods from 2002 to 2010

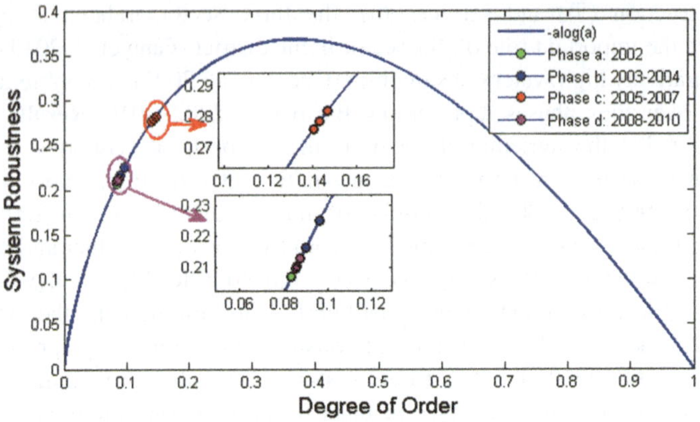


**Fig. 11.6** Robustness of the networks in the four phases from 2002 to 2010. Note that each year is depicted by a circle

## *11.4.3 Relational Analysis*

### 11.4.3.1 Keystone Species Analysis: South Africa

Here we present applications of the ecological relations analyses for ecological and socioeconomic systems, including the application to a perturbed estuarine systems in South Africa and to the Metropolitan area of Beijing, China. In the ecological study of the Mdloti estuary, South Africa, we investigated the impacts of keystone species on a time series of networks representing various seasons. Keystone species were identified using the method described above (Libralato et al. 2006; Ulanowicz and Puccia 1990).

The networks were perturbed to investigate possible temporal changes of their impacts on other network nodes and on the system as a whole. After the keystone species had been identified, its biomass was increased and decreased in 10% intervals to calculate their impacts on other nodes in the system. For five different time steps representing different seasons from March 2002 to March 2003, the identified keystone species were all fish: *Argyrosomus japonicus* (in two of the time steps), *Monodactylus falciformis* (in two of the time steps), and *Caranx sexfasciatus* (in one time step), ranging from trophic level 2.9 to 3.7 as calculated from the respective networks. Keystone species in the five time steps thus differed (the keystone species in one network is not necessarily the keystone in another), however each keystone species in a given network maintained its keystoneness rank over the perturbation scenarios.

The system level impact of the keystone species was at times prominent but not consistent through time (Fig. 11.7). In addition, there was no consistent impact on species belonging to similar trophic levels. Previous studies have shown that this system is very robust in nature (Mukherjee et al. 2015; Scharler 2012) and although variations in the presence of the keystone species affected some components, the resilience of the system as a whole counters these adverse effects and it does not collapse when the keystone species are perturbed or removed. This point is further supported by the change of the keystoneness of the species over time in this system.

From this particular study it could be concluded that although there were significant effects of the keystone species in the ecosystem, that these effects were not consistent throughout the seasonal variation (Fig. 11.7). The ecosystem continues

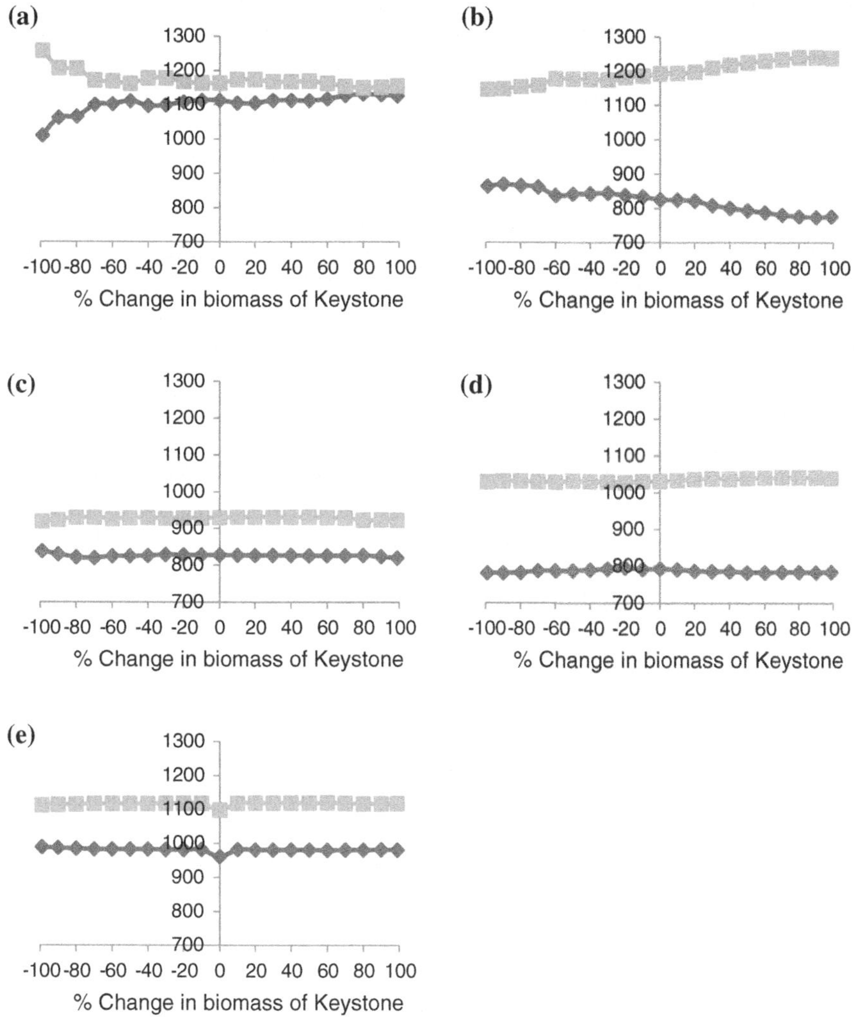


**Fig. 11.7** System-level impacts on the total number of positive (black diamonds) and negative (grey squares) relations (y-axes) after perturbation of keystone species biomass in various seasons of the Mdloti Estuary. **a** March 2002, **b** June 2002, **c** September 2002, **d** December 2002, **e** March 2003

to function under the perturbations where keystone biomass is increased or depleted, but that was within a specific boundary beyond which the system could not persist (i.e. could not be mass balanced).

#### 11.4.3.2 Relational Analysis: China

Another network indicator, related to system resilience, concentrates on the types of interaction between nodes (whether they are positive, negative, or neutral) (Fath 1998; Ulanowicz and Puccia 1990). A novel application of this approach was applied to assess the relation of carbon change due to land use change over time.

The acceleration of urbanisation has greatly changed the morphology of terrestrial surfaces, and about one-third of urban carbon emission results from land use/cover changes (LUCC) such as the replacement of vegetated surfaces with built-up land (Denman et al. 2007). Previous studies have shown a "carbon sink" effect caused by LUCC that promotes carbon sequestration by forests and grasslands after the conversion of farmland to forest or grassland at high latitudes, and a "carbon source" effect caused by the loss of these ecosystems (Kauppi et al. 1992). Beijing is a typical city whose distributions of LUCC vary greatly, which strongly affects the spatial and other characteristics of urban carbon emission and sequestration. Overall, urban resilience will be enhanced with improved carbon emission and sequestration management.

To achieve the research goal, first one needs to characterize the increase or decrease of carbon release and absorption under the existing LUCC conditions. And then, this needs to be analysed in context of the relationships between the social economic systems and the environment that causes these changes. Finally, the direction and size of carbon reduction can be determined based on the spatial distribution of these relationships. In urban systems, these relationships usually include mutualism, competition, exploitation, control, and neutralism (see Zhang et al. 2014).

In contrast to the robustness measure which explored the link distribution between pairs of nodes, utility analysis includes both direct and indirect pathways and how they affect the interaction signs between nodes (or sectors) of a system. For instance, direct feeding relations may reveal positive effects of prey to predator, and negative effects of predators on prey. These represent direct relations, but indirect relations may be classified by exploring the relationship between predators. If they are feeding on the same prey, then their relation is one of competition (negative effects both ways, i.e. −, −) although they are not directly linked through a feeding relation. As such, relations between nodes in a network according to positive and negative impacts describe the type of relation for each pair in a network either as positive (+), negative (−), or neutral (0), via all direct pathways. The relations may also be calculated over indirect pathways, in order to capture any indirect effects that travel over several pathway lengths between nodes (Ulanowicz and Puccia 1990; Fath 1998). The nature of the relations is important to understand the interaction between the nodes that may be mutualistic (+, +), competitive (−, −),

**Table 11.1** Ecological relations specified by binary interactions between components of the ecosystem

| | + | − | 0 |
|---|---|---|---|
| + | (+, +) Mutualism | (+, −) Predation | (+, 0) Commensalism |
| − | (−, +) Altruism | (−, −) Competition | (−, 0) Amesalism |
| 0 | (0, +) Commensalism | (0, −) Amensalism | (0, 0) Neutralism |

absent (0, 0), or any other combination of positive, negative, and neutral (Table 11.1). As positive interactions, especially those configured into so-called autocatalytic loops, are thought to benefit an ecosystem (Ulanowicz and Puccia 1990), in economic sectors these relations may illustrate those sectors that have consistent positive or negative impacts on other sectors, from which may arise management scenarios to either maintain, or improve, certain relations.

The calculation of this particular relational analysis (Utility Analysis: Fath 1998) is very similar to that of the trophic impacts analysis (Ulanowicz and Puccia 1990) described above, and the similarities and differences are treated in Scharler and Fath (2009).

An application of utility analysis to the city of Beijing as a network of the different sectors illustrates the type of interactions between socioeconomic sectors. As urban areas contribute the majority of carbon emissions (IPCC 2007), this study illustrated how the various sectors interact, how each of them expanded or decreased their influence in the network, and which type of interactions were prominent among which sectors. Beijing's urban area expanded to such an extent in recent decades that between 1992 and 2008 about 20% of cultivated land and 28% of the forested land was converted into constructed land (Miao et al. 2011). Within the time period studied here, major changes in land-use and cover occurred before 2000. After 2000, relationships between sectors changed mainly in the central and south-eastern parts of Beijing due to land-use and cover changes in these areas (Xia et al. 2016). Overall, there was a significant shift in the extent of area between sectors, and consequently a difference in the carbon emission and sequestration of each. The various sectors included in this analysis included urban land, rural land (residential area outside of urban land), transportation and industrial land outside urban and rural residential area, cultivated land, bare land, forest and water (reservoirs, rivers). Many of these had subcomponents to further identify the various sectors, resulting in 18 component networks replicated within and between four time periods representing the timespan from 1990 to 2010 (Xia et al. 2016) (Fig. 11.8).

Major changes that occurred during the study period were apparent from the mutualistic relations (+, +) between the natural and socioeconomic sectors. The extent of mutualistic relations decreased substantially by more than 40%, which was mainly disturbed by urban expansion (Table 11.2). Mutualistic relations between the socioeconomic sectors were less easily disturbed. Mutualistic relations were also the most fluctuating in that they changed to a large extent between the

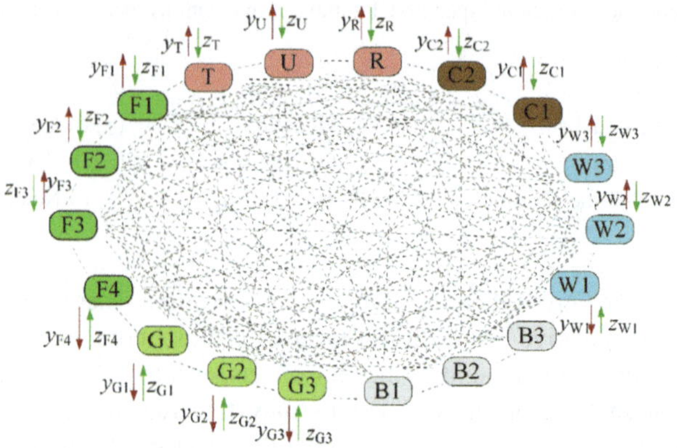


**Fig. 11.8** The 18 component network representing the socioeconomic carbon metabolism network of Beijing. *B1* sand, *B2* barren earth, *B3* bare exposed rock, *C1* irrigated cultivated land, *C2* dry cultivated land, *F1* forest, *F2* shrub land, *F3* open woodland, *F4* other woodland, *G1* high-coverage grassland (vegetation cover >50%), *G2* medium-coverage grassland (vegetation cover 20–50%), *G3* low-coverage grassland (vegetation cover 5–20%), *R* rural, *T* transportation and infrastructure, *U* urban, *W1* rivers, *W2* lakes and reservoirs, *W3* intermittently flooded plain, $z_{ij}$ input flow from *j* to *i*, $y_j$, output flow from *j* to *i*. Reprinted from Xia et al. (2016), with permission from Elsevier

time periods. Exploitative relations (+, −) occurred mainly between two components of the socioeconomic sector, transportation and industrial land. As these sectors are maintained and regulated by managers, they may therefore also provide an opportunity for regulation to reduce the extent of the exploitative relationship. Competitive relationships (−, −) arose mainly between natural components, or between cultivated land and natural components. Such relations were also the most stable over time (Table 11.2), which indicates that once established

**Table 11.2** Changes in the number of each type of relationship during the four parts of the study period

| | 1990–1995 | 1995–2000 | 2000–2005 | 2005–2010 | Mean value | Standard deviation |
|---|---|---|---|---|---|---|
| *M* Index | 1.09 | 1.15 | 0.43 | 0.47 | 0.79 | 0.34 |
| *Number of relationships* | | | | | | |
| Mutualism | 114 | 132 | 38 | 62 | 86.50 | 38.01 |
| Competition | 100 | 110 | 130 | 156 | 12400 | 21.40 |
| Exploitation | 55 | 41 | 30 | 20 | 36.50 | 13.01 |
| Control | 55 | 41 | 30 | 20 | 36.50 | 13.01 |
| Neutralism[a] | 0 | 0 | 99 | 35 | 33.50 | 40.43 |

[a]Neutralism only occurred during the last two periods (between bare land and other components), and because we were mainly concerned with the spatial and temporal variation in relationships that resulted in significant carbon flows, we have not analysed the neutralism relationship further
Reprinted from Xia et al. (2016), with permission from Elsevier

they are difficult to change. Before 2000, more positive than negative relations were found in the entire system, whereas negative relations mostly outweighed positive ones after 2000 (Xia et al. 2016).

Not only did the ecological relations change overall in the entire system, but the spatially explicit analysis revealed a concentration of change in certain areas of Beijing after 2000. Exploitation (+, −), control (−, +), and mutualism (+, +) relationships became concentrated in the southeast of the urban area, whereas competition relations (−, −) dominated its northwest expanse. The increase in competitive relations over time was a consequence of urban expansion competing for land, and increased use of water resources. The results provide an objective basis for planning adjustments to Beijing's land-use patterns to improve its carbon metabolism and reduce carbon emission.

## 11.5 Discussion and Conclusions

Resilience has been an important topic in ecology for many decades, and has garnered attention and application in social and economic systems. In both domains, the same basic system-dynamic question applies in terms of the ability of a system to continue to function in the face of perturbations. Throughout its history, the resilience concept has included the premise of a steady state and return to that state, but recently has been expanded to encompass natural change and adaptation. A resilient system is not one that necessarily stays in the same place. Quite the contrary, the system must remain flexible and adaptive to changing conditions. This refers to either a return to the previous state or the continuation to a new state through a regime change. For example, as referenced above, Fath et al. (2015) frame resilience in terms of the adaptive cycle and a system's ability to survive all phases of growth, conservation, collapse, and reorganization. In addition, similar features are seen as key to develop resilient systems such as incorporating functional diversity and also providing a balanced trade-off between efficiency and redundancy. Analysis of socioeconomic systems, using the information-based measure of balanced trade-off between efficiency and redundancy has shown that these systems have lower resilience (robustness) due to comparatively high redundancy (Kharrazi et al. 2013). In any event, the concern for resilience and the need for better metrics of resilience in socioeconomic systems provides a rich area for further research. Therefore, a main goal of this research was to illustrate the application of systems analysis concepts by providing additional case studies using a suite of resilience metrics derived from ecological network analysis.

The information-theory based analysis was applied to the reservoir ecosystem impacted by the Manwan Dam in Southwest China. This example showed a relatively high level of robustness in the unperturbed state (although not at the optimum which could be because the dam itself is a perturbation from the natural condition). Scenarios to remove functional groups showed that the robustness can be lowered most notably by altering flows and nodes regarding the phytoplankton, which is the key entry point for energy flow in the ecosystem. Application of the methodology to a human designed water supply network, also in China, exhibited a much lower overall robustness measure. The metric was tracked during four phases of water network development from 2002 to 2010 which showed a decrease in the overall robustness during this time. In this case, the goal of the regional water network was delivery for agricultural usage, and while water delivery was achieved it came at the expense of a more vulnerable network. Diversity and redundancy are two features which allow a system to respond effectively to perturbations. These key features must be balanced against efficiency in a trade-off that produces more resilient and robust systems.

These robustness/resilience values used herein have originally been conceived from the analysis of several ecological networks. The networks used to define the "Window of Vitality" as the region of highest robustness and sustainability, in a different approach (Zorach and Ulanowicz 2003) also match a confined region defined by minimum and maximum link density and effective number of trophic levels (Pimm 1982; Ulanowicz 2002; Wagensberg et al. 1990). As the robustness curve is not completely symmetrical, redundancy has a larger influence on resilience than efficiency [although the impact of a marginal change depends on which side of the optimum the system lies (Kharrazi et al. 2017)]. However, the authors (Goerner et al. 2009) concede that the optimum has not yet been defined due to a lack of data. The optimal balance between efficiency and redundancy, as well as the variability that may accompany such a balance is as yet to be explored. This research adds additional case studies employing this method showing that the ecological network of Manwan Dam (unperturbed) lies near this optimum region, and an agricultural water network lies off the optimum on the heavily redundant side, indicating it can improve its efficiency. The methodology allowed for tracking and improvement of resilience or efficiency in the respective networks.

Two additional case studies applied an energy-flow based network analysis that determines the direct and indirect relations between any two nodes in the network. In the South Africa case, this was used to reveal the keystone species in the Mdloti Estuary during five sampling periods from March 2002 until March 2003. Perturbations on the network had variable impacts on the presence of the keystone species, although a general pattern was not obvious. Lastly, the relational analysis was applied to carbon sequestration ability due to land use change around Beijing. Results showed that the number of mutualistic relations decreased over time and since the assessment was spatially explicit, we were able to show that competition relations increased in regions that experienced heavy urban expansion. This is a novel approach to use network analysis for assessing the carbon metabolism due to land use change.

In summary, networks are useful tools to describe the interactions between their constituents, and resilience measures applicable to both ecological and socioeconomic systems have been presented. They are derived from the efficiency of energy and material transfers throughout the networks, as well as the proportion of "detours", or redundant pathways ensuring resilience. Positive and negative relations between sectors of economic systems, or ecosystems, highlight the influence of various economies and species on one another, resulting in a comprehensive picture of relations, impacts and therefore management options to achieve balance between sectors. Both relational analyses are related to resilience by describing the proportions of types of interactions in systems and their response to change and perturbation.

## References


Admiraal, J. F., Wossink, A., De Groot, W. T., & De Snoo, G. R. (2013). More than total economic value: How to combine economic valuation of biodiversity with ecological resilience. *Ecological Economics, 89,* 115–122.

Allesina, S., & Pascual, M. (2008). Network structure, predator—prey modules, and stability in large food webs. *Theoretical ecology, 1,* 55–64.

Carpenter, S., Walker, B., Anderies, J. M., & Abel, N. (2001). From metaphor to measurement: Resilience of what to what? *Ecosystems, 4,* 765–781.

Chapin IIIi, F. S., Zavaleta, E. S., Eviner, V. T., Naylor, R. L., Vitousek, P. M., Reynolds, H. L., Hooper, D. U., Lavorel, S., Sala, O. E., Hobbie, S. E., Mack, M.C., & Díaz, S. (2000). Consequences of changing biodiversity. *Nature 405,* 235–242.

Chen, S., Fath, B. D., Chen, B. (2011). Information-based network environ analysis: A system perspective for ecological risk assessment. *Ecological Indicators, 11*(6), 1664–1672.

Chen, X., & Cohen, J. E. (2001). Transient dynamics and food-web complexity in the Lotka-Volterra cascade model. *Proceedings of the Royal Society of London. Series B, Biological Sciences, 268,* 869–877.

Cheng, G. D., Xiao, H. L., Xu, Z. M., Li, J. X., & Lu, M. F. (2006). Water issue and its counter-measure in the inland river basins of Northwest China—a case study in Heihe River Basin. *Journal of Glaciology and Geocryology, 3,* 406–413.

Christensen, V., & Walters, C. J. (2004). Ecopath with ecosim: Methods, capabilities and limitations. *Ecological Modelling, 172,* 109–139.

Chrystal, R. A., & Scharler, U. M. (2014). Network analysis indices reflect extreme hydrodynamic conditions in a shallow estuarine lake (Lake St Lucia), South Africa. *Ecological indicators, 38,* 130–140.

Denman, K. L., Brasseur, G., Chidthaisong, A., Ciais, P., Cox, P. M., Dickinson, R. E., Hauglustaine, D., Heinze, C., Holland, E., Jacob, D., Lohmann, U., Ramachandran, S., da Silva Dias, P. L., Wofsy, S. C., & Zhang, X. (2007). Couplings between changes in the climate system and biogeochemistry. In S. Solomon, D. Qin, M. Manning, Z. Chen, M. Marquis, K. B. Averyt, M. Tignor, & H. L. Miller (Eds.), *Climate Change 2007: The Physical Science Basis. Contribution of Working Group I to the Fourth Assessment. Report of the Intergovernmental Panel on Climate Change.* Cambridge, United Kingdom and New York, NY, USA: Cambridge University Press.

Dudgeon, D., Arthington, A. H., Gessner, M. O., Kawabata, Z.-I., Knowler, D. J., Lévêque, C., Naiman, R. J., Prieur-Richard, A.-H., Soto, D., Stiassny, M.L.J., Sullivan, C.A. (2006). Freshwater biodiversity : Importance, threats, status and conservation challenges. *Biological reviews of the Cambridge Philosophical Society, 81*(2), 163–182.

Elliott, M., & Quintino, V. (2007). The estuarine quality paradox, environmental homeostasis and the difficulty of detecting anthropogenic stress in naturally stressed areas. *Marine Pollution Bulletin, 54,* 640–646.

Elmqvist, T., Folke, C., Nyström, M., Peterson, G., Bengtsson, J., & Walker, B. (2003). Response diversity, ecosystem change, and resilience. *Frontiers in Ecology and the Environment, 1,* 488–494.

Elton, C. S. (1958). *Ecology of invasions by animals and plants.* London: Chapman & Hall.

Fang, D., Fath, B. D., Chen, B., & Scharler, U. M. (2014). Network environ analysis for socio-economic water system. *Ecological indicators, 47,* 80–88.

Fath, B. (1998). Network synergism: Emergence of positive relations in ecological systems. *Ecological Modelling, 107,* 127–143.

Fath, B. D., Scharler, U., Ulanowicz, R. E., & Hannon, B. (2007). Ecological network analysis: network construction. *Ecological Modelling, 208,* 49–55.

Fath, B. D., Dean, C. A., & Katzmair, H. (2015). Navigating the adaptive cycle: An approach to managing the resilience of social systems. *Ecology and Society, 20*(2), 24.

Folke, C. (2006). Resilience: The emergence of a perspective for social—ecological systems analyses. *Global Environmental Change, 16,* 253–267.

Folke, C., Carpenter, S., Walker, B., Scheffer, M., Elmqvist, T., Gunderson, L., et al. (2004). Regime shifts, resilience, and biodiversity in ecosystem management. *Annual Review of Ecology Evolution and Systematics, 35,* 557–581.

Gamfeldt, L., & Hillebrand, H. (2008). Biodiversity effects on aquatic ecosystem functioning—maturation of a new paradigm. *International Review of Hydrobiology, 93,* 550–564.

Goerner, S. J., Lietaer, B., & Ulanowicz, R. E. (2009). Quantifying economic sustainability: Implications for free-enterprise theory, policy and practice. *Ecological Economics, 69,* 76–81.

Grilli, J., Rogers, T., & Allesina, S. (2016). Modularity and stability in ecological communities. *Nature Communications, 7,* 12031.

Gunderson, L. H., & Holling, C. S. (Eds.). (2002). *Panarchy: Understanding transformations in human and natural systems*. Washington DC: Island Press.

Higashi, M., & Patten, B. C. (1989). Dominance of indirect causality in ecosystems. *The American Naturalist, 133,* 288–302.

Higashi, M., & Nakajima, H. (1995). Indirect effects in ecological interaction networks. I. The chain rule approach. *Mathematical Biosciences, 130,* 99–128.

Holling, C. S. (1973). Resilience and the stability of ecological systems. *Annual Review of Ecology and Systematics, 4,* 1–23.

Holling, C. S. (1986). The resilience of terrestrial ecosystems: local surprise and global change. In W. C. Clark & R. E. Munn (Eds.), *Sustainable Development of the Biosphere*. London: Cambridge University Press.

Hughes, T. P., Baird, A. H., Bellwood, D. R., Card, M., Connolly, S. R., Folke, C., et al. (2003). Climate change, human impacts, and the resilience of coral reefs. *Science, 301,* 929–934.

IPCC—Report of the intergovernmental panel on climate change (2007). Fourth Assessment Report. Climate change 2007: Synthesis report. Cambridge University Press. ISBN 92-9169-122-4.

Jørgensen, S. E., Fath, B. D., Nielsen, S. N., Pulselli, F., Fiscus, D., Bastianoni, S. (2015). *Flourishing within limits to growth: Following nature's way*. Earthscan Publisher. 220 p.

Kharrazi, A., Rovenskaya, E., Fath, B. D., Yarime, M., & Kraines, S. (2013). Quantifying the sustainability of economic resource networks: An ecological information-based approach. *Ecological Economics, 90,* 177–186.

Kharrazi, A., Fath, B. D., & Katzmair, H. (2016). Advancing empirical approaches to the concept of resilience: A critical examination of panarchy, ecological information, and statistical evidence. *Sustainability, 8,* 935.

Kharrazi, A., Rovenskaya, E., & Fath, B. D. (2017). Network structure impacts global commodity trade growth and resilience. *PLoS ONE, 12*(2), e0171184. https://doi.org/10.1371/journal.pone.0171184.

Kauppi, P. E., Mielikainen, K., & Kuusela, K. (1992). Biomass and carbon budget of European forest, 1971 to 1990. *Science, 256,* 70–74.

Libralato, S., Christensen, V., & Pauly, D. (2006). A method for identifying keystone species in food web models. *Ecol. Modell., 195,* 153–171.

Loreau, M. (2000). Biodiversity and ecosystem functioning: Recent theoretical advances. *Oikos, 91,* 3–17.

MacArthur, R. H. (1955). Fluctuations of animal populations and a measure of community stability. *Ecology, 36,* 533–536.

May, R. M. (1972). Will a large complex system be stable? *Nature, 238,* 413–414.

McCann, K. S. (2000). The diversity-stability debate. *Nature, 405,* 228–233.

McNaughton, S. J. (1977). Diversity and stability of ecological communities: A comment on the role of empiricism in ecology. *The American Naturalist, 111*(979), 515–525.

MEA. (2005). Millenium ecosystem assessment, 2005. http://www.millenniumassessment.org.

Meadows, D. H., Meadows, D. L., Randers, J., & Behrens, W. W., III. (1972). *The limits to growth.* New York: Universe Books.

Miao, L. J., Cui, L. F., Luan, Y. B., & He, B. (2011). Similarities and differences of Beijing and Shanghai's land use changes induced by urbanization. *Chinese Journal of Metal Science and Technology, 31*(4), 398–404.

Moore, J. C., & de Ruiter, P. C. (2012). *Energetic Food Webs.* An analysis of real and model ecosystems: Oxford University Press.

Moore, J. C., de Ruiter, P. C., & Hunt, H. W. (1993). Influence of productivity on the stability of real and model ecosystems. *Science, 261,* 906–908.

Mori, A. S., Furukawa, T., & Sasaki, T. (2013). Response diversity determines the resilience of ecosystems to environmental change. *Biological Reviews of the Cambridge Philosophical Society, 88,* 349–364.

Müller, F., Bergmann, M., Dannowski, R., Dippner, J. W., Gnauck, A., Haase, P., et al. (2016). Assessing resilience in long-term ecological data sets. *Ecological indicators, 65,* 10–43.

Mukherjee, J., Scharler, U. M., Fath, B. D., & Ray, S. (2015). Measuring sensitivity of robustness and network indices for an estuarine food web model under perturbations. *Ecological Modelling, 306,* 160–173.

Naeem, S., Chapin III, F. S., Costanza, R., Ehrlich, P. R., Golley, F. B., Hooper, D. U., Lawton, J. H., O'Neill, R. V., Mooney, H. A., Sala, O. E., Symstad, A. J., & Tilman, D. (1999). Biodiversity and ecosystem functioning: Maintaining natural life support processes. *Issues in Ecology, 4,* 1–11. Published by the Ecological Society of America.

Neutel, A.-M., Heesterbeek, J. A. P., & De Ruiter, P. C. (2002). Stability in real food webs: Weak links in long loops. *Science, 296,* 1120–1123.

Pimm, S. L. (1982). *Foodwebs.* London: Chapman and Hall.

Power, M. E., Tilman, D., Estes, J. A., Menge, B. A., Bond, W. J., Mills, L. S., et al. (1996). Challenges in the quest for keystones. *BioScience, 46,* 609–620.

Rist, L., Felton, A., Nyström, M., Troell, M., Sponseller, R. A., Bengtsson, J., et al. (2014). Applying resilience thinking to production ecosystems. *Ecosphere, 5,* 1–11.

Rutledge, R. W., Basore, B. L., & Mulholland, R. J. (1976). Ecological stability: An information theory viewpoint. *Journal of Theoretical Biology, 57,* 355–371.

Salas, A. K., & Borrett, S. B. (2011). Evidence for dominance of indirect effects in 50 trophic ecosystem networks. *Ecological Modelling, 222,* 1192–1204.

Scharler, U. M. (2012). Ecosystem development during open and closed phases of temporarily open/closed estuaries on the subtropical east coast of South Africa. *Estuarine, Coastal Shelf Science, 108,* 119–131.

Scharler, U. M., & Fath, B. D. (2009). Comparing network analysis methodologies for consumer —resource relations at species and ecosystems scales. *Ecological Modelling, 220,* 3210–3218.

Scheffer, M., Carpenter, S., Foley, J. A., Folke, C., & Walker, B. (2001). Catastrophic shifts in ecosystems. *Nature, 413,* 591–596.

Shevtsov, J., & Rael, R. (2015). Indirect energy flows in niche model food webs: Effects of size and connectance. *PLoS ONE, 10*(10), e0137829.

Szyrmer, I., & Ulanowicz, R. E. (1987). Total flows in ecosystems. *Ecological Modelling, 35,* 123–136.

Tilman, D. (1999). The ecological consequences of changes in biodiversity: A search for general principles. *Ecology, 80,* 1455–1474.

Ulanowicz, R. E. (1986). *Growth and development.* New York: Springer.

Ulanowicz, R. E. (2002). Information theory in ecology. *Journal of Computational Chemistry, 25,* 393–399.

Ulanowicz, R. E. (2004). Quantitative methods for ecological network analysis. *Computers and Chemistry, 28,* 321–339.
Ulanowicz, R. E. (2009). The dual nature of ecosystem dynamics. *Ecological Modelling, 220*(16), 1886–1892.
Ulanowicz, R. E., & Puccia, C. J. (1990). Mixed trophic impacts in ecosystems. *COENOSES, 5,* 7–16.
Ulanowicz, R., Goerner, S., Lietaer, B., & Gomez, R. (2009). Quantifying sustainability: Resilience, efficiency and the return of information theory. *Ecological Complexity, 6,* 27–36.
Wackernagel, M., Schulz, N. B., Deumling, D., Linares, A. C., Jenkins, M., Kapos, V., et al. (2002). Tracking the ecological overshoot of the human economy. *PNAS, 99,* 9266–9271.
Wagensberg, J., Garcia, A., & Sole, R. V. (1990). Connectivity and information transfer in flow networks: Two magic numbers in ecology? *Bulletin of Mathematical Biology, 52,* 733–740.
Wootton, J. T. (1994). The nature and consequences of indirect effects in ecological communities. *Annual Review of Ecology and Systematics, 25*(1), 443–466.
Xia, L., Fath, B. D., Scharler, U. M., & Zhang, Y. (2016). Science of the total environment spatial variation in the ecological relationships among the components of Beijing's carbon metabolic system. *Science of the Total Environment, 544,* 103–113.
Zhang, Y., Xia, L. L., & Xiang, W. N. (2014). Analyzing spatial patterns of urban carbon metabolism: A case study in Beijing, China. *Landscape and Urban Planning, 130,* 184–200.
Zorach, A. C., Ulanowicz, R. E. (2003). Quantifying the complexity of flow networks: How many roles are there?. *Complexity, 8*(3), 68–76.

# Chapter 12
# Complexity and Stability of Adaptive Ecological Networks: A Survey of the Theory in Community Ecology

**Pietro Landi, Henintsoa O. Minoarivelo, Åke Brännström, Cang Hui and Ulf Dieckmann**

**Abstract** *Background and Significance of the topic*: The planet is changing at paces never observed before. Species extinction is happening at faster rates than ever, greatly exceeding the five mass extinctions in the fossil record. Nevertheless, human life is strongly based on services provided by ecosystems, thus the responses to global change of the planet's natural heritage are of immediate concern. Understanding the relationship between complexity and stability of ecosystems is of key importance for the maintenance of the balance of human growth and the conservation of all the natural services that ecosystems provide. *Methodology*: The concept of ecological networks and their characteristics are first introduced, followed by central and occasionally contrasting definitions of complexity and stability. The literature on the relationship between complexity and stability in different types of models and few real ecosystems is then reviewed, highlighting the theoretical debate and the lack of consensual agreement. *Application/Relevance to systems analysis*: This chapter uses ecological-network models to study the relationship between complexity and stability of natural ecosystems. *Policy and/or practice implications*: Mathematical network models can be used to simplify the

P. Landi (✉) · H. O. Minoarivelo · C. Hui
Department of Mathematical Sciences,
Stellenbosch University, Stellenbosch, South Africa
e-mail: landi@sun.ac.za

P. Landi · Å. Brännström · U. Dieckmann
Evolution and Ecology Program, International Institute for Applied Systems Analysis, Laxenburg, Austria

H. O. Minoarivelo
Centre of Excellence in Mathematical and Statistical Sciences,
Wits University, Johannesburg, South Africa

Å. Brännström
Department of Mathematics and Mathematical Statistics,
Umeå University, Umeå, Sweden

C. Hui
Mathematical and Physical Biosciences,
African Institute for Mathematical Sciences, Muizenberg, South Africa

 
P. Mensah et al. (eds.), *Systems Analysis Approach for Complex Global Challenges*,
https://doi.org/10.1007/978-3-319-71486-8_12

vast complexity of the real world, to formally describe and investigate ecological phenomena, and to understand ecosystems propensity of returning to its functioning regime after a stress or a perturbation. *Discussion and conclusion*: The chapter concludes by summarising the importance of this line of research for the successful management and conservation of biodiversity and ecosystem services.

## 12.1 Introduction

In the geological era of the Anthropocene, the planet is changing at paces never observed before (Millennium Ecosystem Assessment 2005). Pollution, natural resources exploitation, habitat fragmentation, and climate change are only some of the threats the biosphere is facing. Species extinction is happening at faster rates than ever, largely exceeding the five mass extinctions in the fossil record. Even if it is difficult to realize, human life is strongly based on services provided by ecosystems, thus the responses to global change of the planet's natural heritage are of immediate concern for policy makers. As ecosystems are composed by thousands of interlinked species that interact directly or through their shared environment, such as nutrients, light, or space, a holistic perspective on the system as a whole is normally required to predict ecosystem responses to global changes (Wolanski and McLusky 2011). A systems-analysis approach is thus crucial for acquiring an understanding of all the dynamical feedbacks at the ecosystem level and for accurately managing the biodiversity that humans rely on in terms of ecosystem services. In particular, mathematical network models can be used to simplify the vast complexity of the real world, to formally describe and investigate ecological phenomena, and to understand how ecosystems react to stress and perturbations.

Complex-network models are composed of a set of compartments, describing either species or coarser functional groups, and a set of links that represent interactions (or energy or biomass flows) among compartments (Fig. 12.1, Baird and Mehta 2011; Olff et al. 2009). Thus, such models can describe both biotic and abiotic interactions of species, i.e., both interactions among the species themselves and interactions with their external environment, and consequently they can successfully be used to assess ecosystem stability to perturbations (Saint-Béat et al. 2015). Stability of an ecosystem can be understood as its propensity of returning to its functioning regime after a stress or a perturbation in its biotic components (e.g., decline in species abundances, introduction of alien species, and species extinction) or abiotic components (e.g., exploitation, habitat fragmentation, and climate change). A challenging and central question that has interested ecologists and systems analysts alike for decades is how the stability of an ecosystems depend on its complexity, as roughly measured by the ecosystems' diversity in species and their interactions (D'Alelio et al. 2016; Loreau and de Mazancourt 2013).

In order to appreciate the importance of this question, a classification of the major different functions that ecosystems continuously provide is briefly introduced. Natural ecosystems sustain life and provide services that can be divided into

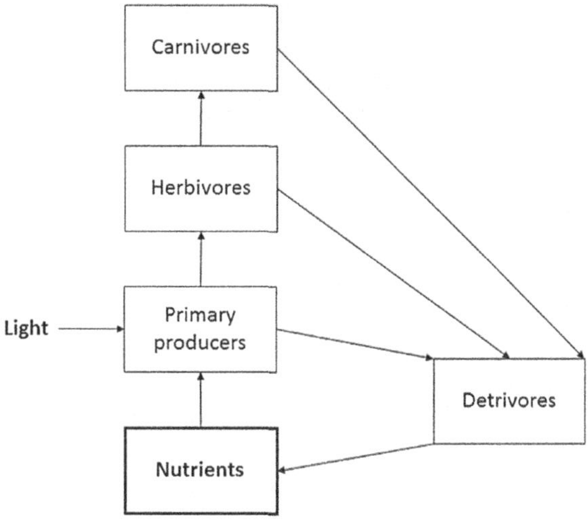


**Fig. 12.1** Schematic representation of an ecosystem using a network model. The abiotic components (light and nutrients, in bold) are transformed through photosynthesis into living biomass of autotrophic primary producers. Such biomass is transferred through trophic interactions to higher heterotrophic levels (herbivores and carnivores), and it returns to the nutrients abiotic compartment through decomposition by detrivore organisms

four areas (Millennium Ecosystem Assessment 2005): *provisioning*, such as the production of food and water; *regulating*, such as the control of climate and disease; *supporting*, such as nutrient cycles and crop pollination; and *cultural*, such as spiritual and recreational benefits. For the management and conservation of ecosystems services it is important to know how the complexity of an ecosystem is related to its stability, thus how the diversity of species in the ecosystem and the network of their interactions can contribute to maintaining a stable supply of services (Kondoh 2005 and Saint-Béat et al. 2015). This is increasingly important in an era in which the pressure exerted on natural ecosystems is strengthening, influencing their structure and functioning, while the services they provide are vital for a continuously increasing number of people. In particular, human activities, directly or indirectly, tend to simplify the composition and the structure of natural ecosystems. Therefore, understanding the relationship between complexity and stability of ecosystems is of key importance for maintaining the balance of human growth with the conservation of natural services that ecosystems provide. Using ecological-network models to study the relationship between complexity and stability of natural ecosystems is the focus of this chapter. The concept of ecological networks and their characteristics are first introduced, followed by central and occasionally contrasting definitions of complexity and stability. After that, a review of the literature on the relationship between complexity and stability in different types of models and in real ecosystems is presented, highlighting the theoretical debate and the lack of consensual agreement. The chapter continues with describing the importance of considering the dynamic adaptation of species behaviour and the resulting changes in ecosystems structure, after which a conclusion summarizing the importance of this line of research for the successful management and conservation of biodiversity and ecosystem services in the current era of the Anthropocene ends the chapter.

This chapter focuses on the theoretical debate in the field of community ecology. A vast research using similar network tools and both stability and complexity indicators is also undertaken in the field of ecosystem ecology (see, e.g., review in Saint-Béat et al. 2015). While community ecology focuses on each species and their interactions, usually studying simplified model versions of real ecosystems, ecosystem ecology groups species together in functional groups (e.g., detritivores, primary producers, herbivores, carnivores, etc.) and quantifies the fluxes (usually of biomass and energy) among such compartments, focusing on emergent quantifiable properties of ecosystems, such as ecosystem growth, organization, development, resilience, redundancy, energy throughput, flow diversity, nutrient (re)cycling, etc. (Baird and Mehta 2011). Although the studied issue is the same, the focus and the methodology have diverged into two distinct fields. This chapter focuses on the theoretical contributions of community ecology, while ecosystem ecology is considered somewhere else in this book.

Although this chapter mainly focus on theoretical issues, ecological networks have also proved to be of extreme practical importance as an applied tool in the fields of restoration, conservation, invasive species, biological control, habitat management, and global warming, and could be very promising and rewarding also in the fields of urban ecology, agroecology, habitat fragmentation, and ecosystem services (Kaiser-Bunbury and Bluthgen 2015; Memmot 2009), especially as a visualisation tool to improve public and stakeholder engagement in the decision-making process of sustainable development and management (Pocock et al. 2016).

## 12.2 Ecological Networks Defined

An ecological network describes interactions among species in a community. There are different types of interactions, e.g., trophic interactions (feeding), mutualistic interactions (pollination, seed dispersal, etc.), and competitive interactions (interference for common resources). Ecological networks can be represented as a set of $S$ *nodes,* characterizing the species, connected by a set of $L$ *links*, characterizing possible interactions among each ordered pair of species (Bonchev and Buck 2007). Links can be described by either a binary variable (0 or 1, absence or presence of interaction) or by a real number characterizing the *weight* (or *strength*) of the interaction. In the first case the network is called *unweighted*, while in the second case it is called *weighted*. Moreover, interactions can be *undirected* (or symmetric), meaning that species $i$ affects species $j$ to a certain amount and equally vice versa, or *directed* (or asymmetric), meaning that species $i$ can affect species $j$ differently from how species $j$ affects species $i$ (Fig. 12.2). Moreover, interactions can be described by their sign (+ or −). For example, trophic networks (food webs) are characterized by the fact that one species is feeding on the other, thus the coefficients $a_{ij}$ (describing the effect of species $j$ on species $i$) and $a_{ji}$ (describing the effect of species $i$ on species $j$) will obviously have opposite signs (thus their product will be negative, $a_{ij}a_{ji} < 0$), i.e., one species is benefiting while the other is suffering from

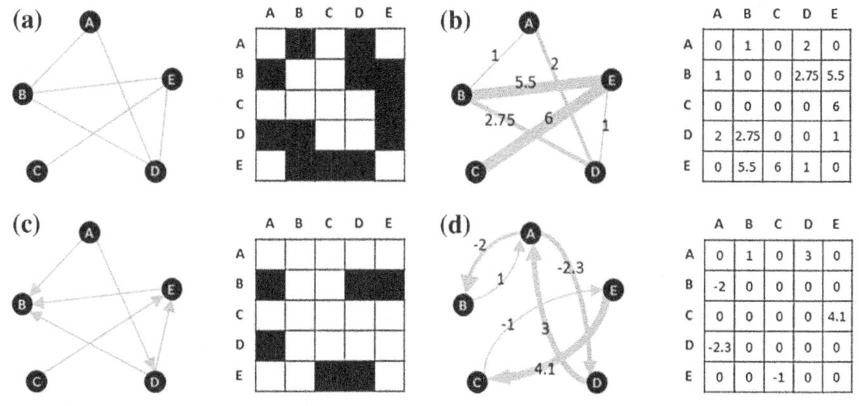


**Fig. 12.2** Categorisation of ecological networks, according to link directionality and weight. Black (white) entries in matrices (**a**) and (**c**) represent presence (absence) of interaction. **a** Unweighted undirected; **b** weighted undirected; **c** unweighted directed; **d** weighted directed. Note that links point to the affected species. For example, species A in panel (**d**) is positively affected by species B and D while negatively affecting species B and D

the interaction. In mutualistic networks both species are benefiting from the interaction, thus both coefficients $a_{ij}$ and $a_{ji}$ will be positive (and so their product, $a_{ij}a_{ji} > 0$), while in competitive networks both species are suffering from the interaction, thus both coefficients $a_{ij}$ and $a_{ji}$ will be negative (thus their product will be again positive, $a_{ij}a_{ji} > 0$) (Fig. 12.3). Notice therefore that trophic networks cannot be undirected (symmetric), since the two coefficients describing the interaction always have opposite sign (and typically also different absolute values).

The structure of the ecological network can be described by the *SxS community matrix* $A = [a_{ij}]$, where each element $a_{ij}$ describes the link between species $i$ and species $j$, i.e., the effect that species $j$ has on species $i$. Such elements of the interaction matrix are called *interaction strengths.* In the particular case of unweighted and undirected network, the interaction matrix is symmetric (i.e., $a_{ij} = a_{ji}$) and its elements are either 0 or $\pm 1$. In the most general case of weighted and directed network the interaction matrix can have any composition of real values. For *bipartite* networks (i.e., those formed by two disjoint groups of respectively $m$ and $n$ species, $S = m + n$, with interactions only between two species of different groups), such as mutualistic networks of plants and their pollinators or antagonistic networks of host-parasite interactions, the interaction matrix $A = [a_{ij}]$ is a $m \times n$ matrix.

Unfortunately, there is no unique quantification of interaction strengths. Depending on the scope (theoretical vs. empirical), several measures and indexes have been used to quantify interaction strengths. For example, theoretical studies mostly refer to $a_{ij}$ as the effect of a perturbation from equilibrium of the abundance of species $j$ on the population growth rate of species $i$ (elements of the Jacobian matrix describing the linearized dynamics of the model ecosystem around equilibrium, see also Sect. 12.5). Unfortunately, such coefficients are well defined in

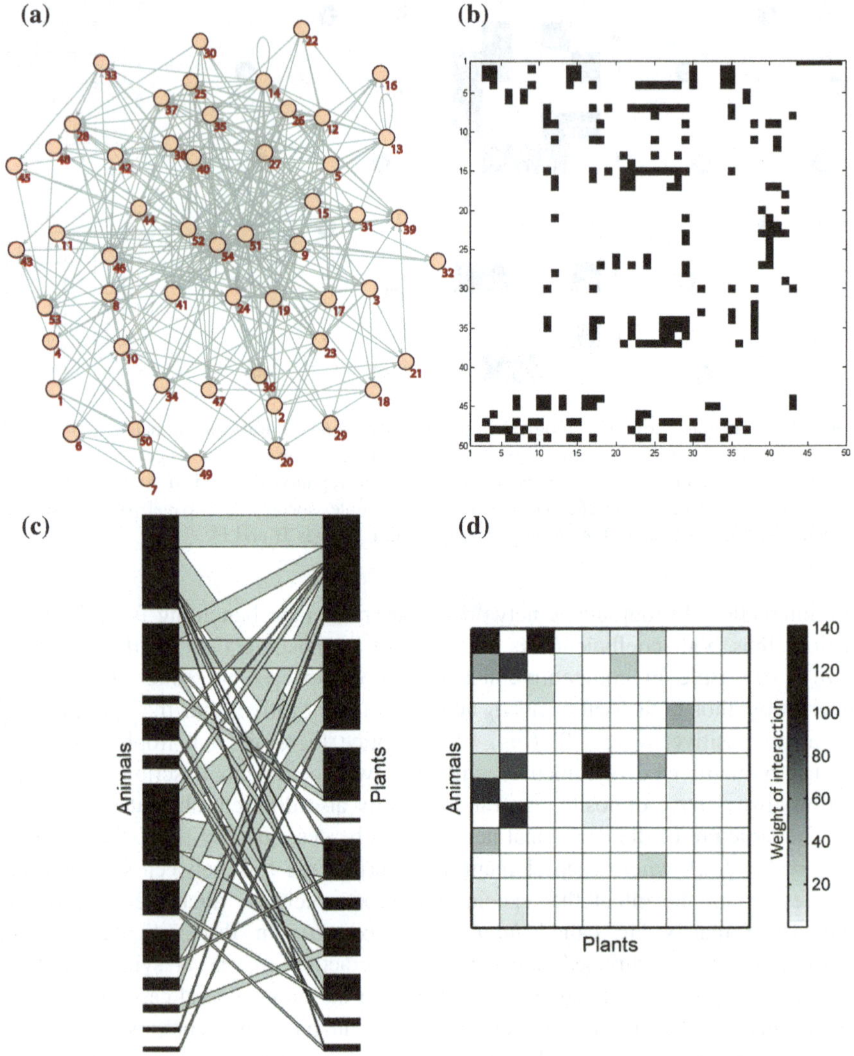


**Fig. 12.3** Examples of real-world ecological networks. **a** and **b** Food web from the estuary river of St. Marks, Florida, USA (Baird et al. 1998). **c** and **d** Mutualistic network of pollination from the Flores Island, one of the Azores oceanic islands (Olesen et al. 2002). Left column: network representation. Second column: matrix representation. The food web is unweighted directed: in (**b**) the black entries in the matrix represent presence of interaction. The mutualistic network is weighted undirected: the link width in (**c**) and the shade of grey in (**d**) are proportional to the weight of the interaction which represents the number of pollinator visits

theory, but very hard to measure in the field or in lab experiments. On the other hand, empirical observations mainly quantify magnitude of energy and biomass flows between compartments in model ecosystems, or consumption rate for resource-consumer and prey-predator interactions, or visiting probabilities in pollination networks. Such quantities are relatively easy to be estimated empirically, but unfortunately they are not directly related to theoretical Jacobian elements as they are independent of species abundances (see Berlow et al. 2004 and Wootton and Emmerson 2005 for reviews on the subject).

Additionally, the network itself can be *adaptive,* i.e., its fundamental characteristics such as the set of nodes and links can evolve through time (Kondoh 2005 and Gross and Sayama 2009). For instance, new species can emigrate or be introduced into the system, originating new interactions in the network, while some other species might emigrate or go extinct, together with all their interactions with the rest of the network. Besides that, adaptive species behaviour—e.g. adaptive consumer foraging and/or adaptive resource defence (Valdovinos et al. 2010)—or phenotypic adaptation—e.g. coevolution of animal and plant traits in pollination systems (Zhang et al. 2013)—can often cause links among species to form, disappear, or change in strength as time progresses. Although some of the complexity and stability indicators that will be described in Sects. 12.3 and 12.5 and used in Sect. 12.6 will partly deal with invasions and extinctions, specific focus to adaptive networks, with particular emphasis on adaptive behaviour and phenotypic traits, will be given in Sect. 12.7.

## 12.3 Network Complexity

*Species richness* $S$, or the total number of interacting species in the network, also known as the network size, has been used as the simplest descriptor of network complexity (MacArthur 1955; May 1973; Pimm 1980a; Bonchev and Buck 2007, and Table 12.1). In the particular case of bipartite networks, species richness $S$ is expressed as $S = m + n$. In food-web studies, the use of *trophic species* (a functional group of species sharing the same set of predators and preys) as a replacement of taxonomic species (i.e., when species are distinguished based on morphological and phylogenetic criteria) is a widely accepted convention (Schoener 1989; Pimm et al. 1991; Goldwasser and Roughgarden 1993; Williams and Martinez 2000; and Dunne et al. 2002b). The use of 'trophic species' has indeed been shown to reduce methodological biases in food web datasets because it reduces scatter in the data and avoids redundancy of interactions (Pimm et al. 1991 and Martinez 1994). Sometimes, the use of *morphospecies* (species distinguished from others by only their morphologies) as a replacement of taxonomic species is also considered because of a lack of taxonomic distinction between species (Olito and Fox 2014). Hence, network size often refers to the number of functional or morphological diversity in the system.

Another commonly-used indicator of complexity is the *connectance* $C$, measuring the proportion of realised interactions among all the possible ones in a network (i.e., the total number of interactions $L$ divided by the square of the number of species $S^2$ or $L$ divided by the product $mn$ in the case of a bipartite network).

**Table 12.1** Measures of network complexity

| Network complexity measure | Definition | References |
|---|---|---|
| Species richness ($S$) | Total number of species in the network | MacArthur (1955), May (1973)<br>*Food webs*: Pimm (1979, 1980a), Cohen and Briand (1984), Cohen and Newman (1985), Havens (1992), Martinez (1992), Haydon (1994); Borrvall et al. (2000), Dunne et al. (2002a, b), Dunne and Williams (2009), Banašek-Richter et al. (2009), Gross et al. (2009), Thébault and Fontaine (2010), Allesina and Tang (2012)<br>*Mutualism*: Okuyama and Holland (2008), Suweis et al. (2015)<br>*Competition*: Lawlor (1980), Lehman and Tilman (2000), Christianou and Kokkoris (2008), Fowler (2009) |
| Connectance ($C$) | Proportion of realized interactions among all possible ones, $L/S^2$ | May (1973)<br>*Food webs*: De Angelis (1975), Pimm (1979, 1980a, 1984), Martinez (1992), Haydon (1994, 2000), Chen and Cohen (2001), Olesen and Jordano (2002), Dunne et al. (2002a, b, 2004), Banašek-Richter et al. (2009), Dunne and Williams (2009), Gross et al. (2009), Thébault and Fontaine (2010), Tylianakis et al. (2010), Allesina and Tang (2012), Heleno et al. (2012), Poisot and Gravel (2014)<br>*Mutualism*: Jordano (1987), Rezende et al. (2007), Okuyama and Holland (2008), Vieira and Almeida-Neto (2015)<br>*Competition*: Fowler (2009) |
| Connectivity ($L$) | Total number of interactions | *Mutualism*: Okuyama and Holland (2008)<br>*Competition*: Fowler (2009) |
| Linkage density ($L/S$) | Average number of links per species | *Food webs*: Pimm et al. (1991), Havens (1992)<br>*Mutualism*: Jordano (1987) |
| Interaction strength ($a_{ij}$) | Weight of an interaction | *Food webs*: De Angelis (1975), Yodzis (1981), Paine (1992), Haydon (1994), de Ruiter et al. (1995), McCann et al. (1998), Berlow (1999), Borrvall et al. (2000), Haydon (2000), Neutel et al. (2002, 2007), Berlow et al. (2004), Emmerson and Yearsley (2004), Wootton and Emmerson (2005), |

(continued)

**Table 12.1** (continued)

| Network complexity measure | Definition | References |
|---|---|---|
| | | Rooney et al. (2006), Allesina and Pascual (2008), Gross et al. (2009), Allesina and Tang (2012), van Altena et al. (2016)<br>*Mutualism*: Jordano (1987), Bascompte et al. (2006), Okuyama and Holland (2008), Suweis et al. (2015)<br>*Competition*: Lawlor (1980), Hughes and Roughgarden (1998), Kokkoris et al. (1999, 2002), Christianou and Kokkoris (2008) |
| Weighted linkage density | Average number of links per species weighted by interaction strength | *Food webs*: Bersier et al. (2002), Tylianakis et al. (2007), Dormann et al. (2009) |
| Weighted connectance | Weighted linkage density divided by species richness | *Food webs*: Haydon (2000), Bersier et al. (2002), Tylianakis et al. (2007), Dormann et al. (2009), van Altena et al. (2016)<br>*Mutualism*: Minoarivelo and Hui (2016) |
| Species degree | Number of interactions the species share with others | *Food webs*: Waser et al. (1996), Memmott (1999), Solé and Montoya (2001), Camacho et al. (2002), Dunne et al. (2002a), Montoya and Solé (2002), Jordano et al. (2003), Vázquez and Aizen (2003), Dunne and Williams (2009) |
| Species strength | Sum of all weighted interactions of a species | *Mutualism*: Bascompte et al. (2006), Okuyama (2008), James et al. (2012), Feng and Takemoto (2014) |
| Dependency of species $i$ on species $j$ | Interaction strength between $i$ and $j$ divided by the total strength of species $i$ | *Mutualism*: Bascompte et al. (2006) |

It accounts intuitively for the probability that any pair of species interact in the network. It is probably one of the earliest and the most popular descriptors of ecological networks structure. Sometimes, a simpler measurement of interactions, known as *connectivity,* has been used instead of connectance. The connectivity of a network is simply its total number of interactions $L$.

In order to understand the average level of specialization of the network, i.e., whether the network is dominated by specialists (species holding few interactions) or generalists (species holding many interactions), food web ecologists have introduced *linkage density*. It is calculated as the average number of links per species, or the connectivity divided by species richness, $L/S$.

By increasing the information value of these network metrics, some theoretical studies have incorporated the strength of interactions. Thus, quantitative counterparts of linkage density and connectance, called respectively *weighted connectance* and *weighted linkage density* have been developed (Bersier et al. 2002; Tylianakis et al. 2007; and Dormann et al. 2009). Weighted linkage density is a measurement based on the Shannon measure (Shannon 1948) of entropy (or uncertainty). Weighted connectance is then computed as the weighted linkage density divided by species richness. There are several reasons for believing that networks metrics incorporating interaction strength are better suited to reflect salient ecosystem properties, among them their ability to give increased weight to strong interactions and the fact that they change continuously as link strength is reduced and the link is eventually removed. The latter can be particularly important in empirical food-web studies in which the sampling resolution typically proves important for the number of links found, with higher resolution leading to many weak links.

As connectance and linkage density are only community-average descriptors of network structure, they do not inform on the relative importance of each species to the overall connectivity. *Node degree* distribution, i.e., the distribution of the number of interactions per species, is another widely used descriptor of network complexity. The degree of a node (or a species) refers to the number of interactions it shares with others. The distribution of node degree in ecological networks have been shown to differ from a Poisson distribution that characterises large random networks (Camacho et al. 2002; Dunne et al. 2002a; Montoya and Solé 2002; and Jordano et al. 2003).

A generalization of the node-degree distribution is the *interaction-strength* distribution, taking into account the weights associated with each link. The strength of each species is computed as the sum of all the weighted interactions of that species. A refinement of species strength is *species dependency* on another species. The dependence of a species $i$ on a species $j$ is defined as the fraction of the interaction strength between $i$ and $j$ ($a_{ij}$) relative to the total strength of species $i$.

## 12.4 Network Architecture

Beyond ecological patterns in interaction and strength distribution, interactions in ecological networks exhibit even more complex topological features, related to the architecture of the network (Table 12.2). Among the most important of these features is the level of modularity or compartmentalization. Modularity depicts the extent to which a network is compartmentalized into delimited modules where species are strongly interacting with species within the same module but not with those from other modules (Olesen et al. 2007). Although a number of metrics have been developed to quantify the level of compartmentalization in a network, *modularity* (developed by Newman and Girvan 2004) has been the most widely accepted. This measure assumes that nodes in the same module have more links between them than one would expect for a random network and form each module by collecting

**Table 12.2** Measures of network architecture

| Network architecture measure | Definition | References |
|---|---|---|
| Modularity | Extent to which a network is compartmentalized into delimited modules | *Food webs*: Moore and Hunt (1988), Ives et al. (2000), Krause et al. (2003), Thébault and Fontaine (2010), Stouffer and Bascompte (2011)<br>*Mutualism*: Olesen et al. (2007), Mello et al. (2011), Dupont and Olesen (2012) |
| Nestedness | When specialists can only interact with subset of the species generalists interact with | *Food webs*: Atmar and Patterson (1993), Neutel et al. (2002), Cattin et al. (2004), Thébault and Fontaine (2010), Allesina and Tang (2012)<br>*Mutualism*: Bascompte et al. (2003), Memmott et al. (2004), Almeida-Neto et al. (2008), Bastolla et al. (2009), Zhang et al. (2011), Campbell et al. (2012), James et al. (2012), Rohr et al. (2014) |

nodes that interact between themselves more frequently than expected by chance. However, see, e.g., Rosvall and Bergstrom (2007) and Landi and Piccardi (2014) for limitations of modularity and other metrics of compartmentalization.

Another important descriptor of ecological network architecture, especially for mutualistic networks, is *nestedness*. It is a pattern of interactions in which specialists can only interact with a subset of species with which more generalists interact. It means that in a nested network, both generalists and specialists tend to interact with generalists whereas specialist-to-specialist interactions are rare (Bascompte et al. 2003). To quantify the nestedness of a network, several metrics have been developed. Among the most commonly used are, for example, the 'temperature' metric by Atmar and Patterson (1993) and the NODF (Nestedness metric based on Overlap and Decreasing fill) metric by Almeida-Neto et al. (2008). Despite the existence of several metrics and algorithms, they are all mainly based on the same assumption: if a species $i$, more specialist than a species $j$, interacts with a species $k$, then the interaction between $i$ and $k$ is rewarded only if species $j$ also interacts with species $k$. Otherwise, the interaction is penalised.

## 12.5 Network Stability

In theoretical studies, each entry $a_{ij}$ of the interaction matrix usually quantifies the change in population growth rate of species $i$ caused by a small perturbation in the abundance of species $j$ around equilibrium abundances (i.e., stationary regime, species abundances are constant in time). Thus the interaction matrix is equivalent to the Jacobian matrix of the dynamical system that describes species abundance dynamics over continuous time, evaluated at equilibrium (Small et al. 2013). Such matrix is very useful to study the (local) asymptotic stability of the equilibrium. In fact, stability is defined by the real part of the leading eigenvalue of the Jacobian matrix (i.e. the eigenvalue with the largest real part). If the real part of the leading eigenvalue is positive, the equilibrium is unstable, i.e., any small perturbation from the equilibrium will be amplified until possible extinction of one (or more) species in the community. Otherwise, if the real part of the leading eigenvalue is negative, then small perturbations around the equilibrium will be dampened and the system will converge back to its stationary regime. Therefore, the sign of the real part of the leading eigenvalue can be a binary indicator of stability. Moreover, if stable, the inverse of the absolute value of the real part of the leading eigenvalue gives an indication of the time needed by the system to return to its equilibrium (the *time constant* of classical system theory). Systems that quickly return to equilibrium after perturbations are called *resilient*. *Resilience* is therefore often measured by the absolute value of the leading eigenvalue (if negative) of the interaction matrix. Notice that resilience is only defined for stable equilibria and it only gives information about the asymptotic behaviour of the system (see Neubert and Caswell 1997 for transient indicators). Global (vs. local) stability implies that any (vs. small) perturbation from the equilibrium will be dampened. Global stability usually refers to the case of a single equilibrium (typical of linear systems).

The notion of *structural* stability of an equilibrium refers to its domain (or probability) of coexistence of all the species in the ecosystem (*feasibility*). A feasible equilibrium for a system is in fact an equilibrium at which all the species $S$ in the system coexist with positive abundances. Structural stability usually refers to perturbations in the system itself (i.e. slightly changing one of its parameters) rather than perturbations in the state of the system (i.e. abundances, see previous paragraph). Assuming that a system is at a feasible equilibrium, a small perturbation in a parameter (e.g. species carrying capacity, intrinsic growth rate, predator conversion efficiency, handling time, …) will generically move the system to a slightly different (in terms of species abundances) feasible equilibrium, unless the system is close to a *bifurcation* point for that parameter. A bifurcation is indeed a qualitative change in the asymptotic behaviour of a system driven by a perturbation in one of its parameters. Such qualitative change could, for example, be a switch to a non-feasible equilibrium (where one or more species go extinct), to a non-stationary (e.g. periodic) orbit, etc. The region in parameter space for which the system has a feasible equilibrium is its domain of stable coexistence, and gives an indication (or probability) of its structural stability. The bigger the domain, the more structurally stable the system.

In addition to this, the number of coexisting species at an equilibrium could trivially be an indicator of stability. This number will be $S$ at a feasible equilibrium, and will be smaller than $S$ at an equilibrium at which some species have gone extinct. If this number is standardized to the total number of species $S$ we obtain the proportion of persistent species once equilibrium is reached, that is, *persistence*.

The notions of asymptotic stability and structural stability can of course be generalized in the case of non-stationary asymptotic regimes (such as cycles, tori, and chaotic attractors), using, e.g., Lyapunov exponents (Lyapunov 1992; Rinaldi et al. 2015). In such cases, or in the study of empirical time series, other stability indicators can however be more useful. For example, *temporal stability* (the reciprocal of *variability*) quantifies the stability of fluctuations in time. It is usually defined as the ratio of the mean value of the variable in time over its standard deviation (the inverse of the Coefficient of Variation). A high mean contributes to temporal stability, as it contributes to values far from 0 (extinction), as well as a low standard deviation that describes fluctuations around the mean.

Another approach to stability considers the effect of removing target species from a system. The *extinction cascade* measures the loss of additional species after the removal of one target species. *Robustness* (or *deletion stability*) is indeed the ability of a system to resist extinction cascades. Species removal can be random or targeted (e.g., the most connected species or species with low or high trophic level).

Instead of removing target species, *invasibility* describes the propensity of a system to be invaded by new species. Non-invadable systems are thought to be more stable than systems that are easily invaded by introduced alien species. Thus, resistance to invasion can be a measure of system stability. Invasion can simply bring the system in a new stable and feasible configuration, or in the worst case it could lead one or more species to extinction (Hui et al. 2016; Hui and Richardson 2017). Moreover, invaded systems tend to become more dominated by generalist interactions because invasive species are usually generalists or super-generalists. Thus, the level of specialization of interactions is a system could be used as a proxy for system resistance to invasion, hence stability.

A summary of the different introduced measures for network stability is given in Table 12.3. See Pimm (1984), Logofet (2005), Donohue et al. (2013), and Borrelli et al. (2015) for additional reviews on different stability concepts.

## 12.6 Complexity-Stability Debate

Before the 1970s, ecologists believed that more diverse communities enhanced ecosystem stability (Odum 1953; MacArthur 1955; Elton 1958). In particular, they believed that natural communities develop into stable systems through successional dynamics. Aspects of this belief developed into the notion that complex communities are more stable than simple ones. A strong proponent of this view was Elton (1958), who argued that "simple communities were more easily upset than richer ones; that is, more subject to destructive oscillations in populations, and more vulnerable to invasions". In fact, both Odum (1953) and Elton (1958) arrived at similar conclusions based on the repeated observation that greatly simplified terrestrial communities are characterized by more violent fluctuations in population density than diverse terrestrial communities. For example, invasions most frequently occur on cultivated land where human influence had produced greatly

**Table 12.3** Measures of network stability

| stability measure | Definition | References |
|---|---|---|
| Asymptotic stability | Perturbations from the ecological regime are dampened. The system return to its ecological regime after a perturbation in the state of the system | May (1973) *Food webs*: De Angelis (1975), Yodzis (1981), De Ruiter et al. (1995), Haydon (2000), Neutel et al. (2002, 2007), Emmerson and Raffaelli (2004), Emmerson and Yearsley (2004), Rooney et al. (2006), Otto et al. (2007), Allesina and Pascual (2008), Gross et al. (2009), Allesina and Tang (2012), Visser et al. (2012), van Altena et al. (2016) *Mutualism*: Feng and Takemoto (2014) *Competition*: Lawlor (1980), Christianou and Kokkoris (2008), Fowler (2009) |
| Resilience | Return time to ecological regime after a small perturbation | *Food webs*: Thébault and Fontaine (2010) *Mutualism*: Okuyama and Holland (2008) *Competition*: Lawlor (1980), Christianou and Kokkoris (2008) |
| Persistence | Proportion of coexisting species (over the total number of species) at ecological regime. In case of a feasible regime the persistence is equal to 1 (i.e., coexistence of all species in the community) | *Food webs*: Haydon (1994), McCann et al. (1998), Krause et al. (2003), Kondoh (2003, 2006, 2007), Thébault and Fontaine (2010), Stouffer and Bascompte (2011), Heckmann et al. (2012) *Mutualism*: Ferrière et al. (2002), West et al. (2002), Bascompte et al. (2006), Bastolla et al. (2009), Olivier et al. (2009), James et al. (2012), Valdovinos et al. (2013), Song and Feldman (2014) *Competition*: Kokkoris et al. (2002), Christianou and Kokkoris (2008) |
| Structural stability | Domain or probability of existence of an ecological regime w.r.t. system perturbations | May (1973) *Food webs*: De Angelis (1975), Haydon (1994), Kondoh (2003, 2006, 2007), Allesina and Tang (2012) *Mutualism*: Rohr et al. (2014) *Competition*: Christianou and Kokkoris (2008) |

(continued)

**Table 12.3** (continued)

| stability measure | Definition | References |
|---|---|---|
| Temporal stability | The reciprocal of temporal variability. It quantifies the stability of fluctuations in time. It is the ratio of the mean value of the variable in time over its standard deviation (the inverse of the coefficient of variation) | Elton (1958)<br>*Food webs*: McCann et al. (1998), Ives et al. (2000), Kondoh (2003, 2006, 2007)<br>*Competition*: Hughes and Roughgarden (1998), Lehman and Tilman (2000), Fowler (2009) |
| Deletion stability (extinction cascade) | Loss of additional species after the removal of one target species | *Food webs*: Pimm (1979, 1980b), Borrvall et al. (2000), Dunne et al. (2002b), Dunne and Williams (2009)<br>*Mutualism*: Memmott et al. (2004), Campbell et al. (2012), Vieira and Almeida-Neto (2015) |
| Robustness | Resistance of a system against additional extinction after species removal | *Food webs*: Dunne et al. (2002b), Dunne and Williams (2009)<br>*Mutualism*: Ramos-Jilibeto et al. (2012) |
| Resistance to invasion | Resistance of a system to be invaded by new species | Elton (1958)<br>*Food webs*: Hui et al. (2016)<br>*Competition*: Kokkoris et al. (1999) |

simplified ecological communities, outbreaks of phytophagous insects occur readily in boreal forests but are unheard of in diverse tropical forests, and the frequency of invasions is higher in simple island communities compared to more complex mainland communities. These observations led Elton (1958) to believe that complex communities, constructed from many predators and parasites (consumers), prevented populations from undergoing explosive growth (e.g., pest outbreaks) and would have fewer invasions. His ideas were closely related to MacArthur's (1955), who hypothesized that "a large number of paths through each species is necessary to reduce the effects of overpopulation of one species." MacArthur (1955) concluded that "stability increases as the number of links increases" and that stability is easier to achieve in more diverse assemblages of species, thus linking community stability with both increased trophic links (e.g., connectance $C$) and increased numbers of species ($S$). In other words, multiplicity in the number of prey and predator species associated with a population freed that population from dramatic changes in abundance when one of the prey or predator species declined in density.

These early intuitive ideas were challenged by the work of May (1973). He used mathematics to rigorously explore the complexity/stability relationship. By using linear stability analysis (asymptotic stability of the Jacobian matrix) on models constructed from a statistical universe (that is, randomly constructed Jacobians with randomly assigned interaction strengths), May (1973) found that complexity tends to destabilize community dynamics. He mathematically demonstrated that network stability decreases with diversity (measured as the number of species *S*), complexity (measured as connectance *C*), and average interaction strength among the species $\sigma$. In particular, he found that more diverse systems, compared to less diverse systems, will tend to sharply transition from stable to unstable behavior as the number of species *S*, the connectance *C*, or the average interaction strength $\sigma$ increase beyond a critical value, i.e., the system is stable if $\sigma\sqrt{SC} < 1$, unstable otherwise.

In his seminal study on food web stability, May (1973) measured asymptotic local stability. In this analysis, it is assumed that the community rests at an equilibrium point where all populations have constant abundances. The stability of this equilibrium is tested with small perturbations. If all species return to the equilibrium —monotonically or by damped oscillations—it is stable. In contrast, if the population densities evolve away from the equilibrium densities—monotonically or oscillatory—they are unstable. In a community of *S* species, this approach is based on the *SxS* Jacobian matrix of species interaction coefficients that describe the perturbation impact of each species *j* on the growth of each species *i* at equilibrium population densities. The *S* eigenvalues of the Jacobian matrix characterize its temporal behavior. Specifically, positive real parts of the eigenvalues indicate perturbation growth, while negative real parts indicate perturbation decay. Accordingly, if any of the eigenvalues has a positive real part the system will be unstable, i.e., at least one of the species does not return to the equilibrium.

### *12.6.1 Food Webs*

The use of random community matrices in May (1973)'s work has attracted much criticism (Table 12.4). It was shown to be extremely unlikely that any of these random communities even remotely resembles ecosystems with a minimum of ecological realism, such as containing at least one primary producer, a limited number of trophic levels, and no consumers eating resources that are two or more trophic levels lower (Lawlor 1978). The non-randomness of ecosystem structure has been demonstrated in detail by more recent food-web topology studies (e.g., Williams and Martinez 2000; Dunne et al. 2002a, b, 2004). Accordingly, subsequent work added more structural realism to those random community matrices by including empirical patterns of food web structure and interaction strength distributions (see Allesina and Tang 2012 for the most recent advance). Several simple models have played an important role characterizing the non-random structure of food webs, including the cascade model (Cohen et al. 1990), the niche model (Williams and Martinez 2000), and the nested-hierarchy model (Cattin et al. 2004). The niche and nested-hierarchy models have been able to capture several structural properties of empirical food webs.

**Table 12.4** Complexity/stability debate in food webs

| References | Complexity/ stability measures | Methods and assumptions | Additional results |
|---|---|---|---|
| *Negative complexity/stability relationship* | | | |
| Haydon (1994) | *S*/asymptotic, feasibility | Lead eigenvalue of random and plausible model Jacobian at feasible equilibria | Stability is reduced by donor control interactions |
| Pimm (1979, 1980b) | *S*, *C*/extinction cascades | Simulation of food web model | If carnivores are removed |
| Gross et al. (2009) | *S*, *C*, interaction strength/ asymptotic | Lead eigenvalue of Jacobian of realistic food web model (niche) | Increasing interaction strength destabilizes large networks |
| Allesina and Pascual (2008), Allesina and Tang (2012) | *S*, *C*/asymptotic | Lead eigenvalue of random, empirical, and model (cascade and niche) Jacobian of antagonistic interactions | |
| Krause et al. (2003), Thébault and Fontaine (2010) | *C*, interaction strength/ resilience, persistence | Simulations of model and real food webs | Stability is enhanced in compartmentalized and weakly connected architectures |
| van Altena et al. (2016) | Interaction strength/ asymptotic | Lead eigenvalue of Jacobian of model obtained from real food web data | Skew toward weak interaction enhances stability |
| Neutel et al. (2002, 2007), Emmerson and Yearsley (2004) | Interaction strength/ asymptotic, resilience, feasibility, persistence | Lead eigenvalue of Jacobian of model (cascade) and real food webs | Weak interactions in long feedback loops of omnivorous species is stabilizing |
| McCann et al (1998) | Interaction strength/ persistence, temporal stability | Nonlinear models away from equilibrium | Weak links and intermediate interaction strengths are stabilizing |
| *Positive complexity/stability relationship* | | | |
| Ives et al. (2000) | *S*/temporal stability | Simulation of model community under environmental variation | Increasing the number of modular subcommunities increases stability |
| Pimm (1979, 1980b) | *S*, *C*/extinction cascades | Simulation of plausible food web model | If herbivores are removed |
| Stouffer and Bascompte (2011) | *S*, *C*/persistence, extinction cascade | Simulation of model (niche) | Compartmentalization increases stability |

(continued)

**Table 12.4** (continued)

| References | Complexity/ stability measures | Methods and assumptions | Additional results |
|---|---|---|---|
| De Angelis (1975) | *C*/asymptotic | Lead eigenvalue of Jacobian of plausible food web model | Stability is increased by donor control interactions |
| Dunne et al. (2002b), Dunne and Williams (2009) | *C*/robustness | Simulation of model obtained from real food webs | Skewness of degree distribution increases robustness |
| Haydon (2000) | Weighted *C*/ Asymptotic | Lead eigenvalue of Jacobian of plausible food web model | |
| van Altena et al. (2016) | Weighted *C*/ asymptotic | Lead eigenvalue of Jacobian obtained from real food webs | No relationship between unweighted *C* and stability |
| Allesina and Pascual (2008), Allesina and Tang (2012) | *S*, interaction strength/ asymptotic | Lead eigenvalue of random, empirical, and model (cascade and niche) Jacobian of antagonistic interactions | Weak interactions are destabilizing |
| Borrvall et al. (2000) | *S* per functional group, Interaction strength/ extinction cascades | Model | Higher risk of extinction if autotrophs (rather than top predators) are removed. Skewness towards weak interactions is destabilizing Omnivory is stabilizing |
| Haydon (1994) | *C*, interaction strength/ asymptotic, feasibility | Lead eigenvalue of random and plausible model Jacobian at feasible equilibria | Stability is reduced by donor control interactions. Stability is increased by increased interaction strengths |
| Yodzis (1981) | Interaction strength/ asymptotic | Lead eigenvalue of Jacobian of empirically inspired food webs | Intraspecific competition is stabilizing whereas interspecific competition tends to be destabilizing |
| de Ruiter et al. (1995), Rooney et al. (2006) | Interaction strength/ asymptotic | Model and time series of real and experimental food webs | Asymmetries in interaction strength (i.e., strong consumer control interactions at lower trophic levels and strong donor control interactions at higher trophic levels) are stabilizing |
| Gross et al. (2009) | Interaction strength/ asymptotic | Lead eigenvalue of Jacobian of realistic food web model (niche) | Increasing interaction strength stabilizes small networks |

**Species Richness** In general, food web features vary with species richness. Although empirical datasets of ecological networks do not display any consistency regarding their size, it has been observed that ecological networks have much smaller size than other published real-world network datasets, such as co-authorships between scientists or the World Wide Web (Dunne et al. 2002a).

Haydon (1994) discussed some of May's hypotheses (such as the measure of stability, the consideration of unfeasible models, and the self-regulatory terms on the diagonal of the interaction matrix describing intraspecific competition) but still found that stability of model ecosystems is reduced by the number of species. Gross et al. (2009) found that small model ecosystems follow other rules than large ecosystems. Indeed, they studied artificial food webs generated by a niche model (Williams and Martinez 2000), well known to be able to recreate realistic food web topology. Then, they used a generalized modelling approach consisting of identifying all parameters that capture the local stability properties of all stationary states of the generated food webs. Thus, adding more details to May's (1973) stability criteria, they showed that the strength of predator-prey links increase the stability of small webs, but destabilize larger webs. They also revealed a new power law describing how food-web stability scales with the number of species. Also Pimm (1979, 1980b) showed that extinction cascades are more likely in model communities with higher number of species, contrasted by Borrvall et al. (2000) that found model food web robustness to increase with network size. Using real food webs, Dunne et al. (2002b) and Dunne and Williams (2009) found the same result, i.e., positive relationship between number of species and robustness. However, a recent paper by Jacquet et al. (2016) disprove the association between species richness and stability in empirical food webs.

**Connectance** Exploring how the number of interactions varies with the number of species has been one of the most basic questions for ecologists trying to find universal patterns in the structure of ecological networks. Contradicting previous works which found that the number of interactions increases linearly with the number of species (Cohen and Briand 1984; Cohen and Newman 1985), Martinez (1992) claimed the constant connectance hypothesis in food webs: trophic links increase approximately as the square of the number of species. However, with the improvement of methodological analysis and datasets, the constant connectance hypothesis has been called into question by later studies (Havens 1992; Dunne et al. 2002a; and Banašek-Richter et al. 2009). One of the most generally accepted rule on food web connectance is that food webs display an average low connectance of about 0.11 (Havens 1992; Martinez 1992; and Dunne et al. 2002a), which is however still relatively high compared to that of other real-world networks (Dunne et al. 2002a).

Since connectance has been used by May (1973) as a descriptor of network complexity, it has become central to early works on the complexity-stability debate (De Angelis 1975; Pimm 1980b, 1984) and continues to be widely used as a descriptor for network structure (Havens 1992; Dunne et al. 2002b; Olesen and Jordano 2002; Tylianakis et al. 2010; Heleno et al. 2012; Poisot and Gravel 2014).

Depending on the way stability is defined, the quality of empirical datasets, or the methods used to generate theoretical networks, contradiction has been observed in the relationship between network stability and connectance. While some studies reinforced May's hypothesis of a negative relationship between connectance and stability (Pimm 1979, 1980b; Chen and Cohen 2001; Gross et al. 2009; Allesina and Tang 2012; Vieira and Almeida-Neto 2015), others found that connectance enhances network stability (De Angelis 1975; Dunne et al. 2002b; Dunne and Williams 2009). For example, using extinction cascade as stability measure, Pimm (1979, 1980b) found that complex model food webs are more likely to lose additional species following the extinction of one species than simple food webs: complexity is negatively correlated with stability. By using different measurements of network stability (resilience and persistence), Thébault and Fontaine (2010) also confirmed the negative relationship between connectance and stability in food webs (however, the opposite holds for mutualistic networks, see next Section). Gross et al. (2009) also revealed a negative power law to describe how food-web stability scales with connectance.

The opposite view is sustained, among others, by De Angelis (1975): using plausible food web models, he showed that the probability of stability can increase with increasing connectance if the food web is characterized by a bias toward strong self-regulation (intraspecific competition) of higher trophic level species, low assimilation efficiencies, or a bias toward donor control. Also Haydon (1994), improving May's assumptions, found stability to increase with connectance. However, in contrast with De Angelis (1975), stability is found to be reduced by the prevalence of donor control interactions. Furthermore, robustness increases with connectance in real food webs (Dunne et al. 2002b; Dunne and Williams 2009). However, a recent paper by Jacquet et al. (2016) disprove the association between connectance and stability in empirical food webs.

**Weighted Connectance** If weighted connectance has been used before in food web studies, only a recent study by van Altena et al. (2016) started its use into the complexity/stability context. Although they contradicted previous studies, finding no relationship between food web stability and unweighted connectance, they emphasized a high level of weighted connectance indeed stabilizes food webs. Following a different perspective, Haydon (2000) focused on communities constructed to be as stable as they could be, and show that communities built in this way require high levels of weighted connectance, in agreement with van Altena et al. (2016). According to those studies, high stability requires high connectance, especially between weakly and strongly self-regulated (intraspecific competition) elements of the community.

**Degree Distribution** Degree distribution in food webs differ from a Poisson distribution (typical of random networks). However, there is no universal shape that fits food webs degree distribution. Most of the webs display exponential degree distribution (Camacho et al. 2002 and Dunne et al. 2002a) and those with high connectance show a uniform distribution. Power-law and truncated power-law with an exponential drop-off in the tail also fit few of food webs degree distribution (mostly those having very low connectance) (Dunne et al. 2002a and Montoya and Solé 2002).

The skewness of degree distribution, especially exponential-type degree distribution (Dunne and Williams 2009) makes food webs more robust to targeted removals (from the most generalists) (Solé and Montoya 2001; Dunne et al. 2002b). However, the hierarchical feeding feature imposes a cost to food web robustness (Dunne and Williams 2009).

**Interaction Strength** In contrast with May's (1973) findings, Haydon (1994) found stability to increase with interaction strength. Yodzis (1981) also found that the networks were far more likely to be stable when interaction strengths are chosen in accord with real food web patterns rather than strictly at random. Neutel et al. (2007) showed how non-random interaction strength patterns in naturally assembling communities explain stability. They used below-ground food webs, whose complexity increased along a vegetational succession gradient. The weight of the feedback loops of omnivorous species characterized stability (omnivory: feeding on more than one trophic level). Low predator–prey biomass ratios (biomass pyramid, a feature common to most ecosystems) in these omnivorous loops were shown to have a crucial role in preserving stability as complexity increased during succession.

Variability in link strengths have also been found to be related with stability, but only for relatively small webs, whereas larger webs are instead destabilized (Gross et al. 2009). Stability is enhanced when species at a high trophic level feed on multiple prey species and species at an intermediate trophic level are fed upon by multiple predator species. Using an energetic approach, de Ruiter et al. (1995) and Rooney et al. (2006) found that structural asymmetry in energy fluxes is key to stability of real food webs. In particular, de Ruiter et al. (1995) showed that simultaneous occurrence of strong top down effects (consumer control) at lower trophic levels and strong bottom up effects (donor control) at higher trophic levels in the patterns of interaction strengths in real food webs is important to ecosystem stability. The patterning is a direct result of the energetic organization of the food web. Rooney et al. (2006) confirm that slow and fast energy fluxes coupled by top-predators in real food webs convey both local and non-local stability to food webs. In conclusion, complexity does not lead to instability.

A skewness of the interaction strength distribution has been widely observed in food webs, i.e., there are many weak interactions and few strong interactions (Paine 1992; Berlow 1999; Berlow et al. 2004; Wootton and Emmerson 2005). This skewness towards weak interactions has been related to stability. For example, McCann et al. (1998) found that weak links and intermediate interaction strength reinforce the stability and the persistence of the community as they dampen the oscillation in predator-prey dynamics. Neutel et al (2002) showed that weak interactions are more likely observed in long loops in real food webs. Specifically, interaction strengths are organized in trophic loops in such a way that long loops contain relatively many weak links. They showed and explain mathematically that this patterning enhances stability, because it reduces the amount of intraspecific interaction needed for matrix stability. On the same line, Thébault and Fontaine (2010) showed that stability of trophic networks is enhanced in weakly connected architectures. van Altena et al. (2016) confirmed the role of weak interactions for stability of real food webs. However, given skewed distributions of interaction strengths towards weak interactions, they found that stability was promoted by even distribution of fluxes over links, in contrast with de Ruiter et al. (1995) and Rooney et al. (2006) who emphasized the role of strong asymmetry. In a recent paper, Jacquet et al. (2016) disprove the association between interaction strength and stability in empirical food webs, but show that the correlation between the effects of predators on prey and those of prey on predators, combined with a high frequency of weak interactions, stabilize food web dynamics. In agreement with Neutel et al. (2002, 2007), Emmerson and Yearsley (2004) showed that a skew towards weak interactions promotes local and global stability only when omnivory is present. A feedback is found between skewness toward weak interactions and omnivory, i.e., skewed interaction strengths are an emergent property of stable omnivorous communities, and in turn this skew creates a dynamic constraint maintaining omnivory. Borrvall et al. (2000) however found that omnivory stabilizes food webs, but the skew towards weak interaction is destabilizing. Omnivory appears to be common in food webs (Sprules and Bowerman 1988; Polis 1991). However, a previous theoretical work (Pimm and Lawton 1978) predicted that it should be extremely rare to find species that feed simultaneously both high and low in real-world food web, and also webs with a large number of omnivores should be rare in real world.

By contrast, Allesina and Pascual (2008) found that stability is highly robust to perturbations of interaction strength, but it is mainly a structural property driven by short and strong predator–prey loops, with the stability of these small modules cascading into that of the whole network. These considerations challenge the current view of weak interaction strength and long cycles as main drivers of stability in natural communities. In addition to that, Allesina and Tang (2012) showed that preponderance of weak interactions decreases the probability of food webs to be stable. In particular, trophic interactions are shown to be stabilizing (as opposed to mutualistic and competitive) but, counterintuitively, the probability of stability for predator–prey networks decreases when a realistic food web structure is imposed or if there is a large preponderance of weak interactions. However, stable predator–prey networks can be arbitrarily large and complex (positive complexity/stability relationship), provided that predator–prey pairs are tightly coupled (i.e., short loops and high interaction strength). Same negative relationship between stability and skewness of interaction strength distribution has been found by Borrvall et al. (2000), although using a different measure of stability (extinction cascade).

**Network Architecture** The effect of network architecture, in particular modular structures, has been observed in food webs and related to their stability. Moore and Hunt (1988) showed that food webs may contain tightly coupled subunits whose numbers may increase with diversity. Communities may be arranged in resource compartments and within them species interaction strength would decline as diversity increased. Same result has been found by Krause et al. (2003) and Thébault and Fontaine (2010), who showed that stability of trophic networks is enhanced in compartmented and weakly connected architectures. Also, Ives et al. (2000) showed that increasing the number of modular subcommunities increases stability through different species reactions to environmental fluctuations (insurance hypothesis). Similarly, Stouffer and Bascompte (2011) demonstrate that compartmentalization increases the persistence of food webs. Compartments buffer the propagation of extinctions through the community and increase long-term persistence. The latter contribution increases with the complexity of the food web, emphasizing a positive complexity/stability relationship. However, the recent study of Grilli et al (2016) shows that the stabilizing effect of modularity is not as general as expected.

Nested diets have been observed in food webs: top predators are very generalists and prey upon all over species, while the next predator exploiting all but the top predators (in the niche model by Williams and Martinez 2000, and the nested-hierarchy model by Cattin et al. 2004). Generalist top predators prey upon intermediate specialist predators also in the results of Neutel et al. (2002).

### *12.6.2 Mutualistic Communities*

As the interaction between a plant and its insect pollinator has often been used as a straightforward illustrative example of a reciprocal coevolution (Darwin 1862), early studies on mutualistic interactions were mainly dedicated to understanding coevolutionary processes (e.g., Ehrlich and Raven 1964; Brown et al. 1978; Wheelwright and Orians 1982; Herrera 1985). However, coevolution is often considered as a diffuse mechanism involving several species. Thus, ecologists started to study mutualism as a whole network of interactions for which the tools provided by complex network theory can be used to address the complexity/stability relationship (Table 12.5).

**Species Richness** Network size (or total number of species in the network) has been considered an important determinant of mutualistic networks stability. By using a theoretical model with empirically informed parameters, Okuyama and Holland (2008) found a positive relationship between community size and community resilience. They mainly attributed this positive relationship to the use of a nonlinear functional response and its saturating positive feedback on population growth. Their finding was later supported by Thébault and Fontaine (2010) who also used a population dynamics model with a nonlinear functional response. They confirmed that a high number of species promotes not only the resilience of mutualistic communities but also their persistence.

**Connectance and Connectivity** Contributing to their complexity, mutualistic networks have been observed to display non-random structural patterns. Motivated by the finding of scale invariance in food webs (Cohen and Briand 1984; Cohen and

**Table 12.5** Complexity/stability debate in mutualistic communities

| References | Complexity/stability measures | Methods and assumptions | Main results |
|---|---|---|---|
| *Negative complexity/stability relationship* | | | |
| Vieira and Almeida-Neto (2015) | *C*/extinction cascade | Stochastic coextinction model applied to a set of empirical networks | Extinction cascades occur more likely in highly connected mutualistic communities |
| Feng and Takemoto (2014) | Heterogeneity of degree distribution, species strength distribution, interaction strength distribution/ asymptotic | Theoretical study based on the analytical expression of the dominant eigenvalue | Heterogeneity of node degrees and interaction strength primarily determines the local stability of mutualistic ecosystems<br>Nestedness additionally affects it |
| Suweis et al. (2015) | *C*/localization | Evaluation of the components of the eigenvalues of a set of empirical networks | Mutualistic communities are localized<br>Localization is negatively correlated with connectance |
| Allesina and Tang (2012) | $S$, $C$, $\sigma$, nestedness/ asymptotic | Analytical analyses of artificial networks with realistic structure | Mutualistic interactions are destabilizing<br>Stability is negatively affected by nestedness |
| Campbell et al. (2012) | Nestedness/extinction cascade | Dynamic boolean network-based model of plant-pollinator community formation | High nestedness may in extreme circumstances promote a critical over-reliance on individual species and enhances extinction cascade |
| Thébault and Fontaine (2010) | Modularity/resilience, persistence | Simulations of model and real pollination networks | A highly connected and nested architecture promotes community stability in mutualistic networks |
| *Positive complexity/stability relationship* | | | |
| Okuyama and Holland (2008) | $S$, $L$, symmetry of interaction strength, nestedness/resilience | Theoretical analysis with empirically informed parameters<br>Non-linear functional response | Community resilience is enhanced by increasing community size and connectivity, and through strong, symmetric interaction strengths of highly nested networks |

(continued)

**Table 12.5** (continued)

| References | Complexity/stability measures | Methods and assumptions | Main results |
|---|---|---|---|
| Thébault and Fontaine (2010) | *S*, nestedness/resilience, persistence | Simulations of model and real pollination networks | A highly connected and nested architecture promotes community stability |
| Memmott et al. (2004) | Nestedness/extinction cascade | Topological extinction model<br>Explore the effects of plant extinction on the preferential removal of the most linked pollinators | Plant species diversity declined most rapidly with l removal of the most-linked pollinators<br>Declines were no worse than linear, because of the nested architecture |
| James et al. (2012) | Species degree, *C*, nestedness/persistence | Use of population dynamics model that incorporates both competition and mutualism | Species degree is a much better predictor of individual species survival and hence, community persistence<br>Nestedness is only of a secondary importance to community persistence |
| Bascompte et al. (2006) | Heterogeneity of species strength distribution, asymmetry of species dependencies/domain of coexistence | Population dynamics model<br>Species dependencies were estimated from empirical quantitative networks | The asymmetry of plant-animal dependences enhance long-term coexistence and facilitate biodiversity maintenance |
| Suweis et al. (2015) | *S*, heterogeneity of species strength distribution/localization | Evaluation of components of the eigenvalues of empirical interaction matrices | Mutualistic communities are localized<br>Localization is positively correlated with the variance of the weighted degree distribution |
| Bastolla et al. (2009) | Nestedness/domain of coexistence | Theoretical model of population dynamics<br>Nestedness of simulated networks informed from empirical networks | Nestedness reduces effective interspecific competition and enhances the number of coexisting species |
| Rohr et al. (2014) | Species degree, species strength, interaction strength, nestedness/structural stability | Population dynamics model<br>Exploration of the range of parameters necessary for a stable coexistence | A maximal level of nestedness, a small trade-off between the number and intensity of interactions a species has, and a high level of mutualistic strength are factors that maximize stability |

Newman 1985). Jordano (1987) studied patterns of connectance and interaction strength observed in a large dataset of pollination and seed-dispersal networks. He found that connectance decreases with species richness but the average number of links per species (or linkage density) is invariant relative to network size. Using empirical mutualistic networks spanning different biogeographic regions (including those used in Jordano 1987). Olesen and Jordano (2002) observed that connectance indeed decreases exponentially with species richness. After controlling for species richness (network size), they also observed that connectance differed significantly between biogeographic regions. On average, mutualistic networks exhibit higher connectance than food webs and other real-world networks. However, mutualistic networks still have low to moderate level of connectance (average of 0.11 in Olesen and Jordano 2002 and 0.18 in Rezende et al. 2007).

The implications of connectance patterns to stability of mutualistic networks has gained attention only recently. When extending the theoretical work of May (1973) to incorporate realistic network structures, Allesina and Tang (2012) found that connectance negatively affects the local stability of mutualistic networks. The analytical study by Suweis et al. (2015) is in agreement with this finding when they looked at the ability of mutualistic networks to reduce the propagation of perturbation (or localization). They found that mutualistic networks are indeed localized and localization decreases with connectance. Moreover, extinction cascades were more likely to happen in highly connected communities (Vieira and Almeida-Neto 2015). However, highly connected communities were also shown to be persistent (James et al. 2012) and resilient (Okuyama and Holland 2008).

**Degree Distribution** Although network size and connectance partially determine the complexity of the network, they neglect important information regarding individual species connectivity as well as the distribution of the overall connectivity among species. Early studies on species connectivity in mutualistic networks mainly concentrated on how interactions are distributed among species. Attentions were mainly focused on the prevalence of either generalists or specialists in mutualistic networks (Waser et al. 1996; Memmott 1999; Vázquez and Aizen 2003). Stimulated by these early studies, Jordano et al. (2003) found generalized patterns in the node degree distribution of a large number of empirical plant-pollinator and plant-frugivore networks. Most of the networks indeed showed a node-degree distribution of that fits a truncated power-law, suggesting the prevalence of specialists and the rarity of super generalists. Few of the networks showed a standard power-law or an exponential distribution in their node degree. Moreover, gamma distribution was also found to best fit the distribution of node degree in mutualistic networks (Okuyama 2008). The heterogeneity of node degree distribution was found years later to be a primary factor negatively affecting the local stability of mutualistic networks (Feng and Takemoto 2014). However, when node degrees are considered individually for each species, they were shown to be a good predictor of species own survival and thus of the community persistence (James et al. 2012).

**Species Strength, Species Dependency and Interaction Strength** Instead of only considering qualitative interactions (presence or absence), quantitative measurement of interaction strength also prevails in mutualistic network studies. In plant-pollinator as well as in plant-frugivory interactions, interaction strength often refers to the relative number of visits of the animal to the plant. Jordano (1987) observed an extremely skewed distribution of interaction strength in empirical mutualistic communities: weak interactions greatly exceed in number strong ones. By including more datasets in their study, Bascompte et al. (2006) confirmed Jordano's (1987) finding of a skewed distribution of interaction strength. The mutual dependence of a species on each one of its mutualistic partners is derived from measurement of interaction strength. Mutualistic networks were also found to be highly asymmetric in terms of species dependences: while animals depend strongly on the plants, plants rely poorly on their animal pollinators or seed dispersers.

Species coexistence, and thus community persistence, was found to be facilitated by the both the heterogeneity of species strength distribution and the asymmetry of species dependences (Bascompte et al. 2006). Localization, or the ability of the system to reduce the propagation of perturbations through the network, has also been shown to be enhanced by the heterogeneity of interaction strengths (Suweis et al. 2015). Contrasting the finding of Bascompte et al. (2006) who used a linear functional response in their model, Okuyama and Holland (2008) argued that the asymmetry of species dependences (implying an asymmetry of interaction strength between animals and plants) has a small negative effect on the resilience of mutualistic communities when a non-linear functional response is used. Feng and Takemoto (2014) also showed that the heterogeneity of species strength distribution indeed impacts negatively on the local stability of mutualistic communities. Moreover, by also using a saturating functional response, Rohr et al. (2014) demonstrated that regardless of the interaction strength distribution, mutualistic communities that have on average a high level of interaction strength are more likely to be structurally stable.

**Network Architecture** Although modularity or compartmentalization is a feature commonly observed in food webs, mutualistic networks also exhibit a certain level of modularity. A test for modularity in a wide dataset allowed Olesen et al. (2007) to affirm that pollination networks with a relatively high number of species are indeed modular. Moreover, the observed level of modularity increases with network size. Modularity patterns, such as the number of modules observed in pollination networks, are found to be invariant to sampling effort at different time (Dupont and Olesen 2012). Mello et al. (2011) also noticed a high level of modularity in seed-dispersal networks. Little is known about the implication of modular structure to mutualistic network stability. Thébault and Fontaine (2010) emphasized that structural patterns favouring stability fundamentally differ in food webs and mutualistic networks: while the modularity pattern enhances food web stability, it has a negative effect on the persistence and resilience of mutualistic networks.

A widely accepted topological feature proper to mutualistic networks is nestedness. Bascompte et al. (2003) started to explore this feature in a meta-analysis of empirical mutualistic communities and found that mutualistic networks are indeed highly nested. They also found that nestedness increases with network complexity expressed in terms of species richness and connectivity. Nestedness has always been believed to be the most important determinant of mutualistic network stability. For example, extinction cascades following the removal of the most generalist pollinator have been shown to happen only linearly because of the stabilizing effect of nestedness (Memmott et al. 2004). The nested structure of mutualistic networks also enhances the number of coexisting species by reducing interspecific competition (Bastolla et al. 2009). Nestedness also has a positive effect on the persistence and resilience of mutualistic communities (Okuyama and Holland 2008; Thébault and Fontaine 2010). Structural stability, or the domain of stable coexistence of species, was shown to be maximized when artificial networks are assumed to have a high level of nestedness (Rohr et al. 2014). However, some recent studies started to discard the importance of nestedness to network stability. James et al. (2012) indeed found that nestedness is, at best, a secondary covariate rather than a causative factor for species coexistence in mutualistic communities and has no significant effect on community persistence. By means of careful analyses of artificial networks with realistic structure, Allesina and Tang (2012) affirmed that local stability is negatively affected by nestedness in mutualistic networks. Campbell et al. (2012) also showed that extreme nestedness facilitates sequential species extinctions (extinction cascades).

### 12.6.3 Competitive Communities

Competitive interactions have sometimes been considered together with trophic interactions in food webs models, and their contribution to stability assessed. There is common agreement that self-regulating interactions due to intraspecific competition (i.e., negative terms on the diagonal of the interaction matrix) increase stability, while the role of interspecific competition is not clear. For example, De Angelis (1975), using plausible food web models, showed that the probability of stability increases if the food web is characterized by a bias toward strong self-regulation (intraspecific competition) of higher trophic level species. Haydon (1994) found a similar result, i.e., that considering intraspecific competition increases food web stability. Again, Haydon (2000) focused on communities constructed to be as stable as they could be, and show that communities built in this way require high connectance between weakly and strongly self-regulated (intra-specific competition) elements of the community. Neutel et al (2002) showed and explain mathematically that the patterning of real food web interaction strengths enhances stability because it reduces the amount of intraspecific interaction needed for matrix stability. Yodzis (1981) also found that the presence of self-regulatory terms (intraspecific competition) in some consumer species stabilizes the network, whereas interspecific competition tends to be destabilizing. By contrast, Allesina and Tang (2012) showed that competitive interactions are destabilizing.

Theoretically speaking, purely competitive communities are simpler compared to food webs and mutualistic networks because they are composed of only one trophic level. This allowed empirical and experimental evidence to be collected in some rare cases (Lawlor 1980; Lehman and Tilman 2000). Moreover, the simplicity of competitive communities makes them also an ideal theoretical framework for studying the relationship between community complexity and stability (Table 12.6).

**Species Richness** Species richness has been reported to affect the stability of competitive communities. Lehman and Tilman (2000) analysed different models of multispecies competition and empirical data, finding that greater diversity increases the temporal stability of the entire communities but decreases the temporal stability of individual populations. Specifically, temporal stability of the entire community increases fairly linearly without saturation with increased diversity. Species composition of each population was also predicted to be as important as diversity in affecting community stability. Lawlor (1980) compared observed communities with analogous random versions of them: he found that stability of observed food webs decreases with the number of species, however, observed communities are generally more stable than randomly constructed communities with the same number of species. The higher stability of observed (compared to random) communities is due to lower similarities among consumer species, suggesting that interspecific competitive processes are very important in shaping communities. The use of a different measurement of stability changes these results. Indeed, Christianou and Kokkoris (2008) reported that increasing the number of species in the community decreases the probability of feasibility of the system. However, they also found that species richness does not significantly affect the probability of local stability and the resilience of competitive communities. Hence, when only feasible systems are taken into account, their finding seems to contradict May (1973)'s finding of a negative correlation between species richness and the probability of local stability. However, their work was contradicted later by Fowler (2009). He demonstrated that increasing the number of species results in an increased probability of local stability in competitive communities.

**Connectance** Fowler (2009) also showed that an increase in network connectance and in the number of competitive links (connectivity) reduces per-capita growth rates through an increase in competitive feedback, thus stabilises oscillating dynamics. Furthermore, he affirmed that these results stay robust to changes in species interaction strengths.

**Interaction Strength** Most studies on competitive communities focused on the implication of competition coefficients, i.e. interactions strengths, on community stability. However, different measurements of stability have been used. For example, Hughes and Roughgarden (1998) studied temporal stability measured as the aggregate community biomass in a two-species competition model. They found

**Table 12.6** Complexity/stability debate in competitive communities

| References | Complexity/stability measures | Methods and assumptions | Additional results |
|---|---|---|---|
| *Negative complexity/stability relationship* | | | |
| Lawlor (1980) | *S*/asymptotic, temporal stability | Lead eigenvalue of random versus observed Jacobian | Observed communities are generally more stable than randomly constructed communities with the same number of species |
| Lehman and Tilman (2000) | *S*/asymptotic, temporal stability | Lead eigenvalue and simulations of three different models (mechanistic, phenomenological, statistical) and empirical time series | Greater diversity decreases the temporal stability of individual populations |
| Christianou and Kokkoris (2008) | *S*, interaction strength/asymptotic, feasibility, persistence, structural stability | Model of competitive community | Asymptotic stability is not affected by the number of species *S*, but structural stability decreases with species richness<br>Weak interaction strengths enhances structural stability |
| Kokkoris et al. (1999) | Interaction strength/ resistance to invasion | Community assembly model from a regional species pool | Weak interaction strengths enhances resistance to invasion |
| Kokkoris et al. (2002) | Interaction strength/ asymptotic, feasibility, persistence, structural stability | Model of competitive community | Weak interaction strengths enhances species coexistence |
| *Positive complexity/stability relationship* | | | |
| Lehman and Tilman (2000) | *S*/asymptotic, temporal stability | Lead eigenvalue and simulations of three different models (mechanistic, phenomenological, statistical) and empirical time series | Greater diversity increases the temporal stability of the entire communities |
| Fowler (2009) | *S*, *C*, connectivity/ asymptotic, temporal stability | Lead eigenvalue and simulation of discrete-time model of competitive community | Result robust to change in interaction strengths |
| Hughes and Roughgarden (1998) | Interaction strength/ temporal stability | Discrete-time two-species competition model | Stability independent on the magnitude but related to asymmetry of interaction strengths |

that stability is relatively independent of the magnitude of interaction strengths but the degree of asymmetry of interactions is the key to community stability. Quantifying the stability of the community by its invulnerability to invasion, Kokkoris et al. (1999) studied the distribution of interaction strengths during the assembly process of theoretical competitive communities. They found that the mean interaction strength drops as assembly progresses and most interactions that are formed are weak. It suggests that communities that are invulnerable to further invasion are those where interspecific interactions are weaker than the average interaction strength between competing species of a regional pool. In a later study (Kokkoris et al. 2002), the same authors explored how the number of coexisting species vary with the average interaction strength. Confirming their previous finding on the importance of weak interactions to community stability, they found that the preponderance of weak interactions indeed allow many species to coexist. Moreover, correlation in the interaction matrix, mainly a result of trade-offs between species characteristics, can increase the probability of species coexistence. Christianou and Kokkoris (2008) even deepened the study on the importance of weak interactions to stability by considering system feasibility of a competitive community. Consistent with previous findings, they showed that the probability of feasibility decreases with increasing interaction strength.

## 12.7 Trait-Mediated Interaction and Adaptive Networks

### *12.7.1 Food Webs*

The discussion thus far has implicitly assumed that network topology and number of nodes and links remain unchanged over time (Kondoh 2005; Gross and Sayama 2009). This is often a simplifying assumption: for instance, new species (nodes) can emigrate or be introduced in the system while some others might go extinct. Additionally, adaptive species behaviour—e.g., adaptive consumer foraging and/or adaptive resource defence (Beckerman et al. 2010; Valdovinos et al. 2010)—can often cause links to form, disappear, or change in strength as time progresses (see Gravel et al. 2016; Strona and Lafferty 2016 for recent studies on the effect of dispersal and environmental change). Adaptive networks describing adaptive species behaviour has been shown to reproduce realistic food-web structures (Nuwagaba et al. 2015), to promote stability, and to allow for positive complexity-stability relationships. For example, Kondoh (2003, 2006) showed that short-term selection on trophic links, arising from a consumer's adaptive food choice, enhances the long-term stability of complex communities. Without adaptive foragers, food-web complexity destabilizes community composition; whereas in their presence, complexity may enhance community persistence through facilitation of dynamical food-web reconstruction that buffers environmental fluctuations (insurance hypothesis). Visser et al. (2012) examined the effect of adaptive foraging behaviour within a tri-trophic food web and demonstrated that adaptive behaviour will always promote stability of community dynamics.

Predator-prey body mass ratio, affecting the interaction strength distribution, contributes largely to food-web stability (Emmerson and Raffaelli 2004; Otto et al. 2007). Heckmann et al. (2012) combined body-size structure and adaptive foraging behaviour. They investigated dynamic random and niche model food webs to evaluate the proportion of persistent species. They showed that stronger body-size structures and faster adaptation stabilise these food webs. Body-size structures yield stabilising configurations of interaction strength distributions across food webs, and adaptive foraging emphasises links to resources closer to the base. Moreover, both mechanisms combined have a cumulative effect. Most importantly, unstructured random food webs evolve via adaptive foraging into stable size-structured food webs, i.e., after adaptation predators tend to focus on prey on lower trophic levels and with smaller body sizes. This offers a mechanistic explanation of how size structure adaptively emerges in complex food webs, thus building a novel bridge between these two important stabilising mechanisms.

Trait adaptation can also be modelled and give rise to complex trophic interaction networks (Brännström et al. 2011; 2012; Fussman et al. 2007; Landi et al. 2013), and their complexity/stability relationship assessed (Kondoh 2007 and Ingram et al. 2009). In particular, Kondoh (2007) studied adaptation in anti-predator defence traits reporting its positive effect on community-level stability (persistence and robustness), while Ingram et al. (2009) studied body size and niche width adaptation, emphasizing a positive relationship between asymptotic stability and resistance to invasion from new phenotypes displacing existing species.

### 12.7.2 Mutualistic Communities

Pioneering studies addressing the effect of mutualistic community structure to community stability often utilized dynamic models of changing population abundance such as those based on the Lotka Voltera model (e.g., Okuyama and Holland 2008; Bastolla et al. 2009; Thébault and Fontaine 2010). Although these models have expanded our knowledge about the structure and dynamics of complex mutualistic systems, they disregarded important biological processes associated with plant-animal interactions. One important biological process is adaptation (Petanidou et al. 2008). Recent studies incorporated adaptation into the foraging behaviour of animal pollinators and seed dispersers. One way to reflect adaptive foraging is through rewiring of interactions. In a study by Zhang et al. (2011), the emergence of nestedness pattern in pollination and frugivory networks has been well-reproduced when species are allowed to switch their mutualistic partner for another one providing higher benefit, as a consequence of adaptive foraging strategy. Going beyond the importance of adaptive rewiring to the emergence of network structure, other studies even explored its implication to network stability. Ramos-Jilibeto et al. (2012) showed that when animal pollinators have the ability to rewire their connections after depletion of host plant abundances, network robustness is enhanced. Moreover, preferential attachment to host plants having higher abundance and few exploiters enhances network robustness more than other rewiring alternatives. Foraging effort of pollinators can also be incorporated directly as an evolving trait affecting pollinator's growth rate. Indeed, Valdovinos et al. (2013) developed a population dynamics model based on pollinator's adaptive foraging and projected the temporal dynamics of three empirical pollinator networks. They found that incorporation of adaptive foraging into the dynamics of a pollination network increases network persistence and diversity of its constituent species. Moreover, Song and Feldman (2014) constructed a mathematical model that integrates individual adaptive foraging behaviour and population dynamics of a community consisting of two plant species and a pollinator species. They found that adaptive foraging at the individual level, complementing adaptive foraging at the species level, can enhance the coexistence of plant species through niche partitioning between conspecific pollinators.

Adaptation in mutualistic networks has also been modelled through the evolution of functional traits determinant of the interactions. Such traits are often those that can impose important constraints on the interactions, such as the proboscis lengths of a pollinator and the flower tube length of a plant. For instance, Olivier et al. (2009) showed that tolerance traits (those responsible for minimising fitness cost but not reducing encounter rate), as opposed to resistance traits (those acting to reduce encounter rate between the interacting partners) are an important factor promoting stability of mutualisms. Moreover, they argued that a tolerance trait such as the phenotypic plasticity in honeydew production can prevent escalation into an antagonistic arms race and led to mutualistic coevolution. Using a theoretical model based on the interplay between ecological and evolutionary processes, Minoarivelo and Hui (2016) studied the evolution of phenotypic traits in mutualistic networks. By assuming that interactions are mediated by the similarity of phenotypic traits between mutualistic partners, they generated certain realistic architectures of mutualistic networks. In particular, they showed that a moderate accessibility to intra-trophic resources and cross-trophic mutualistic support can result in a highly nested web, while low tolerance to trait difference between interacting pairs leads to a high level of modularity.

More abstract traits have also been used in modelling mutualistic coevolution. For instance, Ferrière et al. (2002) defined a trait measured as the per capita rate of commodities trading which represents the probability per unit time that a partner individual receives benefit from a mutualistic interaction. They found that the existence of ‘cheaters’, or individuals that reap mutualistic benefits while providing fewer commodities to the partner species, can lead to the coexistence of mutualistic partners and thus is a key to the persistence of mutualism. In contrast to their study, West et al. (2002) showed that one of the factors that may stabilize mutualistic interactions is when individuals preferentially reward more mutualistic behaviour and punish less mutualistic (i.e. more parasitic) behaviour. Specifically, they explained the stability of the plant-legume mutualism by this cost/reward process:

plants are selected to supply preferentially more resources to nodules that are fixing more $N_2$.

## 12.8 Conclusions

More than 40 years after May's pioneering work, there is still no complete agreement on the complexity/stability relationship in ecosystems. The main issues could be related to the use of different definitions and measures for both complexity and stability, and the use of model vs. real ecosystem data. A joint effort between community ecology and ecosystems ecology, both at the theoretical and experimental/empirical level, should be undertaken to tackle such contemporary questions. Theoretical modelling should improve the realism of described ecological interactions, including spatial and environmental variables with their fluctuations and trends, while experimental and empirical works should monitor species densities and strength of interactions over time to possibly allow dynamic models to be parametrized. Also, species behavioural adaptation and phenotypic evolution has only recently started to be explored, and their contribution in the debate is no doubt important, seen the rapid changes that are affecting our planet and the environment where ecosystems are embedded, that also trigger contemporary plastic phenotypic adaptations. Adaptive and evolutionary models, taking into account adaptive behaviour and phenotypic adaptation, thus seem a promising tool to try to solve the debate: indeed, added ecological realism to community models seems to be reversing May's complexity/stability relationship, reconciling model predictions with ecological first principles and observations. Ecological networks also proved to be a very powerful and effective tool for applied sustainable management of biodiversity and ecosystem services under the threat of alien species invasion and climate change, also favouring stakeholder and decision-makers engagement for the governance of socio-environmental systems in the era of the Anthropocene.

**Acknowledgements** The authors are grateful to the National Research Foundation (NRF) of South Africa and the International Institute for Applied Systems Analysis (IIASA) for organizing the Southern African Young Scientist Summer Program (SA-YSSP). The contribution of two anonymous reviewers is acknowledged. This chapter is based on a review paper by the same authors submitted to Population Ecology.

## References

Allesina, S., & Pascual, M. (2008). Network structure, predator-prey modules, and stability in large food webs. *Theoretical Ecology, 1,* 55–64.

Allesina, S., & Tang, S. (2012). Stability criteria for complex ecosystems. *Nature, 483,* 205–208.

Almeida-Neto, M., Guimarães, P., Guimarães Jr., P. R., et al. (2008). A consistent metric for nestedness analysis in ecological systems: Reconciling concept and measurement. *Oikos, 117,* 1227–1239.

Atmar, W., & Patterson, B. D. (1993). The measure of order and disorder in the distribution of species in fragmented habitat. *Oecologia, 96,* 373–382.

Bascompte, J., Jordano, P., Melián, C. J., et al. (2003). The nested assembly of plant-animal mutualistic networks. *Proceedings of the National Academy of Sciences of the United States of America, 100,* 9383–9387.

Bascompte, J., Jordano, P., & Olesen, J. M. (2006). Asymmetric coevolutionary networks facilitate biodiversity maintenance. *Science, 312,* 431–433.

Bastolla, U., Fortuna, M. A., Pascual-Garcia, A., et al. (2009). The architecture of mutualistic networks minimizes competition and increases biodiversity. *Nature, 458,* 1018–1021.

Baird, D., Luczkovich, J. J., & Christian, R. R. (1998). Assessment of spatial and temporal variability in ecosystem attributes of the St. Marks National Wildlife Refuge, Apalachee Bay, Florida. *Estuarine Coastal Shelf Science, 47,* 329–349.

Baird, D., & Mehta, A. (Eds.). (2011). *Estuarine and coastal ecosystem modeling, Volume 9 in Treatise on estuarine and coastal science.* Amsterdam: Elsevier.

Banašek-Richter, C., Bersier, L. F., Cattin, M. F., et al. (2009). Complexity in quantitative food webs. *Ecology, 90,* 1470–1477.

Beckerman, A., Petchey, O. L., & Morin, P. J. (2010). Adaptive foragers and community ecology: Linking individuals to communities and ecosystems. *Functional Ecology, 24,* 1–6.

Berlow, E. L. (1999). Strong effects of weak interactions in ecological communities. *Nature, 398,* 330–334.

Berlow, E. L., Neutel, A. M., Cohen, J. E., et al. (2004). Interaction strengths in food webs: Issues and opportunities. *Journal of Animal Ecology, 73,* 585–598.

Bersier, L. F., Banašek-Richter, C., & Cattin, M. F. (2002). Quantitative descriptors of food-web matrices. *Ecology, 83,* 2394–2407.

Bonchev, D., & Buck, G. A. (2007). Quantitative measures of network complexity. In *Complexity in chemistry, biology, and ecology.* Berlin: Springer.

Borrelli, J. J., Allesina, S., Amarasekare, P., et al. (2015). Selection on stability across ecological scales. *Trends in Ecology & Evolution, 30,* 417–425.

Borrvall, C., Ebenman, B., & Jonsson, T. (2000). Biodiversity lessens the risk of cascading extinction in model food webs. *Ecology Letters, 3,* 131–136.

Brännström, Å., Loeuille, N., Loreau, M., et al. (2011). Emergence and maintenance of biodiversity in an evolutionary food-web model. *Theoretical Ecology, 4,* 467–478.

Brännström, Å., Johansson, J., Loeuille, N., et al. (2012). Modelling the ecology and evolution of communities: A review of past achievements, current efforts, and future promises. *Evolutionary Ecology Research, 14,* 601–625.

Brown, J. H., Calder III, W. A., & Kodric-Brown, A. (1978). Correlates and consequences of body size in nectar-feeding birds. *American Zoologist, 68,* 687–700.

Campbell, C., Yang, S., Shea, K., et al. (2012). Topology of plant-pollinator networks that are vulnerable to collapse from species extinction. *Physical Review E, 86,* 02192.

Camacho, J., Guimerà, R., & Amaral, L. A. N. (2002). Robust patterns in food web structure. *Physical Review Letters, 88,* 228102.

Cattin, M. F., Bersier, L. F., Banašek-Richter, C., et al. (2004). Phylogenetic constraints and adaptation explain food-web structure. *Nature, 427,* 835–839.

Chen, X., & Cohen, J. E. (2001). Global stability, local stability and permanence in model food webs. *Journal of Theoretical Biology, 212,* 223–235.

Christianou, M., & Kokkoris, G. D. (2008). Complexity does not affect stability in feasible model communities. *Journal of Theoretical Biology, 253,* 162–169.

Cohen, J. E., & Briand, F. (1984). Trophic links of community food webs. *Proceedings of the National Academy of Sciences of the United States of America, 81,* 4105–4109.

Cohen, J. E., & Newman, C. M. (1985). A stochastic theory of community food webs. *Proceedings of the Royal Society of London B, 224,* 421–448.

Cohen, J. E., Briand, F., & Newman, C. M. (1990). *Community food webs: Data and theory*. Biomathematics 20. Berlin: Springer.

D'Alelio, D., Libralato, S., Wyatt, T., et al. (2016). Ecological-network models link diversity, structure and function in the plankton food-web. *Scientific Reports, 6,* 21806.

Darwin, C. (1862). *On the various contrivances by which British and foreign orchids are fertilized by insect*. London: Murray.

De Angelis, D. L. (1975). Stability and connectance in food web models. *Ecology, 56,* 238–243.

De Ruiter, P. C., Neutel, A.-M., & Moore, J. C. (1995). Energetics, patterns of interaction strengths, and stability in real ecosystems. *Science, 269,* 1257–1260.

Donohue, I., Petchey, O. L., Montoya, J. M., et al. (2013). On the dimensionality of ecological stability. *Ecology Letters, 16,* 421–429.

Dormann, C. F., Fründ, J., Blüthgen, N., et al. (2009). Indices, graphs and null models: Analysing bipartite ecological networks. *The Open Ecology Journal, 2,* 7–24.

Dunne, J. A., & Williams, R. J. (2009). Cascading extinctions and community collapse in model food webs. *Philosophical Transactions of the Royal Society of London B, 364,* 1711–1725.

Dunne, J. A., Williams, R. J., & Martinez, N. D. (2002a). Food-web structure and network theory: The role of connectance and size. *Proceedings of the National Academy of Sciences of the United States of America, 99,* 12917–12922.

Dunne, J. A., Williams, R. J., & Martinez, N. D. (2002b). Network structure and biodiversity loss in food webs: Robustness increases with connectance. *Ecology Letters, 8,* 558–567.

Dunne, J. A., Williams, R. J., & Martinez, N. D. (2004). Network structure and robustness of marine food webs. *Marine Ecology Progress Series, 273,* 291–302.

Dupont, Y. L., & Olesen, J. M. (2012). Stability of modularity and structural keystone species in temporal cumulative plant-flower-visitor networks. *Ecological Complexity, 11,* 84–90.

Ehrlich, P. R., & Raven, P. H. (1964). Butterflies and plants: A study in coevolution. *Evolution, 18,* 586–608.

Elton, C. S. (1958). *Ecology of invasions by animals and plants*. London: Chapman and Hall.

Emmerson, M. C., & Raffaelli, D. (2004). Predator-prey body size, interaction strength and the stability of a real food web. *Journal of Animal Ecology, 73,* 399–409.

Emmerson, M. C., & Yearsley, J. M. (2004). Weak interactions, omnivory and emergent food-web properties. *Proceedings of the Royal Society of London B, 271,* 397–405.

Feng, W., & Takemoto, K. (2014). Heterogeneity in ecological mutualistic networks dominantly determines community stability. *Scientific Reports, 4,* 5912.

Ferrière, R., Bronstein, J. L., Rinaldi, S., et al. (2002). Cheating and the evolutionary stability of mutualisms. *Proceedings of the Royal Society of London B, 269,* 773–780.

Fowler, M. S. (2009). Increasing community size and connectance can increase stability in competitive communities. *Journal of Theoretical Biology, 258,* 179–188.

Fussman, G. F., Loreau, M., & Abrams, P. (2007). Eco-evolutionary dynamics of communities and ecosystems. *Functional Ecology, 21,* 465477.

Goldwasser, L., & Roughgarden, J. (1993). Construction of a large Caribbean food web. *Ecology, 74,* 1216–1233.

Gravel, D., Massol, F., & Leibold, M. A. (2016). Stability and complexity in model meta-communities. *Nature Communications, 7,* 12457.

Grilli, J., Rogers, T., & Allesina, S. (2016). Modularity and stability in ecological networks. *Nature Communications, 7,* 12031.

Gross, T., Rudolf, L., Levin, S. A., et al. (2009). Generalized models reveal stabilizing factors in food webs. *Science, 325,* 747–750.

Gross, T., & Sayama, H. (Eds.). (2009). *Adaptive networks: Theory, models and applications*. Berlin: Springer.

Havens, K. (1992). Scale and structure in natural food webs. *Science, 257,* 1107–1109.

Haydon, D. (1994). Pivotal assumptions determining the relationship between stability and complexity: An analytical synthesis of the stability-complexity debate. *American Naturalist, 144,* 14–29.

Haydon, D. (2000). Maximally stable model ecosystems can be highly connected. *Ecology, 81,* 2631–2636.

Heckmann, L., Drossel, B., Brose, U., et al. (2012). Interactive effects of body-size structure and adaptive foraging on food-web stability. *Ecology Letters, 15,* 243–250.

Heleno, R., Devoto, M., & Pocock, M. (2012). Connectance of species interaction networks and conservation value: Is it any good to be well connected? *Ecological Indicators, 14,* 7–10.

Herrera, C. M. (1985). Determinants of plant-animal coevolution: The case of mutualistic dispersal of seeds by vertebrates. *Oikos, 44,* 132–141.

Hughes, J. B., & Roughgarden, J. (1998). Aggregate community properties and the strength of species' interactions. *Proceedings of the National Academy of Sciences of the United States of America, 95,* 6837–6842.

Hui, C., & Richardson, D. M. (2017). *Invasion dynamics*. Oxford University Press.

Hui, C., Richardson, D. M., Landi, P., et al. (2016). Defining invasiveness and invasibility in ecological networks. *Biological Invasions, 18,* 971–983.

Ingram, T., Harmon, L. J., & Shurin, J. B. (2009). Niche evolution, trophic structure, and species turnover in model food webs. *American Naturalist, 174,* 56–67.

Ives, A. R., Klug, J. L., & Gross, K. (2000). Stability and species richness in complex communities. *Ecology Letters, 3,* 399–411.

Jacquet, C., Moritz, C., Morissette, L., et al. (2016). No complexity-stability relationship in empirical ecosystems. *Nature Communications, 7,* 12573.

James, A., Pitchford, J. W., & Plank, M. J. (2012). Disentangling nestedness from models of ecological complexity. *Nature, 487,* 227–230.

Jordano, P. (1987). Patterns of mutualistic interactions in pollination and seed dispersal: Connectance, dependence asymmetries, and coevolution. *American Naturalist, 129,* 657–677.

Jordano, P., Bascompte, J., & Olesen, J. M. (2003). Invariant properties in coevolutionary networks of plant animal interactions. *Ecology Letters, 6,* 69–81.

Kaiser-Bunbury, C. N., & Blutghen, N. (2015). Integrating network ecology with applied conservation: A synthesis and guide to implementation. *AoB Plants, 7,* plv076.

Kokkoris, G. D., Troumbis, A. Y., & Lawton, J. H. (1999). Patterns of species interaction strength in assembled theoretical competition communities. *Ecology Letters, 2,* 70–74.

Kokkoris, G. D., Jansen, V. A. A., Loreau, M., et al. (2002). Variability in interaction strength and implications for biodiversity. *Journal of Animal Ecology, 71,* 362–371.

Kondoh, M. (2003). Foraging adaptation and the relationship between food-web complexity and stability. *Science, 299,* 1388–1391.

Kondoh, M. (2005). Is biodiversity maintained by food-web complexity? The adaptive food-web hypothesis. In *Acquatic food webs: An ecosystem approach* (pp. 130–142). Oxford University Press.

Kondoh, M. (2006). Does foraging adaptation create the positive complexity-stability relationship in realistic food-web structure? *Journal of Theoretical Biology, 238,* 646–651.

Kondoh, M. (2007). Anti-predator defence and the complexity-stability relationship of food webs. *Proceedings of the Royal Society of London B, 274,* 1617–1624.

Krause, A. E., Frank, K. A., Mason, D. M., et al. (2003). Compartments revealed in food-web structure. *Nature, 426,* 282–285.

Landi, P., Dercole, F., & Rinaldi, S. (2013). Branching scenarios in eco-evolutionary prey-predator models. *SIAM Journal on Applied Mathematics, 73,* 1634–1658.

Landi, P., & Piccardi, C. (2014). Community analysis in directed networks: In-, out-, and pseudocommunities. *Physical Review E, 89,* 012814.

Lawlor, L. R. (1978). Comment on randomly constructed model ecosystems. *American Naturalist, 111,* 445–447.

Lawlor, L. R. (1980). Structure and stability in natural and randomly constructed competitive communities. *American Naturalist, 116,* 394–408.

Lehman, C. L., & Tilman, D. (2000). Biodiversity, stability, and productivity in competitive communities. *American Naturalist, 156,* 534–552.

Logofet, D. O. (2005). Stronger-than-Lyapunov notions of matrix stability, or how "flowers" help solve problems in mathematical ecology. *Linear Algebra and its Applications, 398,* 75–100.
Loreau, M., & de Mazancourt, C. (2013). Biodiversity and ecosystem stability: A synthesis of underlying mechanisms. *Ecology Letters, 16,* 106–115.
Lyapunov, A. M. (1992). *The general problem of the stability of motion.* London: Taylor & Francis.
May, R. M. (1973). *Stability and complexity in model ecosystems.* Princeton University Press.
MacArthur, R. H. (1955). Fluctuations of animal populations and a measure of community stability. *Ecology, 36,* 533–536.
Martinez, N. D. (1992). Constant connectance in community food webs. *American Naturalist, 139,* 1208–1218.
McCann, K., Hastings, A., & Huxel, G. R. (1998). Weak trophic interactions and the balance of nature. *Nature, 395,* 794–798.
Martinez, N. D. (1994). Scale-dependent constraints on food-web structure. *American Naturalist, 144,* 935–953.
Mello, M. A. R., Marquitti, V. M. D., Guimarães Jr., P. R., et al. (2011). The modularity of seed dispersal: Differences in structure and robustness between bat– and bird–fruit networks. *Oecologia, 167,* 131–140.
Memmott, J. (1999). The structure of a plant-pollinator food web. *Ecology Letters, 2,* 276–280.
Memmott, J., Waser, N. M., & Price, M. V. (2004). Tolerance of pollination networks to species extinctions. *Proceedings of the Royal Society of London B, 271,* 2605–2611.
Memmot, J. (2009). Food webs: A ladder for picking strawberries or a practical tool for practical problems? *Philosophical Transactions of the Royal Society of London B, 364,* 1693–1699.
Millennium Ecosystem Assessment. (2005). *Ecosystems and human well-being: Synthesis.* Washington: Island Press.
Minoarivelo, H. O., & Hui, C. (2016). Trait-mediated interaction leads to structural emergence in mutualistic networks. *Evolutionary Ecology, 30,* 105–121.
Montoya, J. M., & Solé, R. V. (2002). Small world patterns in food webs. *Journal of Theoretical Biology, 214,* 405–4012.
Moore, J. C., & Hunt, H. W. (1988). Resource compartmentation and the stability of real ecosystems. *Nature, 333,* 261–263.
Neubert, M. G., & Caswell, H. (1997). Alternatives to resilience for measuring the responses of ecological systems to perturbations. *Ecology, 78,* 653–665.
Neutel, A.-M., Heesterbeek, J. A. P., & de Ruiter, P. C. (2002). Stability in real food webs: Weak links in long loops. *Science, 296,* 1120–1123.
Neutel, A.-M., Heesterbeek, J. A. P., van de Koppel, J., et al. (2007). Reconciling complexity with stability in naturally assembling food webs. *Nature, 449,* 599–603.
Newman, M., & Girvan, M. (2004). Finding and evaluating community structure in networks. *Physical Review E, 69,* 026113.
Nuwagaba, S., Zhang, F., & Hui, C. (2015). A hybrid behavioural rule of adaptation and drift explains the emergent architecture of antagonistic networks. *Proceedings of the Royal Society of London B, 282,* 20150320.
Odum, E. P. (1953). *Fundamentals of ecology.* Philadelphia: Saunders.
Okuyama, T. (2008). Do mutualistic networks follow power distributions? *Ecological Complexity, 5,* 59–65.
Okuyama, T., & Holland, J. N. (2008). Network structural properties mediate the stability of mutualistic communities. *Ecology Letters, 11,* 208–216.
Olesen, J. M., & Jordano, P. (2002). Geographic patterns in plant-pollinator mutualistic networks. *Ecology, 83,* 2416–2424.
Olesen, J. M., Eskildsen, L. I., & Venkatasamy, S. (2002). Invasion of pollination networks on oceanic islands: Importance of invader complexes and endemic super generalists. *Diversity and Distributions, 8,* 181–192.

Olesen, J. M., Bascompte, J., Dupont, Y. L., et al. (2007). The modularity of pollination networks. *Proceedings of the National Academy of Sciences of the United States of America, 104,* 19891–19896.

Olff, H., Alonso, D., Berg, M. P., et al. (2009). Parallel ecological networks in ecosystems. *Philosophical Transactions of the Royal Society of London B, 364,* 1755–1779.

Olito, C., & Fox, J. W. (2014). Species traits and relative abundances predict metrics of plant-pollinator network structure, but not pairwise interactions. *Oikos, 124,* 428–436.

Olivier, T. H., Leather, S. R., & Cook, J. M. (2009). Tolerance traits and the stability of mutualism. *Oikos, 118,* 346–352.

Otto, S. B., Rall, B. C., & Brose, U. (2007). Allometric degree distributions facilitate food-web stability. *Nature, 450,* 1226–1229.

Paine, R. T. (1992). Food-web analysis through field measurement of per capita interaction strength. *Nature, 355,* 73–75.

Petanidou, T., Kallimanis, A. S., Tzanopoulos, J., et al. (2008). Long-term observation of a pollination network: Fluctuation in species and interactions, relative invariance of network structure and implications for estimates of specialization. *Ecology Letters, 11,* 564–575.

Pimm, S. L. (1979). Complexity and stability: Another look at MacArthur's original hypothesis. *Oikos, 33,* 251–257.

Pimm, S. L. (1980a). Properties of food webs. *Ecology, 61,* 219–225.

Pimm, S. L. (1980b). Food web design and the effect of species deletion. *Oikos, 35,* 139–149.

Pimm, S. L. (1984). The complexity and stability of ecosystems. *Nature, 307,* 321–326.

Pimm, S. L., & Lawton, J. H. (1978). On feeding on more than one trophic level. *Nature, 275,* 542–544.

Pimm, S. L., Lawton, J. H., & Cohen, J. E. (1991). Food web patterns and their consequences. *Nature, 350,* 669–674.

Pocock, M. J. O., Evans, D. M., Fontaine, C., et al. (2016). The visualization of ecological networks, and their use as a tool for engagement, advocacy and management. *Advances in Ecological Research, 54,* 41–85.

Poisot, T., & Gravel, D. (2014). When is an ecological network complex? Connectance drives degree distribution and emerging network properties. *PeerJ, 2,* e251.

Polis, G. (1991). Complex trophic interactions in deserts: An empirical critique of food web theory. *American Naturalist, 138,* 123–155.

Ramos-Jiliberto, R., Valdovinos, F. S., de Espanés, P. M., et al. (2012). Topological plasticity increases robustness of mutualistic networks. *Journal of Animal Ecology, 81*, 896–904.

Rezende, E. L., Jordano, P., & Bascompte, J. (2007). Effects of phenotypic complementarity and phylogeny on the nested structure of mutualistic networks. *Oikos, 116,* 1919–1929.

Rinaldi, S., Della Rossa, F., Dercole, F., et al. (2015). *Modeling love dynamics*. Singapore: World Scientific.

Rohr, R. P., Saavedra, S., & Bascompte, J. (2014). On the structural stability of mutualistic systems. *Science, 345,* 1253497.

Rosvall, M., & Bergstrom, C. T. (2007). An information-theoretic framework for resolving community structure in complex networks. *Proceedings of the National Academy of Sciences of the United States of America, 104*, 7327–7331.

Rooney, N., McCann, K., Gellner, G., et al. (2006). Structural asymmetry and the stability of diverse food webs. *Nature, 444,* 265–269.

Saint-Béat, B., Baird, D., Asmus, H., et al. (2015). Trophic networks: How do theories link ecosystem structure and functioning to stability properties? A review. *Ecological Indicators, 52,* 458–471.

Shannon, C. E. (1948). A mathematical theory of communication. *AT&T Technology Journal*, *27*, 379–342.

Schoener, T. W. (1989). Food webs from the small to the large. *Ecology, 70,* 1559–1589.

Small, M., Judd, K., & Stemler, T. (2013). The stability of networks—Towards a structural dynamical systems theory. ArXiv.

Solé, R. V., & Montoya, J. (2001). Complexity and fragility in ecological networks. *Proceedings of the Royal Society of London B, 268*, 2039–2045.
Song, Z., & Fledman, M. W. (2014). Adaptive foraging behaviour of individual pollinators and the coexistence of co-flowering plants. *Proceedings of the Royal Society of London B, 281*, 20132437.
Sprules, W. G., & Bowerman, J. E. (1988). Omnivory and food chain length in zooplankton food webs. *Ecology, 69*, 418–426.
Strona, G., & Lafferty, K. D. (2016). Environmental change makes robust ecological networks fragile. *Nature Communications, 7*, 12462.
Stouffer, D. B., & Bascompte, J. (2011). Compartmentalization increases food-web persistence. *Proceedings of the National Academy of Sciences of the United States of America, 108*, 3648–3652.
Suweis, S., Grilli, J., Banavar, J. R., et al. (2015). Effect of localization on the stability of mutualistic ecological networks. *Nature Communications, 6*, 10179.
Thébault, E., & Fontaine, C. (2010). Stability of ecological communities and the architecture of mutualistic and trophic interactions. *Science, 329*, 853–856.
Tylianakis, J. M., Tscharntke, T., & Lewis, O. T. (2007). Habitat modification alters the structure of tropical host-parasitoid food webs. *Nature, 445*, 202–205.
Tylianakis, J. M., Laliberte, E., Nielsen, A., et al. (2010). Conservation of species interaction networks. *Biological Conservation, 143*, 2270–2279.
Valdovinos, F. S., de Espanés, P. M., Flores, J. D., et al. (2013). Adaptive foraging allows the maintenance of biodiversity of pollination networks. *Oikos, 122*, 907–917.
Valdovinos, F. S., Ramos-Jiliberto, R., Garay-Narvaez, L., et al. (2010). Consequences of adaptive behaviour for the structure and dynamics of food webs. *Ecology Letters, 13*, 1546–1559.
van Altena, C., Hemerik, L., & de Ruiter, P. C. (2016). Food web stability and weighted connectance: The complexity stability debate revisited. *Theoretical Ecology, 9*, 49–58.
Vázquez, D. P., & Aizen, M. A. (2003). Null model analyses of specialization in plant–pollinator interactions. *Ecology, 84*, 2493–2501.
Vieira, M. C., & Almeida-Neto, M. (2015). A simple stochastic model for complex coextinctions in mutualistic networks: Robustness decreases with connectance. *Ecology Letters, 18*, 144–152.
Visser, A. W., Mariani, P., & Pigolotti, S. (2012). Adaptive behaviour, tri-trophic food-web stability and damping of chaos. *Journal of the Royal Society, Interface, 9*, 1373–1380.
Waser, N. M., Chittka, L., Price, M. V., et al. (1996). Generalization in pollination systems, and why it Matters. *Ecology, 77*, 1043–1060.
West, S. A., Kiers, E. T., Pen, I., et al. (2002). Sanctions and mutualism stability: When should less beneficial mutualists be tolerated? *Journal of Evolutionary Biology, 15*, 830–837.
Williams, R. J., & Martinez, N. D. (2000). Simple rules yield complex food web. *Nature, 404*, 180–183.
Wheelwright, N. T., & Orians, G. H. (1982). Seed dispersal by animals: Contrasts with pollen dispersal, problems of terminology, and constraints on coevolution. *American Naturalist, 119*, 402–413.
Wolanski, E., & McLusky, D. (Eds.). (2011). *Treatise on estuarine and coastal science*. Amsterdam: Elsevier.
Wootton, J. T., & Emmerson, M. (2005). Measurement of interaction strength in nature. *Annual Reviews of Ecology and Systematics, 36*, 419–444.
Yodzis, P. (1981). The stability of real ecosystems. *Nature, 289*, 674–676.
Zhang, F., Hui, C., & Terblanche, J. S. (2011). An interaction switch predicts the nested architecture of mutualistic networks. *Ecology Letters, 14*, 797–803.
Zhang, F., Hui, C., & Pauw, A. (2013). Adaptive divergence in Darwin's race: how coevolution can generate trait diversity in a pollination system. *Evolution, 67*, 548–560.

# Chapter 13
# Aggregation Methods in Analysis of Complex Multiple Scale Systems

**Jacek Banasiak, Aleksandra Falkiewicz and Milaine S. S. Tchamga**

**Abstract** *Background and Significance of the topic:* Complexity of many advanced models practically precludes its robust analysis. Fortunately, in many cases the models involve multiple time or size scales and thus yield themselves to asymptotic analysis that allows for significant simplifications of them without losing essential features of their dynamics. *Methodology:* We apply various methods of asymptotic analysis, such as the Tikhonov-Vasilieva theory, geometric singular perturbation theory, asymptotic expansions, or the degenerate convergence theory. *Application/Relevance to systems analysis*: The presented theory allows for significant simplifications of multiscale complex systems in a way that preserves the main features of their dynamics. *Policy and/or practice implications:* In many cases a smart aggregation of the equations of a model leads to a complete solution of the problem at a lower computational cost. *Discussion and conclusion:* We presented two models and two ways of their asymptotic analysis. It is important to note that not always does the naive approach work–applicability of the asymptotic theories often requires a subtle mathematical analysis. Sometimes only the approximation of macroscopic variables is available. Whenever possible, the analysis of complex multiscale systems should be preceded by assessing the possibility of their simplification through the aggregation of equations and variables. This often requires application of advanced analytic methods but, if successful, leads to significantly simpler systems that can be solved at a lower computational cost.

J. Banasiak (✉)
Department of Mathematics and Applied Mathematics,
University of Pretoria, Pretoria, South Africa
e-mail: jacek.banasiak@up.ac.za

J. Banasiak · A. Falkiewicz
Institute of Mathematics, Łódź University of Technology, Łódź, Poland
e-mail: aleksandrafalkiewicz@gmail.com

M. S. S. Tchamga
School of Mathematics, Statistics and Computer Science,
University of KwaZulu-Natal, Durban, South Africa
e-mail: seuneumilaine@gmail.com

 
P. Mensah et al. (eds.), *Systems Analysis Approach for Complex Global Challenges*,
https://doi.org/10.1007/978-3-319-71486-8_13

## 13.1 Introduction

The Universe is an evolving collection of interacting subsystems that are connected both vertically, that is, in the direction of increasing complexity, and horizontally, in the sense of ensembles of entities of comparable complexity. Clearly, its exact description would require a model as big as itself. However, we usually do not need the information on every single entity in the Universe at any time; even if given, it would not be manageable. Thus, to describe a particular process, we take advantage of the fact that most of the subsystems in the Universe are loosely connected and thus we can focus on the ones that, in our view, are relevant to the process we are interested in. This choice, like the whole modelling process itself, is highly subjective in the sense that different modellers can arrive at widely different models of the same phenomenon. Even if the models well fit the data, they may provide a diverse assessment of the relevance of particular subsytems of the model, hence leading to inconsistent predictions (Wearing et al. 2005). The existence of various models is not something inherently wrong. It comes from the parsimony principle (Gauch 2003); that is, we build models that should offer good value for money. For instance, often we are interested in a crude description of the most relevant features of the process, but available in a short time, or for a low computational cost, and thus we only need a simple model of it. At other times we need to understand fine details of the process, with time or cost being of secondary importance, and we opt for a more complex model.

To give a concrete example, in epidemiology, since the pioneering work of Ross and MacDonald on malaria (MacDonald 1957; Ross 1911), and Kermack and McKendric on the Bombay plague (Kermack and McKendrick 1991), there have been an abundance of relatively simple, so-called compartmental, models involving a small amount of state variables and parameters. In such models, the populations is subdivided into classes relevant to the progress of the disease (e.g. susceptibles, infectives and recovered in the so-called SIR model) and the state variables are typically chosen to be the population densities in each class. Such models can be extended in various ways by including, for instance, the vector populations in vector borne diseases, or the reservoir in the diseases caused by environmental factors such as cholera. They are given by systems of ordinary differential, or difference, equations that typically yield themselves to a comprehensive analysis. Since, however, they only are structured with respect to the relevance of each class to the disease, they provide a global, macroscopic, view of the evolution of the disease. Nevertheless, even in compartmental models, the increase of the dimensionality of the system of equations, coupled with the occurrence of many parameters that often widely differ in magnitude, may negatively impact on the robustness of their analysis.

Recent advances in the experimental and observational techniques, and the available computational power, have made it possible and necessary to develop more realistic models by taking into account other structures in the population. These could include differentiation of individuals with respect to their position in space, or age, but also could involve its interactions with other communities and entities of the food web (Martcheva 2015). Also, in contrast to the assumptions of the compartmental models, individuals are not homogeneous but differ with respect to the changing metabolism, physiology and behavior. If we assume that the additional structure is discrete; that is, the modelled entities can occur in one of the denumerable states, we create an expanded compartmental model of high complexity that, as discussed above, may prove to be difficult to analyse. Otherwise, if the attribute defining the structure is continuous, we arrive at partial differential equations, or integro-differential equations, the analysis of which brings yet other challenges.

Similarly, if we describe a metapopulation being a network of many interacting subpopulations, a crude model would consist of a system of ordinary differential equations with the averaged states of each subpopulation as unknowns. Such a model would correspond to a compartmental model in epidemiology, see e.g. (Bellomo et al. 2015; Borsche et al. 2014). The state of the subpopulation occupying a particular node of the network is given by a number, for instance, the size or density of the subpopulation, that changes in time. However, to get a more realistic description of the metapopulation, one should recognize that each subpopulation has its own structure that should be modelled. If, for example, we are interested in the age structure of the subpopulations, then each node would become a one dimensional manifold parametrized by the age of individuals and the model would become a system of McKendric equations on these manifolds coupled by an appropriate rule of exchange of individuals between the subpopulations by migrations, or by the gene exchange in the reproduction process (Banasiak et al. 2016; Lebowitz and Rubinov 1974; Rotenberg 1983). In addition, the communication between the subpopulations may be modelled by some dynamical process such as transport or diffusion. It is also important to note that for large, for instance ecological, networks that are described by systems of ODEs, there is a somewhat reverse process of lumping together functional subpopulations; that is, subpopulations having roughly the same duties in the system (Bellomo et al. 2015; Lewis 1977). This results in a simpler, more crude description of the network by a lower dimensional system of equations.

Incorporating more details in models is necessary to provide a more realistic description of the processes occurring in nature and to better understand the interdependence of their drivers. The drawback is that adding more and more details makes the model less clear and more difficult to analyse. Indiscriminately increasing the size of the model often results in incorporating redundant information and thus obscuring its essential features. For instance, for some diseases with quick turnover, such as flu or common cold, the demographic processes are not as relevant as for HIV/AIDS, that often lasts the whole human life. Similarly, the age of individuals is important in modelling sexually transmitted diseases or, for instance, the measles, but not that relevant for a flu. It is thus important to find the details that are important and must be incorporated in the model but still keep the model tractable.

The problem we address in this chapter can be thus phrased as follows. Let us assume that we have a detailed model of a certain process, that we call a micro

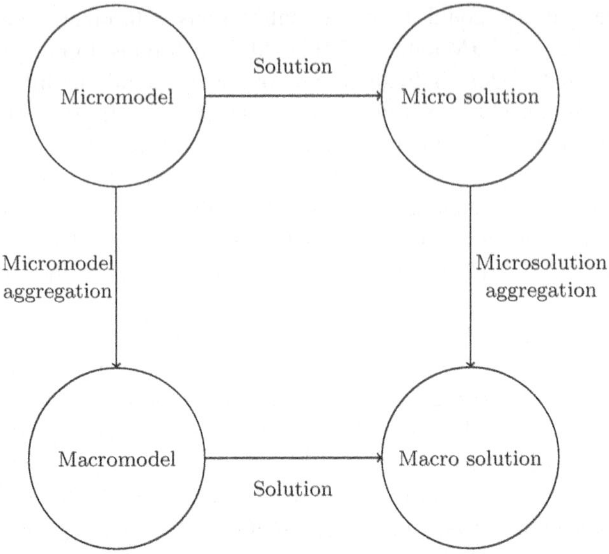


**Fig. 13.1** Commutativity of the aggregation diagram

model. It may involve a large quantity of variables to accurately describe even minute interactions between subpopulations. Is there a systematic way, called aggregation, of finding a reduced set of variables, termed macro-variables, whose dynamics is close to the dynamics of the original process at the required level of accuracy? What are the relations between the micro model and the macro model describing the evolution of the reduced variables? If there are other models of the process, derived at this level of accuracy from first principles, is our macro model consistent with them?

For further discussion we must precisely understand the objects we are dealing with. We have a detailed micro model of some process with its solution giving the micro variables. To obtain its macro description we can aggregate this solution to define appropriate global, or macro, variables. This, however, is of little interest, as to obtain the global variables in this way we need to first solve the detailed model which may be costly. To avoid dealing with the detailed model, we can first aggregate the micro model without solving it to derive a system of equations governing the global variables. The problem is how close the global variables, obtained by solving the aggregated micro model (that is, the macro model) are to the global variables obtained from aggregating the solutions of the original micro model. In other words, we ask to what extend the diagram in Fig. 13.1 commutes.

To give an example, if we consider an epidemiological SI model with continuous age structure as a micro model, the age specific densities of susceptibles and infectives are our micro variables, while the total numbers of susceptibles and infectives individuals are our macro variables. Clearly, once we have the micro solution, we can recover the macro solution by integrating the former with respect to age. On the other hand, we can try to derive the equation governing the total numbers of susceptibles and infectives either from first principles, or by integrating the micro model with respect to age and using some additional argument to obtain a closed system of ODEs.

Here it is important to note that some of the original variables in the system may already be the global variables, determining the dynamics of the system and 'dragging' the other variables with them in the so-called slave mode. This is often the case in models written as systems of ODEs in the so-called Tikhonov form, discussed in Sect. 13.2.

If the dynamics of the aggregated system is consistent with the original dynamics of the microsystem in the sense that the solutions of the aggregated system behave exactly as the aggregated micro variables (that is, the diagram in Fig. 13.1 commutes), we say that the aggregation is perfect. Otherwise, we say that the aggregation is approximate (Auger et al. 1936; Iwasa et al. 1987; Iwasa et al. 1989). Perfect aggregation occurs in very particular situations (Atay and Roncoroni 2016), and here we only are concerned with Approximate aggregation, where the agreement between the micro and macro dynamics is required to by only approximate.

In many cases it is only the modeller's intuition that guides us through the process of selecting the relevant components of the model and rejecting the insignificant ones. However, in many cases the detailed model describes several processes occurring with widely different intensities; that is, having different reference (characteristic) times. Then we are dealing with a multiple scale model that often allows for a more systematic approach to (approximate) aggregation. In such a case the aggregation process begins with selecting the time scale we are interested in and nondimensionalizing the system using the characteristic time of this scale as the unit of time. For instance, when modelling a human population often we take the average lifespan of an individual as the unit of time. With this choice, we indicate that we focus on demographic processes occurring in the population, such as the birth or death processes, which are important for an individual in his/her lifetime. On the other hand, we could consider a 1000 years as the unit of time if we were interested in the evolution of civilizations, or an hour or a day if our main interest was to model the changes in the distribution of the population due to the daily movements of the individuals, such as going to work, school, returning home, etc.

If we nondimensionalize the problem using the characteristic time of a particular process as the time unit, we rescale all parameters of the problem by writing them in this new unit. Hence, the numerical values of some of the parameters significantly increase, or decrease. This separates the model into the part evolving at the time scale we are interested in (the terms multiplied by O(1) parameters), the fast part (the terms multiplied by large parameters) and the slow part (the terms multiplied by small parameters). On physical or biological grounds we expect that it should be possible to discard the slow part and to aggregate fast variables; that is, the variables describing the fast processes, creating thus a macro model in which the macro variables are that aggregated fast variables.

For instance, consider a disease in an age-structured population. If we focus on short-lasting illnesses such as flu or common cold and model their long-term impact on the population, it is natural to define the small parameter to be the ratio of the recovery rate and the life expectancy in the population. In this case often the sizes of susceptibles and infectives reach equilibrium in a time that is short in comparison with the chosen time scale and thus they can be aggregated, or averaged over time, leaving the equations for the age specific densities of the population as the macro model. The macro model does not explicitly involve the variables related to the disease, see Banasiak and M'pika Massoukou (2014), Banasiak et al. (2013). On the other hand, if we tried to model the effect of HIV/AIDS on this population, the ratio of the characteristic times of the vital and epidemiological processes would be close to 1, as the latter disease is terminal and its duration is comparable to the average lifespan of individuals in the population. Hence the aggregation described in the previous paragraph would not be possible.

The intuition that we can discard small terms in the model is, in general, correct. On the other hand, approximate aggregation of fast variables is faced with many difficulties and not always does it amount to just replacing the reciprocal of the large parameters by zero. One of the reason is that the equations describing the system in micro and macro regimes often are of a completely different type—for instance partial differential equations on the micro level are expected to be aggregated to ordinary differential equations and the macro level. Clearly it is neither possible, nor desirable, that all initial and boundary conditions imposed on solutions of the former be satisfied by the solutions of the latter. Thus, not only proving that the aggregated model is a good approximation of the original one, but also just constructing the former presents a formidable challenge.

While the modelling process and the identification of the time scales does not belong to mathematics (though requires a mathematical mind), once the model is constructed and the small and large parameters identified, we can ask several mathematically legitimate questions: what is the limit of the solutions of the model as these parameters tend, respectively, to zero and infinity, and what equations are satisfied by this limit? That limit gives the approximately aggregated (macro) variables and the equations form the macro model. Note that typically the macro variables and the macro model are uniquely defined for the given scaling and, as we shall see, not always correspond to results obtained by an ad hoc aggregation of variables.

There are several mathematical techniques facilitating approximate aggregation of variables such as the Tikhonov–Vasilieva theory (Banasiak and Lachowicz 2014; Hoppensteadt 1967; Tikhonov et al. 1985), geometric singular perturbation theory of Fenichel (Fenichel 1979; Jones and Khibnik 2001; Kuehn 2015), Trotter-Kato-Sova-Kurtz theory (Banasiak and Bobrowski 2009; Bobrowski 2016) or asymptotic expansion methods (Banasiak and Lachowicz 2014; Mika and Banasiak 1995; Verhulst 2005). It is impossible to give even a brief survey of all of them in one chapter. Thus here we have decided to describe two techniques and illustrate them of simplified models.

In Sect. 13.2 we consider compartmental models of vector born disease and of a flu in a population with vital dynamics. In the first case different time scales are due to different life-spans of mosquitoes and humans, while in the other they are offered by the difference in the humans' lifespan and the disease's duration. A specific feature of the second model is the so-called delayed stability switch whereby the set of aggregated variables changes at a certain moment in time but the solution continues to be approximated for some time by the 'wrong' set before eventually jumping to the 'correct' one, see Banasiak and Kimba Phongi (2015, Banasiak et al. (2013), Banasiak and Tchamga (2017). In Sect. 13.3 we look at fast transport along the edges of a network and at a similar model describing the spread of a genotype in a population of cells with age structure (Lebowitz and Rubinov 1974; Rotenberg 1983). We use the Trotter-Kato-Sova-Kurtz theory to find the conditions under which such a model can be approximated by an ODE model describing evolution of the macro variables which are the total masses of the substance (or, respectively, the sizes of the cell populations) along the edges. An interesting feature of this model is that the ODE can only approximate the aggregated micro solution but, contrary to many similar case, it cannot approximate the whole micro solution, see Banasiak and Falkiewicz (2017, Banasiak et al. (2016).

## 13.2 Multiple Scales in Epidemiological Models

As we mentioned in the Introduction, basic epidemiological models are given by systems of ordinary differential equations. In many cases, however, to get a better insight into the progress of the disease we need to introduce additional structure, for instance age, and thus to combine the pure epidemiological model with demographic processes. Also, often we need to combine the dynamics of the disease in humans or other hosts with that of the vector such as mosquitoes. It is clear that the vital dynamics of mosquitoes occur on a much faster time scale (days) than that of humans (years) hence any model combining host and vector dynamics necessarily is a two scale model. Similarly, modelling the effect of quick diseases such as flu or measles, on a population, we combine time scales of the disease (days) and of humans (years). In the Introduction we explained that by nondimensionalizing a multiple scale model using a relevant reference time unit, we introduce a small parameter given by the ratio of the typical time scales of different processes in the model. In many cases it is possible to describe such a model; that is, the micro model in this theory, by the system of differential equations in the so-called Tikhonov form

$$\begin{cases} \dot{x} = f(t,x,y,\epsilon), \ x(0) = x_0, \ x \in \mathbb{R}^n, \\ \epsilon\dot{y} = g(t,x,y,\epsilon), \ y(0) = y_0, \ y \in \mathbb{R}^p, \end{cases} \tag{13.1}$$

where "." denotes the differentiation with respect to time $t \in \mathbb{R}$, $\epsilon > 0$ is a small positive parameter representing the ratio of typical time scales and $f, g$ are known functions. Here it is clear that $x$ should be considered as the global variable in the sense described in the Introduction since the dynamics of $y$ occurs on a much faster time scale.

The presence of the small parameter multiplying the derivative makes the problem singularly perturbed in the sense that the aggregated system

$$\begin{cases} \dot{x} = f(t,x,y,0), \ x(0) = x_0, \\ 0 = g(t,x,y,0), \ y(0) = y_0, \end{cases} \tag{13.2}$$

obtained by setting $\epsilon = 0$, though certainly makes the model simpler, completely changes its structure and therefore it is not clear what is the relation of (13.2) later to (13.1).

Here we describe the Tikhonov–Vasilieva theory that is one of the theories that provide conditions under which the micromodel (13.1) can be approximated by the aggregated model (13.2), (Tikhonov et al. 1985). The precise assumptions of the theory are intertwined and quite technical (Banasiak and Lachowicz 2014), so that we only describe its essential features. Let us denote by $\bar{y}(t,x)$ the solution to

$$0 = g(t,x,y,0), \tag{13.3}$$

that is often called the *quasi steady state*, and by $\bar{x}(t)$ the solution of the equation

$$\dot{x} = f(t,x,\bar{y}(t,x),0), \qquad x(0) = x_0, \tag{13.4}$$

obtained from the first equation of (13.1) by substituting the unknown $y$ by the known quasi steady state $\bar{y}$. The first essential assumption of the Tikhonov theorem is that the quasi-steady states are isolated in some set $[0,T] \times \overline{\mathcal{U}}$, where $T > 0$ and $\overline{\mathcal{U}}$ is a compact subset of the $x$-domain of (13.1) and that they have well defined basins of attraction, explained below. Crucial roles are played by the *auxiliary equation*, describing the fast dynamics,

$$\frac{d\tilde{y}}{d\tau} = g(t,x,\tilde{y},0), \tag{13.5}$$

obtained from the second equation in (13.1) by the change of the time variable $\tau = t/\epsilon$ and setting $\epsilon = 0$ in the resulting equation (here $t$ and $x$ play the role of parameters) and by the *initial layer problem*

$$\frac{d\hat{y}}{d\tau} = g(0, x_0, \hat{y}, 0), \qquad \hat{y}(0) = y_0, \tag{13.6}$$

obtained from (13.5) by setting $t = 0$ and $x = x_0$, where $(x_0, y_0)$ are the initial conditions for (13.1). We note that for each fixed $t$ and $x$, the quasi steady state solution $\bar{y}(t)$ of (13.3) is an equilibrium of (13.5). Then we require that $\bar{y}(t, x)$ be an asymptotically stable equilibrium of (13.5) uniformly for $(t, x) \in [0, T] \times \overline{\mathcal{U}}$. Further, we assume that $\bar{x}(t) \in \mathcal{U}$ for $t \in [0, T]$ provided $x_0 \in \overline{\mathcal{U}}$ and that $y_0$ belongs to the basin of attraction of the stationary point $\bar{y}(0, x_0)$ of (13.6). Then the following theorem is true.

**Theorem 1** *Let the above assumptions be satisfied. Then there exists $\epsilon_0 > 0$ such that for any $\epsilon \in [0, \epsilon_0]$ there exists a unique solution $(x_\epsilon(t), y_\epsilon(t))$ of Problem (1) on $[0, T]$ and*

$$\begin{aligned} \lim_{\epsilon \to 0} x_\epsilon(t) &= \bar{x}(t), \qquad t \in [0, T], \\ \lim_{\epsilon \to 0} y_\epsilon(t) &= \bar{y}(t), \qquad t \in [0, T], \end{aligned} \tag{13.7}$$

where $\bar{x}(t)$ is the solution of (13.4) and $\bar{y}(t) = \bar{y}(t, x)$ is the solution of (13.3).

The convergence in $(7)_1$ is uniform with respect to $t \in [0, T]$, but in $(7)_2$ it is not uniform close to $t = 0$. However, in the latter case the convergence is uniform on any interval $[\zeta, T]$, $\zeta > 0$. This is the so-called initial layer effect and one can include the initial layer term to obtain the uniform convergence on $[0, T]$.

**Proposition 1** *Under the assumption of Theorem* 13.1 *we have*

$$\lim_{\epsilon \to 0} \left( y_\epsilon(t) - \bar{y}(t) - \hat{y}\left(\frac{t}{\epsilon}\right) + \bar{y}(0, x_0) \right) = 0, \tag{13.8}$$

*uniformly for $t \in [0, T]$, where $\hat{y}(\tau) - \bar{y}(0, x_0)$ is the correction in initial layer (the initial layer term).*

A serious drawback of the original formulation of the Tikhonov theorem is the fact that the convergence is only ensured on finite time intervals $[0, T]$. Some improvement in this direction can be found in (Hoppensteadt 1966)—it requires additional stability features of (13.5).

### 13.2.1 Application of the Tikhonov Theorem to Vector Borne Diseases

Vector borne diseases and, in particular, diseases caused by mosquito transmitted viruses, have proved to be both deadly and able to evolve and invading new habitats. A recent example of the Zica virus shows the widening of the distribution of its carrier due to global trade and travel as well as its ability to adapt to cold climate (Kraemer et al. 2015). The interplay of the host, vector and, possibly, pathogen dynamics leads to models with very rich dynamical features, from bifurcation up to deterministic chaos, see e.g. (Aguiar et al. 2011; Fischer and Halstead 1970).

In this section we shall show how the Tikhonov theorem can be used to aggregate a simple model of a vector borne disease, based on the Dengue fever, that was studied by the centre manifold theory in Rocha et al. (2013). Every year between 50 and 100 million people are infected by the Dengue fever (Aguiar and Stollenwerk 2007; Rocha et al. 2013). Unlike malaria, the Dengue fever in general is not fatal but, without proper medical care, the disease can develop into *dengue hemorrhagic fever* which causes a high number of deaths (Aguiar and Kooi 2009). This disease is transmitted by the mosquito *Aedes aegypti* and up to four immunologically distinct dengue serotypes (Gibbons et al. 2007; Gubler 1998) can coexist in an endemic area. While the infection with one serotype leads to life-long immunity with respect to that sereotype, no immunity is induced towards the others ones (Gibbons et al. 2007; Gubler 1998; Kumar et al. 2010).

In Rocha et al. (2013) the authors considered a simplified model of the Dengue fever, obtained by combining an SIR model with vital (Malthusian) dynamics for humans and mass action dynamics for mosquitoes,

$$\begin{aligned}
\dot{S}_h &= \mu(N - S_h) - \frac{\beta}{M} S_h I_v,\\
\dot{I}_h &= \frac{\beta}{M} S_h I_v - (\gamma + \mu) I_h,\\
\dot{R}_h &= \gamma I_h - \mu R_h,\\
\dot{S}_v &= \psi - \nu S_v - \frac{\vartheta}{N} S_v I_h,\\
\dot{I}_v &= \frac{\vartheta}{N} S_v I_h - \nu I_v,
\end{aligned} \tag{13.9}$$

where $S_h, I_h, R_h$ are, respectively, the number of susceptible, infective and recovered humans and $S_v$ and $I_v$ are the numbers of susceptible and infective of (female) mosquitoes. We observe that the parameters are chosen so that the human total population size $N$ is constant. Typical values of the parameters, Rocha et al. (2013), are as follows: the host death rate $\mu = \frac{1}{65}$ year$^{-1}$, the recovery rate

$\gamma = \frac{1}{10}$ day$^{-1}$ $= \frac{365}{10}$ year$^{-1}$, infection rate $\beta = 2\gamma$, the death rate of mosquitoes $\nu = \frac{1}{10}$ day$^{-1}$. Further, the recruitment rate for a stable mosquito population $M$ is $\psi = \nu M$ and, since a female mosquito has a fixed number of blood meals in her lifetime, the biting rate is assumed to be $\vartheta = 2\nu$.

The authors of Rocha et al. (2013) considered an even more simplified model of the Dengue fever. They argued that the time scale of the human disease expansion is given by the slow transition of the infected individuals through the recovered class $R$ back to the susceptible class. Thus, the SIR models for humans can be replaced by the simpler SIS model and thus the combined dynamics can be modelled by

$$\begin{aligned}
\dot{S}_h &= \alpha I_h - \frac{\beta}{M} S_h I_v, \\
\dot{I}_h &= \frac{\beta}{M} S_h I_v - \alpha I_h, \\
\dot{S}_v &= \psi - \nu S_v - \frac{\vartheta}{N} S_v I_h, \\
\dot{I}_v &= \frac{\vartheta}{N} S_v I_h - \nu I_v,
\end{aligned} \tag{13.10}$$

where the new parameter, $\alpha$, describes the return of the infectives to the susceptible class. It is assumed that $\alpha = \frac{1}{10}$ year$^{-1}$ and $\beta = 2\alpha$.

Using the fact that the populations of humans and mosquitoes are constant, $N$ and $M$, respectively, we further simplify (13.10) to

$$\begin{aligned}
\dot{I}_h &= \frac{\beta}{M} I_v (N - I_h) - \alpha I_h \\
\dot{I}_v &= \frac{\vartheta}{N} I_h (M - I_v) - \nu I_v.
\end{aligned} \tag{13.11}$$

Introducing 1 year as the reference time unit, we see that the parameters for the mosquito dynamics are much larger than that for human dynamics. Then, defining $\epsilon = \frac{1}{365}$, the scaled system (13.10) takes the form

$$\begin{aligned}
\dot{I}_h &= \frac{\beta}{M} I_v (N - I_h) - \alpha I_h \\
\epsilon \dot{I}_v &= \left( \frac{\tilde{\vartheta}}{N} I_h (M - I_v) - \tilde{\nu} I_v \right),
\end{aligned} \tag{13.12}$$

where $\tilde{\vartheta} = \frac{\vartheta}{\nu}\alpha$ and $\alpha = \tilde{\nu}$.

To check the assumptions of the Tikhonov theorem, we see that the quasi steady state is given by

$$\bar{I}_v(I_h) = \frac{\tilde{\vartheta} M I_h}{\tilde{\vartheta} I_h + \tilde{v} N} \tag{13.13}$$

and the auxiliary equation takes the form

$$\frac{d}{d\tau}\tilde{I}_v = \tilde{\vartheta}\frac{I_h}{N}(M - \tilde{I}_v) - \tilde{v}\tilde{I}_v = \tilde{\vartheta}\frac{I_h M}{N} - \left(\tilde{\vartheta}\frac{I_h}{N} + \tilde{v}\right)\tilde{I}_v.$$

Clearly, the quasi steady state is unique and, since the auxiliary equation is linear with negative leading coefficient, each point of the quasi steady state is uniformly asymptotically stable. Then, according to the Tikhonov theorem, the solution to (13.12) can be approximated by $(\bar{I}_h, \bar{I}_v)$, where $\bar{I}_h$ is the solution of

$$\dot{I}_h = \beta I_h \frac{N(\tilde{\vartheta} - \alpha\tilde{v}) - \tilde{\vartheta}(1-\alpha)I_h}{\tilde{\vartheta} I_h + \tilde{v} N}, \tag{13.14}$$

with $\bar{I}_h(0) = I_h(0)$ and

$$\bar{I}_v = \frac{\tilde{\vartheta} M \bar{I}_h}{\tilde{\vartheta}\bar{I}_h + \tilde{v} N}. \tag{13.15}$$

In the terminology of the Introduction, system (13.10) describes our original macro model. Under our assumptions, it is possible to make a partial perfect aggregation of the variables that leads to (13.11) and then, utilising the existence of the time scales, to approximately aggregate the model in the macro model (13.14). As we said earlier, the fact that the system is in the Tikhonov form immediately identifies $I_h$ as the global variable, with $\bar{I}_h$ being its approximation and with $I_v$ approximated by $\bar{I}_v$, given by (13.15), $S_h(t) \approx N - \bar{I}_h(t)$ and $S_v(t) \approx M - \bar{I}_v(t)$ following $\bar{I}_h(t)$ in the slave mode.

To illustrate the results, we used the parameters from (Rocha et al. 2013) that we listed above and $(I_h(0), I_v(0)) = (100, 2500)$. The result presented in Fig. 13.2 shows the simulation of the infective vectors versus infective hosts for the micro-model (13.12) as $\epsilon$(eps) tends to 0 and the quasi steady state (green).

### 13.2.2 Application of Tikhonov's Theorem to the Case of a Quick Disease

SIS models describing the evolution of a nonlethal disease are among the simplest epidemiological models. The mathematical form of the SIS model is

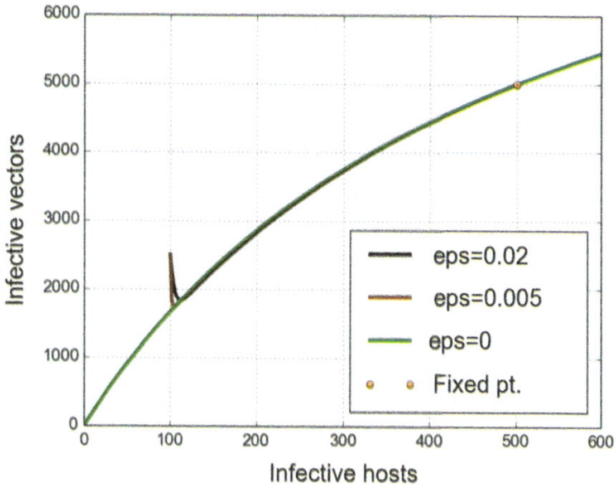


**Fig. 13.2** Infective vectors versus infective hosts for the micromodel (13.12) as $\epsilon$(eps) tends to 0 and the quasi steady state (green)

$$\begin{aligned} \dot{S} &= \lambda SI + \gamma I, \quad S(0) = S_0, \\ \dot{I} &= \lambda SI - \gamma I, \quad I(0) = I_0, \end{aligned} \tag{13.16}$$

where $S$ and $I$ are, respectively, the size of susceptible and infected population; the total population $N$ is given by $N = S + I$. The force of infection is denoted by $\lambda$ and the recovery rate by $\gamma$. Here we aim to model quick diseases, such as common cold or influenza, with the recovery time of several days. Hence it is natural to take 1 day as the unit of time for the disease related parameters $\gamma$ and $\lambda$. However, if we consider such a disease in an evolving population with the birth rate $\beta$ and the death rate $\mu$, then the characteristic time of birth and death processes can be taken to be 10 years—then the numerical values of, say, $\gamma$ and $\mu$ in the respective time units are of the same order of magnitude. In such a case, the ratio of the characteristic times is $\epsilon = 3650^{-1}$. Hence, a quick disease (13.16) in a population evolving according to the malthusian law can be modelled by

$$\begin{aligned} \dot{S}_\epsilon &= \beta N_\epsilon - \mu S_\epsilon - \frac{1}{\epsilon}(\lambda S_\epsilon I_\epsilon - \gamma I_\epsilon), \quad S_\epsilon(0) = S_0, \\ \dot{I}_\epsilon &= -(\mu + \mu^*) I_\epsilon + \frac{1}{\epsilon}(\lambda S_\epsilon I_\epsilon - \gamma I_\epsilon), \quad I_\epsilon(0) = I_0, \end{aligned} \tag{13.17}$$

where the time unit is 10 years and the numerical values of the parameters are of the same order of magnitude. In (13.17) we have introduced the additional parameter $\mu^*$ that gives the disease induced death rate.

The dynamics of the total population $N_\epsilon = S_\epsilon + I_\epsilon$ is given by

$$\dot{N}_\epsilon = (\beta - \mu) N_\epsilon - \mu^* I_\epsilon, \quad N_\epsilon(0) = N_0, \tag{13.18}$$

and (13.17) is equivalent to

$$\dot{N}_\epsilon = (\beta - \mu)N_\epsilon - \mu^* I_\epsilon, \quad N_\epsilon(0) = N_0,$$
$$\epsilon \dot{I}_\epsilon = -\epsilon(\mu + \mu^*)I_\epsilon + I_\epsilon(\lambda(N_\epsilon - I_\epsilon) - \gamma), \quad I_\epsilon(0) = I_0. \tag{13.19}$$

Setting $\epsilon = 0$ in the second equation of (13.19), we find two quasi steady states

$$I_1 = 0 \quad \text{and} \quad I_2 = N - \nu,$$

where $\nu = \frac{\gamma}{\lambda}$. Thus we have two possible reduced equations:

$$\dot{N} = (\beta - \mu)N, \quad N(0) = N_0, \tag{13.20}$$

or

$$\dot{N} = (\beta - \mu - \mu^*)N + \mu^* \nu, \quad N(0) = N_0. \tag{13.21}$$

The quasi steady states cross each other at $N = \nu$ and, since $N$ changes with time, the assumption of the Tikhonov theorem concerning isolated quasi steady states may be not satisfied. In such a case, we can adopt the following strategies:

(a) find conditions for $N$ to stay below $\nu$ for all time;
(b) find conditions for $N$ to stay above $\nu$ for all time;
(c) investigate what happens if $N$ brakes through the threshold $\nu$.

It is easy to check that quasi steady states switch stability at the point of intersection. Precisely, $I_1$ is attracting and $I_2$ is repelling for $N < \nu$, while $I_2$ becomes attracting and $I_1$ repelling for $N > \nu$.

One of the advantages of the Tikhonov theorem is that it is enough to check the conditions in (a) and (b) for $N$ that is the solution of the respective reduced equation, (13.20) or (13.21). First we consider $I_1 = 0$. Then it is easy to see that the solution $\bar{N}_1 = N_0 e^{(\beta-\mu)t}$ of (13.20) satisfies $\bar{N}(t) < \nu$ for all $t \in \mathbb{R}_+$ provided $N_0 < \nu$ and $\beta - \mu \leq 0$. Then the assumptions of the Tikhonov theorem are satisfied and

$$\lim_{\epsilon \to 0} N_\epsilon(t) = N_0 e^{(\beta-\mu)t}, \quad t \in [0, T], \quad \lim_{\epsilon \to 0} I_\epsilon(t) = 0, \quad t \in (0, T] \tag{13.22}$$

for any $T < +\infty$, see the left picture in Fig. 13.3, where we present simulations for $\mu = 0.038$, $\beta = 0.03$, $\mu^* = 0.04$, $\gamma = 0.14$, $\lambda = 0.018$ and $I_0 = 20$, $N_0 = 70$.

On the other hand, if we consider $I_2 = N - \nu$, then the solution of the reduced Eq. (13.21) is given by

$$\bar{N}_2(t) = \frac{((\beta - \mu - \mu^*)N_0 + \mu^* \nu)e^{(\beta-\mu-\mu^*)t} - \mu^* \nu}{\beta - \mu - \mu^*}$$

and we see that if $N_0 > \nu$ and $\beta - \mu \geq 0$, then $\bar{N}_2(t) > \nu$ for all $t \in \mathbb{R}_+$. Hence, again the assumptions of the Tikhonov theorem are satisfied and

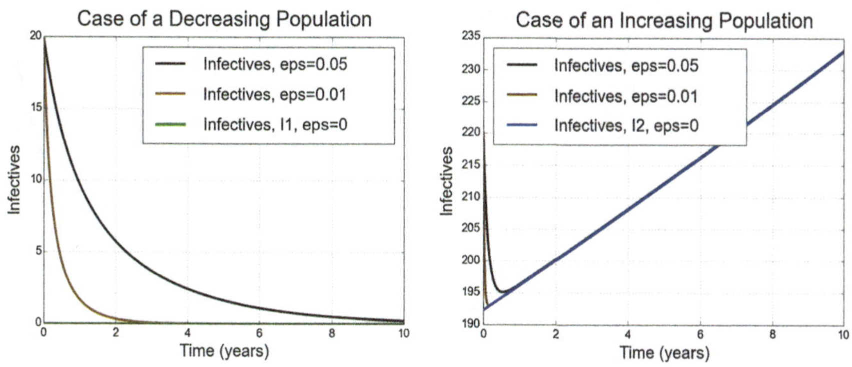


**Fig. 13.3** Case of an decreasing (left) and increasing population (right) staying within the domain of attraction on one quasi steady state

$$\lim_{\epsilon\to 0} N_\epsilon(t) = \bar{N}_2(t), \quad t \in [0,T], \quad \lim_{\epsilon\to 0} I_\epsilon(t) = \bar{N}_2(t) - v, \quad t \in (0,T] \tag{13.23}$$

for any $T < +\infty$, see the right picture in Fig. 13.3, where the values $\mu = 0.013$, $\beta = 0.038$, $\mu^* = 0.015$, $\gamma = 0.14$, $\lambda = 0.018$, $I_0 = 220$ and $N_0 = 270$ were used.

Now let us consider the case when the solution passes close to intersection of the quasi steady states. To simplify considerations, we assume that $\mu^* = 0$ so that, denoting $r = \beta - \mu$, (13.19) becomes

$$\dot{I}_\epsilon = -\mu I_\epsilon + \frac{1}{\epsilon}(\lambda I_\epsilon(N_0 e^{rt} - v) - \gamma I_\epsilon), \quad I_\epsilon(0) = I_0, \tag{13.24}$$

which is the Bernoulli equation, whose solution is

$$I_\epsilon(t) = \frac{I_0 e^{\frac{1}{\epsilon}\left(\frac{N_0\lambda}{r}(e^{rt}-1)-\gamma t\right)-\mu t}}{1 + \frac{\lambda I_0}{\epsilon}\int\limits_0^t e^{\frac{1}{\epsilon}\left(\frac{N_0\lambda}{r}(e^{rs}-1)-\gamma s\right)-\mu s} ds}. \tag{13.25}$$

As before, the quasi steady states are $I_1 = 0$ and $I_2 = N(t) - v = N_0 e^{rt} - v$ and they intersect at

$$t_c = \frac{1}{r} log(\frac{v}{N_0}). \tag{13.26}$$

We see that $t_c > 0$ if $r > 0$ and $N_0 < v$ (growing population starting below the threshold), or $r < 0$ and $N_0 > v$ (decreasing population starting above the threshold). In the first case $I_1$ is attractive for $0 < t < t_c$ and repelling for $t > t_c$, while $I_2$ is

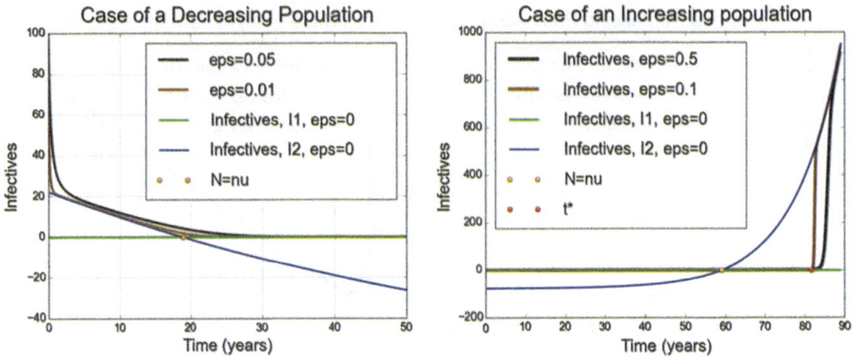


**Fig. 13.4** Decreasing (left) and increasing (right) population in the case of stability switch. One can observe an immediate switch on the left picture and a delayed one the right picture

attractive for $t > t_c$ (and negative, hence irrelevant, for $0<t<t_c$). In the latter, the stabilities are reversed.

**Case** $r > 0$. It is easy to see that $I_\epsilon$ converges to $I_1 = 0$ provided the function

$$G(t, \epsilon) = \frac{N_0\lambda}{r}(e^{rt} - 1) - \gamma t - \epsilon\mu t,$$

that appears in the exponents in (13.25), is positive. By graphing $G$, we find that there is a unique $t^* > 0$ such that $G(t^*) = 0$. However, $t^* > t_c$. This means that $I_\epsilon(t)$ stays close to $I_1 = 0$ also on $(t_c, t^*)$, where $I_1$ is already repelling. It can be proved that $I_\epsilon$ converges to $I_2(t) = N_0e^{rt} - \nu$ only for $t > t^*$. Thus

$$\lim_{\epsilon\to 0} I_\epsilon(t) = 0, \quad t \in (0, t^*), \quad \lim_{\epsilon\to 0} I_\epsilon(t) = N_0e^{rt} - \nu, \quad t \in (t^*, T]$$

for any $T < +\infty$, see the right picture in Fig. 13.4. This is the so-called delayed stability switch that is related to canard solutions (Kuehn 2015). In the presented example we observe a delay of 20 years. A failure to recognize this phenomenon while approximating the original problem by the quasi steady states may lead to significant error in prediction.

**Case** $r<0$. It is interesting to note that in this case, though $t^* > t_c$ as above, there is no delay in the switch between the quasi steady states

$$\lim_{\epsilon\to 0} I_\epsilon(t) = N_0e^{rt} - \nu, \quad t \in (0, t_c), \quad \lim_{\epsilon\to 0} I_\epsilon(t) = 0, \quad t \in (t_c, T]$$

for any $T < +\infty$, see the left picture in Fig. 13.4.

Here we used $\lambda = 0.0018$, $\gamma = 0.14$, $\beta = 0.1$, $\mu = 0.0143$, $N_0 = 0.5$ and $I_0 = 0.4$ for the increasing case and $\lambda = 0.0018$, $\gamma = 0.14$, $\beta = 0.001$, $\mu = \frac{1}{70}$, $I_0 = 100$ and $N_0 = 100$, when the population is decreasing.

These results are valid also for the full problem (13.19), as well as for other planar problem of similar structure (Banasiak and Tchamga 2017), and in higher dimensions (Banasiak and Kimba Phongi 2015).

## 13.3 Asymptotic State Lumping in Network Models

### *13.3.1 Introduction to Network Transport and Related Models*

Let us consider transport of a material along a one dimensional channel (edge). We are interested in characterizing a certain feature of this material in terms of its position along the edge and time. For instance, we can think about the density of a substance in blood at a given position in a blood vessel, or about the density of car traffic on a road network in a specific area. Mathematically, we can formulate such a problem as a system of transport equations along the edges of a (directed) graph with appropriate transmission conditions specified at its nodes.

To be more precise, we consider a directed graph $G$ without multiple edges that consists of $n$ nodes $\{w_1, w_2, \ldots, w_n\}$ and $m$ edges $\{e_1, e_2, \ldots, e_m\}$. We identify each edge with the unit interval $[0, 1]$ and thus the graph with the product of the unit intervals. Then the distribution of a material on the graph can be described by a vector function $\mathbf{u} = [u_1, u_2, \ldots, u_n]$, where each $u_i$, with the domain $[0, 1]$, describes the density of the material flowing along the edge $e_i$. The velocity of the flow can be different on each edge $e_i$ and hence it is denoted by $v_i$. The wave that propagates along each edge is described by a linear scalar transport equation and the whole system takes a form

$$\begin{cases} \partial_t u_i(t,x) + v_i \partial_x u_i(t,x) = 0, & x \in (0,1), \quad t \geq 0, \\ u_i(0,x) = \mathring{u}_i(x), \\ v_i u_i(t,0) = \sum\limits_{j=1}^{m} k_{ij} v_j u_j(t,1), \end{cases} \tag{13.27}$$

The second condition gives the initial distribution of the material along the edges, while the third one governs the transition between the edges. More precisely, the coefficient $k_{ij}$ describes the part of the material flowing from the edge $e_j$ that is directed $e_i$ provided that $e_j$ and $e_i$ are somehow connected. In particular, in the transport models mentioned above, the end of $e_j$ is the beginning of $e_i$. Then the matrix $\mathbb{K} = (k_{ij})_{i,j=1,2,\ldots,m}$ must have a special form—it must be the weighted adjacency matrix of the line digraph of $G$ (Banasiak and Falkiewicz 2015). It follows, however, that the above model is well defined for arbitrary entries $k_{ij}$,

$i,j = 1,2,\ldots,m$, and includes, in particular, the Lebovitz–Rubinov–Rotenberg model of cell maturation (Lebowitz and Rubinov 1974; Rotenberg 1983).

The well-posedness of (13.27) is studied within the framework of the semigroup theory, see e.g. (Engel and Nagel 2000; Pazy 1983). We write the system in the form called an Abstract Cauchy Problem,

$$\begin{cases} \mathbf{u}_t(t) = & [\mathcal{A}\mathbf{u}](t), \\ \mathbf{u}(0) = & \mathring{\mathbf{u}}, \end{cases} \tag{13.28}$$

where the operator $\mathcal{A}$ is defined by $\mathcal{A} = -\mathbb{V}\partial_x$, with $\mathbb{V} = \mathrm{diag}(v_i)_{i=1,2,\ldots,m}$, on the domain

$$D(\mathcal{A}_\epsilon) = \{\mathbf{u}, \mathbf{u}' \in L^1([0,1])^m : \quad \mathbf{u}(0) = \mathbb{K}\mathbf{u}(1)\}. \tag{13.29}$$

We note that the basic space has been chosen to be the space of integrable functions $\mathbb{X} = L^1([0,1])$ since the integral of the density $u_i$ along the edge $e_i$ gives the total amount of the substance present on this edge; it is expected to be the macro variable in this problem.

By Banasiak et al. (2016), (13.28)–(13.29) is well-posed for any $\mathbb{K}$ and there exists a a strongly continuous semigroup $(e^{t\mathcal{A}})_{t\geq 0}$ such that the solution is given by $\mathbf{u}(x,t) = [e^{t\mathcal{A}}\mathring{\mathbf{u}}](x)$.

In order to explain in detail how such a model fits into the theory of micro and macro characterization of complex systems, presented in the Introduction, we consider a specific model.

### 13.3.2 *Model of a Spread of Genotype in a Population of Cells with Age Structure*

Let us consider a group of cells divided into $m$ classes according to one feature of the genetic code, say $\mu_i$. For instance, $\mu_i = i$ can be the number of a specific gene responsible for drug resistance of cancer cells (Kimmel and Stivers 1994). The feature $\mu_i$ may change due to mutations and, typically, the spread of this feature is modelled at the macro scale, by a system of ordinary differential equations

$$\mathbf{y}'(t) = \mathbb{C}\mathbf{y}(t), \qquad \mathbf{y}(0) = \mathbf{y}, \tag{13.30}$$

where the matrix $\mathbb{C} = (c_{ij})_{i,j=1,2\ldots,m}$ gives the rates of mutations between classes $j$ and $i$ and $\mathbf{y} = (y_1,\ldots,y_m)$, gives the sizes of the classes $i = 1,\ldots,m$ that are unknown global variables.

However, the cells have their own vital dynamics and also the mutation process has a complex internal structure. To provide a more detailed model of the dynamics resulting from the coupling of these two processes, we describe the aging of the cells with feature $\mu_i$ as the transport along the edge $e_i$ of a graph and assume that each cell has the same lifespan so that the cells are born at the end-point of the edge with $x = 0$ and divide at the end-point with $x = 1$. By introducing the velocity of the flow on the interval $[0, 1]$ that, in this context, gives the rate of aging, we may distinguish between groups with different lifespans. When a cell reaches the end of the edge, it divides into (possibly) several offspring that may have the genetic code of their parent but also, due to mutations, may change it according to some transition matrix $\mathbb{K}$.

Our aim is to show that, indeed, (13.30) is an approximately aggregated version of an appropriately scaled version of (13.27). To achieve this we identify two time scales in the latter by assuming that the lifespan of a singe cell is much shorter than the time scale of the whole population, which will be taken as the reference time unit. It turns out, Banasiak et al. (2016), that this assumption is not sufficient to derive (13.30) and it should be supplemented by the assumption that in most cases the offspring will have the same genetic code as their parents and the mutations will change it only in a small number of daughter cells.

Incorporating the above assumptions into (13.27), we consider it with the operator $\mathcal{A}$ replaced by

$$\mathcal{A}_\epsilon = \epsilon^{-1}\mathcal{A} = \epsilon^{-1}\mathbb{V}\partial_x, \qquad \mathbb{K} = \mathbb{I} + \epsilon\mathbb{B}, \tag{13.31}$$

where $\mathbb{B} = (b_{ij})_{i,j=1,2\dots,m}$ gives the fractions of the offspring moving from $j$ to $i$ due to replication errors upon the division of the mother cell. Since for a fixed $\epsilon > 0$ such a rescaled problem has the same structure as (13.27), there is a semigroup $(e^{tA_\epsilon})_{t\geq 0}$ solving it.

First we note that the problem is singularly perturbed, as setting $\epsilon = 0$ leads to a substantially different dynamics. Thus, to derive the aggregated problem, we need to use some advanced techniques. Here we use the Trotter-Kato theorem (Bobrowski 2016), that gives conditions under which there exists a subspace $\mathbb{X}_0$ of the original state space, that we call the space of regular convergence, where the solutions $e^{tA_\epsilon}\mathring{\mathbf{u}}$, $\mathring{\mathbf{u}} \in \mathbb{X}_0$, converge to solutions to the Abstract Cauchy Problem generated by some limit operator $\mathcal{A}_0$ on $\mathbb{X}_0$. The latter operator will describe the approximately aggregated dynamics.

To apply the Trotter–Kato theorem, we prove that the operators $\mathcal{A}_\epsilon$ generate $C_0$-semigroups that are bounded independently of $\epsilon$ and that for appropriately large $\lambda$

$$\lim_{\epsilon\to 0^+} (\lambda - \mathcal{A}_\epsilon)^{-1}\mathbf{u} = (\lambda - \mathbb{V}\mathbb{B})^{-1}\mathcal{P}\mathbf{u}, \quad \mathbf{u} \in L^1([0,1])^m,$$

where $\mathcal{P}$ is a projection onto $\mathbb{R}^m$, that can be identified with the space of regular convergence $\mathbb{X}_0$, given by

$$\mathcal{P}\mathbf{u} = \int_0^1 \mathbf{u}(x)dx \in \mathbb{R}^m. \tag{13.32}$$

Details of the calculations can be found in Banasiak and Falkiewicz (2017) for a more general case including a bounded perturbation of the operator $\mathcal{A}_\epsilon$. Then the Trotter-Kato theorem gives us the following result.

**Theorem 2** *If the initial condition is independent of x on each edge; that is,* $\mathbf{u} \in \mathbb{R}^m$, *then*

$$\lim_{\epsilon \to 0^+} e^{t\mathcal{A}_\epsilon}\mathring{\mathbf{u}} = e^{t\mathbb{V}\mathbb{B}}\mathring{\mathbf{u}}$$

*uniformly on compact subsets of* $[0, +\infty)$.

We note that since $\mathbb{V}\mathbb{B}$ is a finite dimensional matrix, $(e^{t\mathbb{V}\mathbb{B}})_{t\geq 0}$ is the solution of the system of ordinary linear differential equations with the coefficient matrix $\mathbb{V}\mathbb{B}$. Thus, we have shown that the dynamics described by $(e^{t\mathcal{A}_\epsilon})_{t\geq 0}$ can be approximately described by solutions to (13.30) with $\mathbb{C} = \mathbb{V}\mathbb{B}$, albeit only for a very specific initial conditions of the original, microscopic problem. We note that the projection operator (13.32) indeed provides the expected aggregation of micro variables (the age specific density in each class) to macro variables (the total amount of cells in each class).

Translating this result into the biological language, we proved that the population of cells with uniform distribution of organisms according to age structure, can be modeled at the macro scale without loss of generality, provided the maturation rates are large. A natural question to ask is what happens if the initial distribution of the cells has a non-constant age structure? In mathematical terms, can we obtain the convergence of $(e^{t\mathcal{A}_\epsilon})_{t\geq 0}$ for any $\mathring{\mathbf{u}} \in L_1([0,1])^m$ and not only for $\mathring{\mathbf{u}} \in \mathbb{X}_0$? Although there are many models, such as a model of diffusion on the graph, see Banasiak et al. (2016), or the McKendrick model with fast migrations discussed in Sect. 13.3.3, for which the convergence can be extended to the whole space, we shall see in the examples below that in transport problems one cannot expect the convergence of the semigroup outside the regularity space.

**Example 1** In order to have an intuition about the dynamics outside the space of regular convergence for the considered system, on Fig. 13.5 we present numerical simulations of the problem with

$$\mathbb{B} = \begin{pmatrix} 1 & 0 \\ 1 & -2 \end{pmatrix}, \quad \mathring{\mathbf{u}}(x) = (e^{-x}, \sin \pi x) \tag{13.33}$$

see (13.31). We see that both components of the solution outside the space $\mathbb{X}_0$ oscillate around zero for parameters $\epsilon = 0.1$ and $\epsilon = 0.05$. This suggests that there

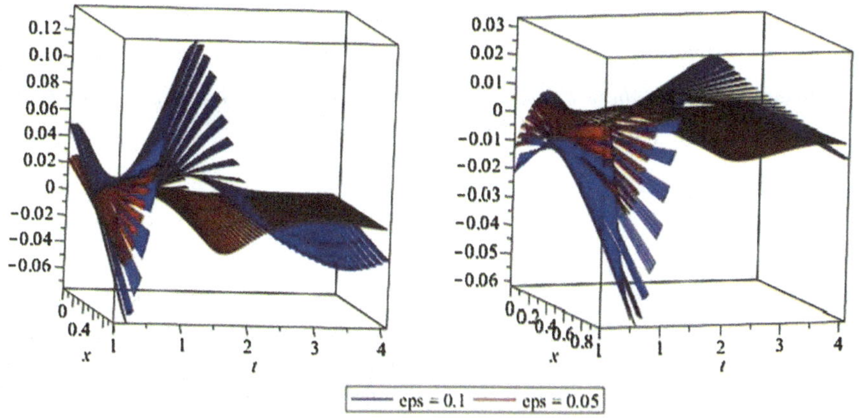


**Fig. 13.5** Components of the solution $u_1$ (left) and $u_2$ (right) to the problem in Example 13.1 for two values of $\epsilon$

may be no convergence of $(e^{tA_\epsilon})_{t\geq 0}$ for any initial condition in $\mathbb{X}$. In the next example we shall provide an analytical illustration of this fact.

**Example 2** In order to ease the calculations we consider a graph that consists of two edges $e_1$, $e_2$ with one possible direction of flow, from $e_2$ to $e_1$, with the rate equal to one. We consider problem (13.28)–(13.29) with (13.31), where

$$\mathbb{V} = \mathbb{I}, \quad \mathbb{B} = \begin{pmatrix} 0 & 1 \\ 0 & 0 \end{pmatrix}.$$

The solution for $-k \leq x - \frac{t}{\epsilon} \leq -k+1$, where $k = -\lfloor x - \frac{t}{\epsilon} \rfloor \in \mathbb{N}$, takes the form

$$\mathbf{u}_\epsilon(t,x) = \begin{pmatrix} 1 & \epsilon k \\ 0 & 1 \end{pmatrix} \mathring{\mathbf{u}}\left(x - \frac{t}{\epsilon} + k\right).$$

Let us take $x \in \left(0, \frac{1}{2}\right)$ and $t = 1$. Then, for $\epsilon_k = \frac{1}{k}$ and $\epsilon'_k = \frac{2}{2k+1}$ we obtain, respectively,

$$\lim_{k\to\infty} \mathbf{u}_{\epsilon_k}(1,x) = \begin{pmatrix} 1 & 1 \\ 0 & 1 \end{pmatrix} \mathring{\mathbf{u}}(x), \qquad \lim_{k\to\infty} \mathbf{u}_{\epsilon'_k}(1,x) = \begin{pmatrix} 1 & 1 \\ 0 & 1 \end{pmatrix} \mathring{\mathbf{u}}\left(x + \frac{1}{2}\right).$$

This indicates that there is no limit semigroup for $(e^{tA_\epsilon})_{t\geq 0}$ as $\epsilon \to 0$. The above example leads to the conclusion that a nontrivial initial age structure of the population has a significant impact on the micro picture of the dynamics. It is interesting that, in spite of this, the aggregated system (13.30) still provides a good approximation of $(e^{tA_\epsilon})_{t\geq 0}$ for arbitrary $\mathring{\mathbf{u}} \in L_1([0,1])^m$ if we only are interested in the approximation of the aggregated micro solutions—see Fig. 13.1. For this we need, however, a technical assumption on the lifespan in each class. Denote by $\tau_j = 1/v_j$ the expected life span in class $j$.

**Theorem 3** *Assume there exists $\tau \in \mathbb{R}_+$ such that each $\tau_j, j = 1, \ldots, m$, is a natural multiple of $\tau$. Then, for any $\mathbf{f} \in L^1([0,1])^m$ we have*

$$\lim_{\epsilon \to 0^+} \mathcal{P} e^{tA_\epsilon} \mathring{\mathbf{u}} = e^{t\mathbb{V}\mathbb{B}} \mathcal{P} \mathring{\mathbf{u}}, \tag{13.34}$$

*where $\mathcal{P}$ was defined in (13.32).*

The result presented above is called the asymptotic state lumping: the aggregation operator (13.32) takes the cells that in the micro model are further subdivided into groups of individuals differing from each other in age, and lumps them into the macro variables that just provide the total numbers of individuals at each edge. Equation (13.34) shows in particular that for this model the diagram in Fig. 13.1 is (approximately) commutative.

In the following example we illustrate the results of Theorem 13.3 for the problem introduced in Example 13.1 (Fig. 13.6).

**Example 3** The error of the approximation at each coordinate is presented in Fig. 13.7. It is visible that the approximations quickly converge to the solution of the aggregated model.

The result presented in Theorem 13.3 confirms that using the procedure of aggregation from the Introduction, we obtain the model describing the same dynamics as the macro model of the phenomena.

It is, however, important to observe that, in general, applying this strategy may lead to a micro model whose singular limit is completely different from the original macro model. An example of such situation is offered by the McKendrick model with geographical structure and fast migrations between the patches.

### *13.3.3 Structured McKendrick Model*

We consider a population of organisms in a fragmented environment and we divide it into $m$ groups with respect to the patch where they live. Further, we consider migrations between the patches. Such a process often is modelled at the macro level by the system of Eqs. (13.30), where this time the entries $k_{ij}$ of $\mathbb{K}$ give the rates of migrations between patches $j$ and $i$ and the solution represents the sizes of the population in each patch.

As before, we may provide a detailed micro model of the process by introducing the demographic processes in each patch. Since, in general, we consider more complex organisms, the demography is described by the McKendrick model (Kot 2001). The population is described by $\mathbf{u}(x) = (u_1(x), \ldots, u_m(x))$, where $u_i(x)$ is the age specific density of the population in patch $i$, with $x \in [0, \infty)$. The number of newborns is given by

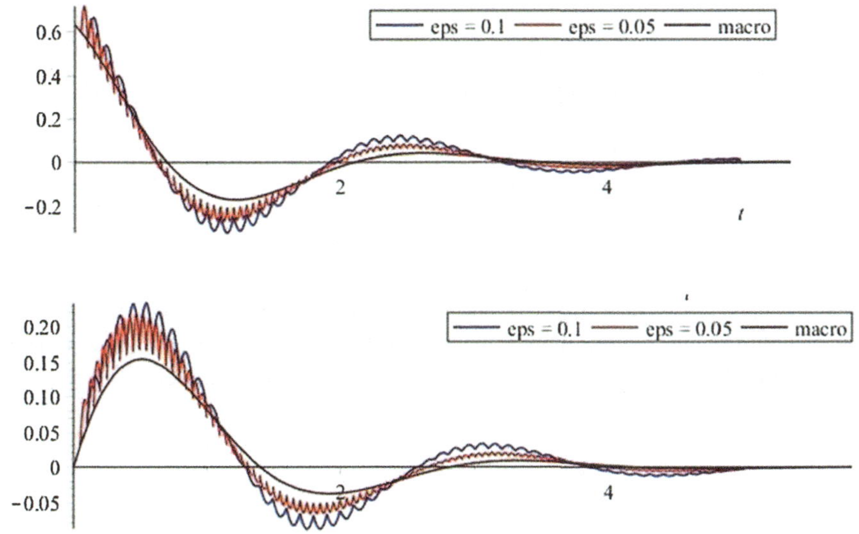


**Fig. 13.6** The macro solution (black) and two aggregated micro solutions for $\epsilon = 0.1$ (blue) and for $\epsilon = 0.05$ (red) at the two edges of the network

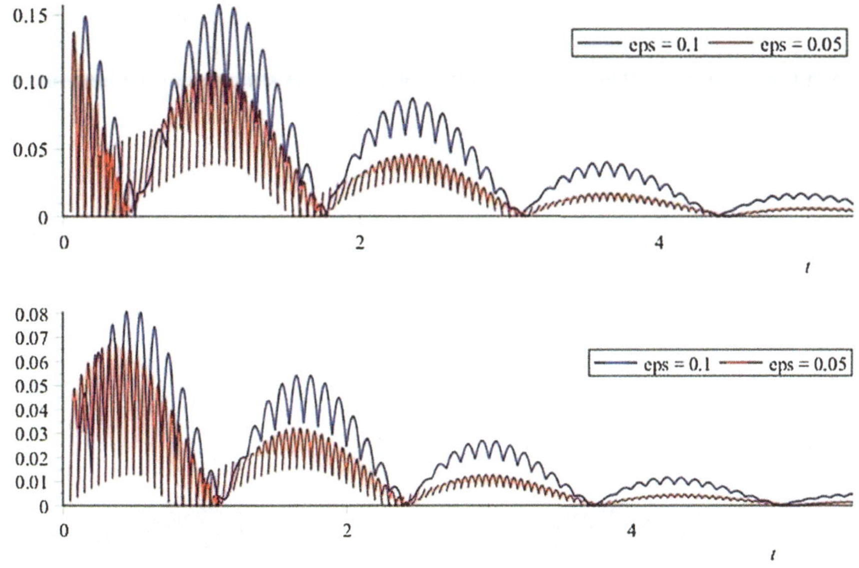


**Fig. 13.7** The error of the approximation of aggregated micro solution by macro solution on each edge

$$\mathbf{u}(0) = \int_0^\infty \mathbb{B}(x)\mathbf{u}(x)dx, \tag{13.35}$$

where $\mathbb{B}(x) = \operatorname{diag}(\beta_i(x) : i = 1, 2, \ldots, m)$ gives the patch and age specific birth rates; mortality described by a matrix $\mathbb{M} = \operatorname{diag}(m_i : i = 1, 2, \ldots, m)$. The migration between the patches is described as in the macro model, by a Kolmogorov transition matrix $\mathbb{K} = (k_{ij})_{i,j=1,2,\ldots,m}$; that is, $\mathbb{K}$ is a non-negative off-diagonal matrix whose columns sum up to 0. Assuming that the migrations between the patches occur much faster then the demographic processes (for instance, commuting between home, work, school etc.) and taking the time unit to be the characteristic time of the demographic processes, we obtain the following model

$$\begin{cases} \mathbf{u}_t(t) = & [\mathcal{B}_\epsilon \mathbf{u}](t), \\ \mathbf{u}(0) = & \mathring{\mathbf{u}}, \end{cases}$$

where the operator $\mathcal{B}_\epsilon = \operatorname{diag}(\partial_x \ldots, \partial_x) + \mathbb{M} + \frac{1}{\epsilon}\mathbb{K}$ is defined on the domain

$$D(\mathcal{B}_\epsilon) = \left\{ \mathbf{u} \in L^1([0,\infty))^m : \quad \mathbf{u}(0) = \int_0^\infty \mathbb{B}(x)\mathbf{u}(x)dx \right\}.$$

Biological heuristics suggests that the macro model should not depend on the parameters related to the geographical structure as we consider a very large inter-patch transition rates. The precise result depends on the structure of the network of migrations described by the matrix $\mathbb{K}$. In particular, if $\mathbb{K}$ is irreducible then, subject to some technical assumptions, see Banasiak et al. (2011), the total size the population $u_\epsilon = \sum_{i=1}^m u_{i,\epsilon}$ converges as $\epsilon$ goes to 0 to the solution of the following scalar system

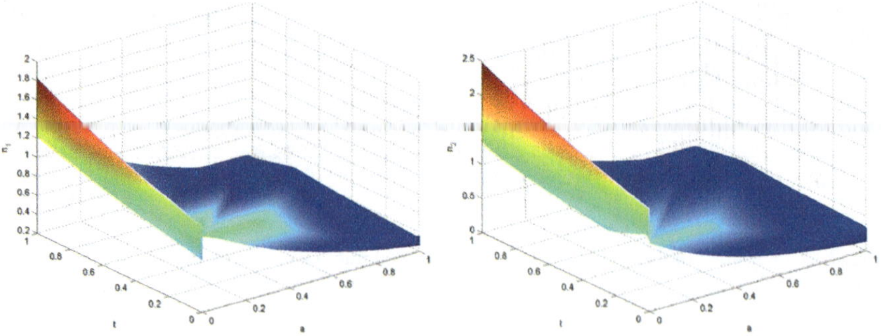

**Fig. 13.8** The components $u_1$ and $u_2$ of the solution in Example 4

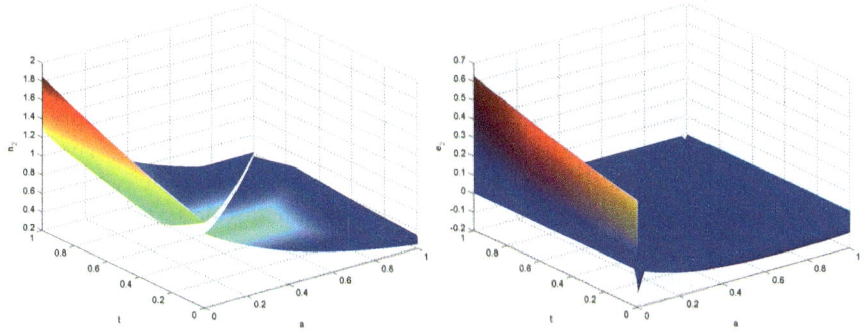

**Fig. 13.9** The aggregated density $u$ (right) given by (13.38) and the error of the approximation of $(u_1, u_2)$ by $u\mathbf{N}$ (left)

$$\partial_t u = -\partial_x u - \mu^* u, \quad u(t,0) = \int_0^\infty \beta^* u(t,x)dx, \quad u(0,x) = \sum_{i=1}^m \mathring{u}_i(x), \qquad (13.36)$$

where $\mu^* = \sum_{i=1}^m \mu_i N_i$, $\beta^* = \sum_{i=1}^m \beta_i N_i$ and $\mathbf{N} = (N_1, N_2, \ldots, N_m)$ gives the stable population distribution; that is, the approximate fractions of the total population residing in different patches (Kot 2001), $\mu^*$ and $\beta^*$ are the death and birth rates averaged with weights corresponding to the relative size of each patch.

In contrast to the Lebovitz–Rubinov–Rotenberg model, here we can also prove the convergence of solutions $\mathbf{u}_\epsilon(t,x) = [e^{t\mathcal{B}_\epsilon}\mathring{\mathbf{u}}]$ to $u(t)\mathbf{N}$ in the norm of $L_1([0,\infty))^m$ for any initial condition from the state space and for any $t > 0$.

To provide a numerical illustration, we consider a simple two dimensional problem with parameters

$$\mathbb{M} = \operatorname{diag}(1,1), \quad \mathbb{B} = \operatorname{diag}(1,2), \quad \mathbb{C} = \begin{bmatrix} -1 & 1 \\ 1 & -1 \end{bmatrix}, \quad \mathring{\mathbf{u}}(x) = (e^{-x}, e^{-2x}). \qquad (13.37)$$

In this case $\mathbf{N} = (1/2, 1/2)$ and the aggregated equation is given by

$$u_t + u_x = -u, \quad u(x) = e^{-x} + e^{-2x}, \quad u(t,0) = \frac{3}{2}\int_0^\infty u(t,x)dx. \qquad (13.38)$$

The solution $(u_1, u_2)$ to (13.36) with the data introduced in this example and $\epsilon = 10^{-3}$ is illustrated in Fig. 13.8, while the solution of the aggregated problem and the error of the approximation is given on Fig. 13.9. We can see that the approximation is poor close to $t = 0$ or $a = 0$. This is the so-called initial and boundary layer effect that can be eliminated by introducing appropriate correctors, see Banasiak et al. (2011).

## 13.4 Conclusion

Two models and their corresponding asymptotic analyses were presented in this chapter. It is important to note that naïve approaches do not always work and therefore applicability of the asymptotic theories often require a subtle mathematical analysis. Sometimes only the approximation of macroscopic variables is available. Whenever possible, the analysis of complex multiscale systems should be preceded by an assessment of the possibility of their simplification through the aggregation of equations and variables. This often requires application of advanced analytic methods but, if successful, leads to significantly simpler systems that can be solved at a lower computational cost.

## References

Aguiar, M., Ballesteros, S., Kooi, B., & Stollenwerk, N. (2011). The role of seasonality and import in a min-imalistic multi-strain dengue model capturing differences between primary and secondary infections: Complex dynamics. *Jounal of Theoretical Biology, 289,* 181–196.

Aguiar, N. S. M., & Kooi, B. (2009). Torus bifurcations, isolas and chaotic attractors in a simple dengue model with ade and temporary cross immunity. *International Journal of Computer Mathematics, 86,* 1867–1877.

Aguiar, M., & Stollenwerk, N. (2007). A new chaotic attractor in a basic multi-strain epidemiological model with temporary cross-immunity. arXiv:0704.3174v1.

Atay, F. M., & Roncoroni, L. (2016). Lumpability of linear evolution equations in Banach spaces. *Evolution Equation and Control Theory, 6*(1), 15–34.

Auger, P., Bravo de la Parra, R., Poggiale, J.-C., Sánchez, E., Nguyen-Huu, T. (2008). Aggregation of variables and applications to population dynamics. In *Structured population models in biology and epidemiology*, volume 1936 of *Lecture Notes in Mathematics* (pp. 209–263). Berlin: Springer.

Banasiak, J., & Bobrowski, A. (2009). Interplay between degenerate convergence of semigroups and asymptotic analysis: A study of a singularly perturbed abstract telegraph system. *Journal of Evolution Equations, 9*(2), 293–314.

Banasiak, J., & Falkiewicz, A. (2015). Some transport and diffusion processes on networks and their graph realizability. *Applied Mathematical Letter, 45,* 25–30.

Banasiak, J., & Falkiewicz, A. (2017). A singular limit for an age structured mutation problem. *Mathematical Bioscience Engineering, 14*(1), 17–30.

Banasiak, J., Falkiewicz, A., & Namayanja, P. (2016a). Asymptotic state lumping in transport and diffusion problems on networks with applications to population problems. *Mathematcial Models Methods Applied Science, 26*(2), 215–247.

Banasiak, J., Falkiewicz, A., & Namayanja, P. (2016b). Asymptotic state lumping in transport and diffusion problems on networks with applications to population problems. *Mathematical Models Methods Applied Science, 26*(2), 215–247.

Banasiak, J., Falkiewicz, A., & Namayanja, P. (2016c). Semigroup approach to diffusion and transport problems on networks. *Semigroup Forum, 93*(3), 427–443.

Banasiak, J., Goswami, A., & Shindin, S. (2011). Aggregation in age and space structured population models: An asymptotic analysis approach. *Journal of Evolution Equations, 11*(1), 121–154.

Banasiak, J., & Kimba Phongi, E. (2015). Canard-type solutions in epidemiological models. *Discrete and Continuous Dynamical System* (Dynamical systems, differential equations and applications. 10th AIMS Conference. Suppl.), 85–93.

Banasiak, J., & Lachowicz, M. (2014). *Methods of small parameter in mathematical biology. Engineering and Technology. Modeling and Simulation in Science*. Cham: Birkhäuser/ Springer.

Banasiak, J., & M'pika Massoukou, R. Y. (2014). A singularly perturbed age structured SIRS model with fast recovery. *Discrete and Continuous Dynamical System Series B, 19*(8), 2383–2399.

Banasiak, J., Phongi, E. K., & Lachowicz, M. (2013). A singularly perturbed SIS model with age structure. *Mathematcial Bioscience and Engineering, 10*(3), 499–521.

Banasiak, J., & Tchamga, M. S. S. (2017). Delayed stability switches in singularly perturbed predator-prey models. *Nonlinear Analaysis Real World Applications, 35,* 312–335.

Bellomo, N., Elaiw, A., Althiabi, A. M., & Alghamdi, M. A. (2015). On the interplay between mathematics and biology: Hallmarks toward a new systems biology. *Physics of Life Reviews, 12,* 44–64.

Bobrowski, A. (2016). *Convergence of one-parameter operator semigroups in models of mathematical biology and elsewhere. New Mathematical Monographs*. Cambridge: Cambridge University Press.

Borsche, R., Göttlich, S., Klar, A., & Schillen, P. (2014). The scalar Keller-Segel model on networks. *Mathematical Models Methods Applied Science, 24*(2), 221–247.

Engel, K.-J., & Nagel, R. (2000). *One-parameter semigroups for linear evolution equations*, volume 194 of *Graduate Texts in Mathematics*. New York: Springer. With contributions by S. Brendle, M. Campiti, T. Hahn, G. Metafune, G. Nickel, D. Pallara, C. Perazzoli, A. Rhandi, S. Romanelli & R. Schnaubelt.

Fenichel, N. (1979). Geometric singular perturbation theory for ordinary differential equations. *Journal of Differential Equations, 31*(1), 53–98.

Fischer, D., & Halstead, S. (1970). Observations related to pathogenesis of dengue hemorrhagic fever. v. examination of age specific sequential infection rates using a mathematical model. *Yale Journal of Biology and Medecine, 42*(5), 329–349.

Gauch, H. G. (2003). *Scientific method in practice*. Cambridge: Cambridge University Press.

Gibbons, R., Kalanarooj, S., Jarman, R., Nisalak, A., Vaughn, D., et al. (2007). Analysis of repeat hospital admissions for dengue to estimate the frequency of third or fourth dengue infections resulting in admissions and dengue hemorrhagic fever, and serotype sequences. *The American Journal of Tropical Medicine and Hygiene, 77*(5), 910–913.

Gubler, D. (1998). Dengue and dengue hemorrhagic fever. *Clinical Microbiology Reviews, 11,* 480–496.

Hoppensteadt, F. C. (1966). Singular perturbations on the infinite interval. *Transactions of the American Mathematical Society, 123,* 521–535.

Hoppensteadt, F. (1967). Stability in systems with parameter. *Journal of Mathematical Analysis and Applications, 18,* 129–134.

Iwasa, Y., Andreasen, V., & Levin, S. (1987). Aggregation in model ecosystems. I. Perfect aggregation. *Ecological Modelling, 37,* 287–302.

Iwasa, Y., Andreasen, V., & Levin, S. (1989). Aggregation in model ecosystems. II. Approximate aggregation. *IMA Journal of Mathematics Applied in Medicine* and *Biology, 6,* 1–23.

Jones, C. K. R. T., & Khibnik, A. I. (Ed.). (2001). *Multiple-time-scale dynamical systems*, volume 122 of *The IMA Volumes in Mathematics and its Applications*. New York: Springer.

Kermack, W., & McKendrick, A. (1991). Contributions to the mathematical theory of epidemics - I. *Bulletin of Mathematical Biology, 53*(1–2), 33–55.

Kimmel, M., & Stivers, D. N. (1994). Time-continuous branching walk models of unstable gene amplification. *Bulletin of Mathematical Biology, 56*(2), 337–357.

Kot, M. (2001). *Elements of mathematical ecology*. Cambridge: Cambridge University Press.

Kraemer, M. U. G., Sinka, M. E., Duda, K. A., et al. (2015). The global distribution of the arbovirus vectors aedes aegypti and ae. albopictus. *ELife, 4,* e08347.

Kuehn, C. (2015). *Multiple time scale dynamics*, volume 191 of *Applied Mathematical Sciences*. Cham:Springer.
Kumar, K., Singh, P., Tomar, J., & Baijal, S. (2010). Dengue: Epidemiology. *Asian Pacific Journal of Tropical Medicine, 3*, 997–1000.
Lebowitz, J. L., & Rubinov, S. I. (1974). A theory for the age and generation time distribution of a microbial population. *Journal of Theoretical Biology, 1*, 17–36.
Lewis, E. R. (1977). *Network models in population biology*. Berlin: Springer.
MacDonald, G. (1957). *The epidemiology and control of malaria*. London: Oxford University Press.
Martcheva, M. (2015). *An introduction to mathematical epidemiology*, volume 61 of *Texts in Applied Mathematics*. New York: Springer.
Mika, J. R., & Banasiak, J. (1995). *Singularly perturbed evolution equations with applications to kinetic theory*, volume 34 of *Series on Advances in Mathematics for Applied Sciences*. River Edge, NJ: World Scientific Publishing Co., Inc.
Pazy, A. (1983). *Semigroups of linear operators and applications to partial differential equations*, volume 44 of *Applied Mathematical Sciences*. New York: Springer.
Rocha, F., Aguiar, M., Souza, M., & Stollenwerk, N. (2013). Time scale separation and centre manifold analysis describing vector-borne disease dynamics. *International Journal of Computer Mathematics, 90*(10), 2105–2125.
Ross, R. (1911). *The prevention of malaria* (2nd ed.). London: Murray.
Rotenberg, M. (1983). Transport theory for growing cell population. *Journal of Theoretical Biology, 103*, 181–199.
Tikhonov, A. N., Vasil eva, A. B., & Sveshnikov, A. G. (1985). *Differential equations. Springer Series in Soviet Mathematics*. Berlin: Springer.
Verhulst, F. (2005). *Methods and applications of singular perturbations*, volume 50 of *Texts in Applied Mathematics*. Boundary layers and multiple timescale dynamics. New York: Springer.
Wearing, H. J., Rohani, P., & Keeling, M. J. (2005). Appropriate models for the management of infectious diseases. *PLoS Medicine, 2*(7), 0621–0627.

# Part IV
# Population, Health and Aging

# Chapter 14
# Introduction to Part IV: People-Based Systems Analysis of Health, Education and Institutions

**Wolfgang Lutz**

**Abstract** People are the agents of change and people are empowered to improve their own lives and those of others primarily through education and health.

People are the agents of change and people are empowered to improve their own lives and those of others primarily through education and health. This view also lies at the foundations of the Sustainable Development Goals as they were specified in Agenda 2030 with health and education being explicitly covered in SDG3 and SDG4. These goals lie at the heart of all SDGs because without the empowerment of people none of the other SDGs relating to economic development, quality of institutions and the environment will be possible. There is also serious concern that rapid population growth in Sub-Saharan Africa—the population is expected to increase by a factor of 2–4 over the course of this century—will make it much more difficult, if not impossible to reach the poverty, food security and economic SDGs. This consequently will have serious negative implications for all the other SDGs, including the ones relating to the environment, quality of institutions and peace. It has recently been shown that if the SDG goals relating to education and health are being met, this will—because of effect of female education on lower desired family size—lead to a strictly voluntary decline of birth rates which results in a considerably lower trajectory of world population growth (Abel et al. 2016).

In preparation for the 2002 Johannesburg Earth Summit on Sustainable Development, IUSSP (International Union for the Scientific Study of Population) and IIASA (International Institute for Applied Systems Analysis) formed a "Global Science Panel on Population in Sustainable Development" under the joint patronage of Maurice Strong and Nafis Sadik and with support from UNFPA. The conclusions of this high level panel (Lutz and Shah 2002) stressed the need to broaden the focus of population analysis to complement the head count by a consideration of what is in the heads, i.e. a focus of human capital including health, or in economic terms, adding quality to quantity.

W. Lutz (✉)
Wittgenstein Centre for Demography and Global Human Capital
(IIASA, ÖAW and WU), Vienna, Austria
e-mail: lutz@iiasa.ac.at



P. Mensah et al. (eds.), *Systems Analysis Approach for Complex Global Challenges*,
https://doi.org/10.1007/978-3-319-71486-8_14

The three papers in this section make contributions to this important new line of research from a South African perspective. They address the role of education and health related interventions for improving the quality of life of Africans. The implications of antiretroviral interventions in rural areas that are significantly enhancing survival chances is addressed by Ledwaba et al. (2018). A review of the importance that indigenous knowledge can play for maternal and child health programs is provided by Nikodem and Silaigwana (2018) and Shoko (2018) addresses population aging and the still significant differentials among the races. All three papers show how a systems analytical perspective can help to broaden the picture and gain deeper insights. They further demonstrate how concern about human wellbeing should combine a people-based approach with a proper consideration of the role of institutions in a systemic way.

## References

Abel, G., Barakat, B., KC, S., & Lutz, W. (2016). Meeting the sustainable development goals leads to lower world population growth. *Proceedings of the National Academy of Sciences (PNAS) 113*(50), 14294–14299. Available from: http://www.pnas.org/content/early/2016/11/28/1611386113.

Ledwaba, J., & Wotela, K. (2018). Evaluating Outcomes of the Antiretroviral Intervention in South Africa: A Systems Thinking Research Framework. In P. Mensah, S. Hachigonta, D. Katerere and A. Roodt (Eds.), *Systems Analysis Approach for Complex Global Challenges*. Heidelberg: Springer. Forthcoming 2018.

Lutz, W., & Shah, M. (2002). Population should be on the Johannesburg agenda (letter to the editor). *Nature 418*, 17 (4 July 2002). Other signatories are: R. E. Bilsborrow, J. Bongaarts, P. DasGupta, B. Entwisle, G. Fischer, B. Garcia, D. J. Hogan, A. Jernelov, Z. Jiang, R. W. Kates, S. Lall, F. L. MacKellar, P. K. Makinwa-Adebusoye, A. J. McMichael, V. Mishra, N. Myers, N. Nakicenovic, S. Nilsson, B. C. O'Neill, X. Peng, H. B. Presser, N. Sadik, W. C. Sanderson, G. Sen, M. F. Strong, B. Torrey, D. van de Kaa, H. J. A. van Ginkel, B. Yeoh, H. Zurayk.

Nikodem, C., & Silaigwana, B. (2018). Integrating indigenous knowledge into maternal and child health programs in Southern Africa. In P. Mensah, S. Hachigonta, D. Katerere and A. Roodt (ed), *Systems Analysis Approach for Complex Global Challenges*. Heidelberg: Springer. Forthcoming 2018.

Shoko, M. (2018). New approaches to measuring ageing in South Africa. In P. Mensah, S. Hachigonta, D. Katerere and A. Roodt (Eds.), *Systems Analysis Approach for Complex Global Challenges*. Heidelberg: Springer. Forthcoming 2018.

# Chapter 15
# Integrating Indigenous Knowledge into Maternal and Child Health Programs in Southern Africa

**Cheryl V. Nikodem and Blessing Silaigwana**

**Abstract** *Background and Significance of the topic:* Indigenous knowledge (IK) can be defined as cultivated knowledge encompassing personal experiences, observations and trial and error experiments that evolved over centuries, and that are transferred from one generation to the next. This chapter presents a systematic review on the efficacy of traditional medicinal plants (TMPs) used for the treatment of soil-transmitted helminthiasis (STH) with the intention of encouraging western health professionals to acknowledge cultural practices, and to support the use of IK, as well as the IK users' quest for justice in accepting their health belief systems. *Methodology:* The methodology utilised involved a systematic review comparing the efficacy of antihelmintic plants with albendazole and/or mebendazole, and any other antihelmintic TMPs or placebo. *Application/Relevance to systems analysis:* A systematic review is an effective way to systematically analyse the effectiveness of traditional medicinal plants (TMPs) for use as an antihelmintic. *Policy and/or practice implications:* This analysis highlights the absence of high quality randomised controlled trials (RCTs) that are so critical to evidence-based healthcare and we therefore identified a need for more professionally executed studies with larger samples: results from such superior quality trials could provide sufficient, conclusive evidence to influence policy regarding the integration of plant-based antihelmintic treatment. *Discussion and conclusion:* This study demonstrated that TMPs was just as effective as routine treatment with albendazole or mebendazole, but slightly more adverse events were reported with the use of TMPs (19.1% vs. 16.8%). It is concluded that plant-based, antihelmintic treatment has the potential to treat intestinal helminthic infections effectively in humans. However, it must be acknowledged that

**Disclaimer**: The opinions, conclusions and recommendations expressed in this review are strictly those of the authors and do not necessarily reflect those of DST, NRF or IIASA.

C. V. Nikodem (✉)
School of Health Sciences, University of Fort Hare, East London, South Africa
e-mail: CNikodem@ufh.ac.za

B. Silaigwana
School of Applied Human Sciences, University of KwaZulu-Natal, Scottsville, Pietermaritzburg, South Africa

 
P. Mensah et al. (eds.), *Systems Analysis Approach for Complex Global Challenges*,
https://doi.org/10.1007/978-3-319-71486-8_15

not one of the studies was a true, high quality RCT, and hence there is a need for research clinicians to conduct high quality, large-scale multicentre RCTs to investigate the efficacy and safety of antihelmintic TMPs in treating helminthic infections.

## 15.1 Introduction

Indigenous knowledge (IK) of the use of traditional medicinal plants (TMPs) is unique and usually transferred, verbally, from one generation to the next (Peter and Babu 2012: 145; Saslis-Lagoudakis et al. 2012; Tapfumaneyi and Rupande 2013: 579). Indigenous knowledge can be defined as cultivated knowledge encompassing personal experiences, observations and trial and error experiments, that evolved over centuries, and that are transferred from one generation to the next (Tapfumaneyi and Rupande 2013: 579). The element of secrecy surrounding the verbal transfer makes the IK transfer process vulnerable to alteration and, occasionally, some of its received wisdom can be lost at each point where knowledge transfer occurs (Tapfumaneyi and Rupande 2013: 579). As far as is known, innovative integration of the uses of TMPs for the treatment of helminthiasis in humans has not yet been explored in South Africa. Written records of the scientific effectiveness of using TMPs to treat helminthiasis are scarce; hence, there is a real need for systematic documentation to determine whether TMPs can safely be used for the treatment of soil-transmitted helminthiasis (STH). The challenge is that very few randomised controlled trials have been published on the efficacy of using TMPs to treat STH, so the aim of this systematic review is to synthesise primary research that evaluated the safety and effectiveness of TMPs for the treatment of STH in humans.

This chapter will briefly discuss the burden of helminthiasis, as well as the value and importance of conserving IK related to TMPs. Additionally, we are reporting on a systematic review on the efficacy of TMPs used for the treatment of STH with the intention of encouraging conventional health professionals to acknowledge cultural practices, and to support the use of IK, as well as the IK users' quest for justice in accepting their health belief systems.

## 15.2 Background

There are approximately 20 major helminthic infections of humans: *Ascaris lumbricoides* (roundworm), *Necator americanis* (hookworm) and *Trichuris trichiura* (whipworm) are the most common helminths found in humans (Centers for Disease Control and Prevention [CDC] 2013; WHO 2017). Pullan and co-workers (Pullan et al. 2014) estimated that, globally, about 819 million people are infected with roundworms, 464.6 million people with whipworms and 438.9 million people with hookworms.

The first stage of the basic life cycle of STH commences when the adult worms inhabit the intestines of their host. Roundworms and hookworms live inside small intestines, whilst whipworms inhabit the colon. Once inside, the adult worms mate to produce eggs, which are then disseminated into the environment through human faecal matter (Saboya et al. 2013). Victims are infected through ingestion of soil, food, vegetation, or water contaminated with these parasitic eggs (roundworms and whipworms); but, infection can also be transmitted through direct contact with the larvae or penetration into the skin (hookworms).

Traditional medicinal plants date back to time immemorial, and indigenous people have been the custodians of TMPs and the use thereof for thousands of years. Yaniv (2014: 1) claimed that the Assyrians and the Egyptians were amongst the first populations to document the use of TMPs, whilst Leyel (2013: 1) states that the practice of TMPs can be traced as far back as 60,000 years ago. There is literature to show that over 400 plants were used by the Assyrian people, and twice as many (850) by the Egyptians (Sumner 2000: 17; Yaniv 2014: 2). It is estimated that today there are about 350,000 species of plants used for indigenous medicine (Pan et al. 2013), with a global market value of USD 61 billion, which is expected to grow yearly (Chaudhary and Singh 2011: 179).

Soil transmitted helminthiasis is a global public health challenge and is currently classified as one of the seven most common, neglected tropical diseases (NTD) in the world (Sabin 2014). The highest reported incidence of helminthiasis is found in South America, China, East Asia, and Sub-Saharan Africa (World Health Organization [WHO] 2017.) Moreover, it is estimated that over 1.5 billion individuals and more than 270 million preschool age children are infected with soil-transmitted helminths globally (WHO 2017). Pullan and colleagues (Pullan et al. 2014) claim that intestinal worms cause nearly 5 million disability-adjusted life years to be lost, of which 65% are attributable to hookworm, 22% to *A. lumbricoides* and the remaining 13% to *T. trichiura.* "Helminthiasis is particularly prevalent in children who live in poverty, in overcrowded housing, or have poor sanitation and inadequate health care" (Taylor-Robinson et al. 2015; Nxasana et al. 2013). Pregnant women are also at a higher risk of developing morbidities such as anaemia and preterm delivery brought on by helminthiasis (Mofid et al. 2015). Although not a significant cause of mortality, STH remains a serious, and often silent, neglected health problem. It is ranked as the greatest cause of morbidity in children aged between five and fifteen years when compared with any other infectious disease (Taylor-Robinson et al. 2015). Children who suffer from STH may experience stunting, anaemia or impaired cognitive development and behaviour (Nxasana et al. 2013).

The World Health Organization (WHO) recommends that children living in areas where helminthiases are high should be dewormed at regular intervals (WHO 2017). However, concerns have been raised about the sustainability of periodic deworming with albendazole or mebendazole where there is evidence of low susceptibility, as this could well create resistance to the drugs (Soukhathammavong et al. 2012). The use of TMPs as a vermicide may potentially be an alternative treatment where there is evidence of such drug resistance (Adams et al. 2005; Tandon et al. 2011: 351).

South Africa was a signatory to the World Health Assembly Resolution 54, 19 in 2001, which urges member states to "ensure access to essential drugs against schistosomiasis and soil-transmitted helminth infections in all health services in endemic areas" (World Health Assembly [WHA] 2001). The resolution pronounced that all clinical cases, and groups at high risk of morbidity, such as women and children, should be treated with antihelmintic medication. Its aim was for signatory countries to attain a minimum target for the regular administration of antihelmintic medication to at least 75%, and up to 100%, of all school-age children at risk of morbidity (Nxasana et al. 2013). However, South Africa discontinued the regular deworming of school-aged children, even though it had successfully reduced the burden of STH disease in South Africa in 2003 (Nxasana et al. 2013). Subsequent studies in the Western and Eastern Cape showed that the overall infection rate of STH amongst school-going children now range between 50 and 79.6%, with about 56.2% having poly-parasitism, and it is possible that this high prevalence could be due to the discontinuation of the regular school deworming programme (Adams et al. 2005; Nxasana et al. 2013).

The initial manifestations of STH infections are subclinical and often remain undetected by clinicians for extended periods, only being recognised when severe clinical symptoms become present. Victims with low infections usually show no symptoms. Morbidity is linked to the quantity of worms hosted within a victim. Heavier STHs are known to cause major complications such as intestinal obstruction, anaemia, malnutrition, dysentery syndrome, fever, dehydration, vomiting, and colitis, and have been shown to affect physical fitness and cognitive performance of the infected children adversely (Adams et al. 2005: 276; Uneke et al. 2006; Ullah et al. 2009: 132; WHO 2017). Taylor-Robinson and associates (Taylor-Robinson et al. 2015) reported on the association between STH, on the one hand, and malnutrition, poor growth, and anaemia in children, on the other. Soil-transmitted helminths impair the nutritional status of the people they infect in multiple ways: the WHO reported that STHs affect the nutritional status and cognitive functioning of more than two billion people in more than 100 countries (WHO 2017). Mireku and colleagues (Mireku et al. 2015) reported that children at one year of age demonstrated poor gross motor functions when their mothers had been infected with hookworms during pregnancy.

The worms feed on host tissues, including blood, which leads to a loss of iron and protein. The worms increase malabsorption of nutrients. In addition, roundworms may possibly compete for vitamin A in the intestine. STH also cause loss of appetite and, therefore, a reduction of nutritional intake and physical fitness. In particular, *T. trichiura* can cause diarrhoea and dysentery which leads to malnutrition; stunting usually occurs through complications associated with long-term malnutrition over long periods, as affected by recurrent and chronic illness (Shaw and Friedman 2011). Children and pregnant women may suffer from anaemia, which reduces the blood's ability to transport oxygen (Shaw and Friedman 2011).

In most parts of the world, TMPs have been used over many years for the treatment of human diseases, and remain an important source of medicine in rural parts of developing countries (Tandon et al. 2011: 352; Ukwubile 2012). Globally, many people use TMPs for ailments, including gastrointestinal infections caused by parasitic worms (Vyagusa et al. 2013). India is one of the countries well known for its indigenous medicinal system, using its enormous resources of TMPs (Tandon et al. 2011: 351). Plants have constitute phytochemicals or biologically active substances which can be used for therapeutic purposes (Tandon et al. 2011: 351). To date, several plants used by different ethnic groups in indigenous systems of medicine have been reported to have good antihelmintic properties (Mali and Mehta 2008).

The species *Dysphania ambrosioides* (L.) Mosyakin and Clemants, also known as Wormseed (US), Epazoite (Latin America), or Paico (Peru), is a well-recognised remedy for intestinal worms (Sá et al. 2016: 533). The plant is widely distributed in the Caribbean regions, and in central and southern America; the fresh or dried parts of the plant are usually used by indigenous people to combat intestinal parasites (Sá et al. 2016: 533). The plant remedy is often prepared from leaf and root decoctions, and 300 mg of dry plant material per kg body weight has shown positive results against STH (Yadav et al. 2007: 133). *Dysphania ambrosioides* has been tested in a few small-scale, quasi-randomised clinical trials, one in U.S and two in Peru (Nakazowa 1996; de Guimaraes et al. 2001). *Dysphania ambrosioides* contains a rich source of monotepernes, whilst the fruit and seeds are rich in essential oil, which contains a bioactive chemical, ascaridole. The 'oil' of *Dysphania ambrosioides* has been used worldwide over many centuries to treat intestinal STH; phytochemicals in the plant-based antihelmintics may have both vermifugal (expel worms from the host intestine) or vermicidal (killing of worms) effects (Yadav et al. 2007: 132).

A study by Ukwubile (2012) investigating antihelminthic properties of: *Annona senegalensis* Persoon (Annonaceae); *Cucurbita ficifolia* Bouché; *Calotropis procera* (Aiton) W. T. Aiton; *Allium sativum* L. and *Seriphidium brevifolia.* Wall. ex DC. in Nigeria demonstrated that these plants had very strong potency against STH. A study done on 60 worm-infected children in Nigeria showed that seeds of *Carica papaya* L. (pawpaw) have high efficacy rates (70–100%) against STH (Okeniyi et al. 2007: 194). The authors claim that *Carica papaya* contains antihelmintic phytochemicals, namely papain and benzyl isothiocynate (Okeniyi et al. 2007: 194).

Medicinal plants contain phytochemicals which numerous studies have shown to be effective against intestinal STH (Tandon et al. 2011: 352).

## 15.3 Methodology for the Systematic Review

### *15.3.1 Criteria for Considering Studies for the Review*

We included randomised controlled trials (RCTs), cluster RCTs, quasi-RCTs, and before-and-after trials in children and adults, comparing the efficacy of anti-helmintic plants with albendazole and/or mebendazole, and any other antihelmintic TMPs or placebo. All studies reported on original data, and all eligible studies were included, regardless of language, country of origin or publication status.

### *15.3.2 Types of Participants*

Children on their own, and/or children and adults.

### *15.3.3 Types of Interventions*

The main comparison was between the use of antihelmintic TMPs (regardless of type of plant species, dosage, or method of plant preparation, i.e. infusions or decoctions of whole plants, leaves, roots, dried material, etc.) and a standard pharmaceutical antihelmintic drug or other antihelmintic TMPs or placebo. Before-and-after studies were not included in this systematic review.

### *15.3.4 Types of Outcomes*

The expected outcomes were the presence of a parasitic burden, egg-reduction rate percentage, and the reporting of any adverse effects of treatment.

### *15.3.5 Searching for Studies*

We searched the Cochrane central register of controlled trials (CENTRAL), Cochrane Infectious Diseases Group Specialized Register, the databases MEDLINE, EMBASE, EBSCO, and the website, www.ClinicalTrials.gov, using a combination of terms such as helminth, worms, random* clinical trial, and specific names of antihelmintic plants, such as: *Dysphania ambrosioides;* Paico; Wormseed; and others. In addition, we scrutinised reference lists of identified studies, and enlisted the librarians' help to search for the relevant studies.

## 15.4 Data Collection and Analysis

The authors assessed the eligibility of the articles independently by first reading the title and abstract. Thereafter, the full versions of potentially eligible articles were retrieved and assessed for inclusion, using the pre-specified criteria. Any disagreements were resolved by discussion and consensus.

The selected studies were then independently evaluated by the reviewers to assess quality and the risk of bias (selection bias, performance bias, attrition bias, reporting bias and other potential threats to validity). Bias risk was graded according to Higgens (2011) as low, high or unclear. It is very important to note that all of these studies showed very poor methodology: none stated what type of process had been used to generate a random sequence to exclude possible selection bias; none reported on the allocation concealment; and one study, Okeniyi (2007), even stated that the laboratory technician had not known the group allocation.

The results of the meta-analysis are presented as a summary risk ratio with 95% confidence intervals for dichotomous data. The $I^2$ statistic was used to measure heterogeneity among the trials, and heterogeneity was regarded as substantial if the $I^2$ statistic was above 50%. When we suspected that heterogeneity between studies was sufficient to suggest a difference in treatment effects between trials, we used random-effects meta-analysis.

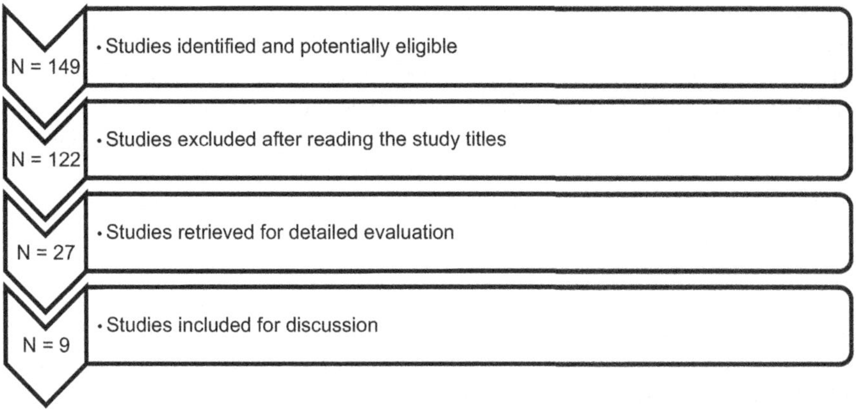


**Fig. 15.1** Flow diagram of the inclusion of studies for the systematic review

## 15.5 Results

We decided to include the studies despite the high degree of heterogeneity and the very poor quality of methodologies. Meta-analysis and statistical findings are presented but need to be interpreted with caution. A table of results, with a comprehensive narrative summary of the results for each outcome, are provided below. A total of nine studies were included for discussion (Fig. 15.1).

## 15.6 Included Studies

The first study was done in Peru and investigated the effect of *Ficus glabrate* latex mixed with aguardiente and sugar in different strengths, and mebendazole, on STH (Hansson et al. 1986). The plant medication was administered on an empty stomach for three consecutive days to 174 children and adults, between the ages of 2 and 21 years.

The second trial also took place in Peru under the supervision of de Guimaraes et al. (2001). The researchers randomly allocated 30 children between the ages of 3 and 14 to receive *Chenopodium ambriosioides* (Paico) leaf extract, diluted in juice, at 1 ml/kg/day for children under 10 kg, and 2 ml/kg/day for children weighing more than 10 kg, for three consecutive days before breakfast. The control group also consisted of 30 children between the ages of 3 and 14. Children under five received a single dose of 200 mg Albendazole, and those of five and older, 400 mg.

The third study was conducted in Nigeria on 60 children aged between 3 and 6 years (Okeniyi et al. 2007). The 30 children who were allocated to the intervention group were given 20 ml of an elixir of air dried *Carica papaya* seeds (4 g) and honey, whilst the 30 children in the control group were given only 20 ml of honey.

The fourth study was conducted in China on children and adults between the ages of 10 and 74 years (Li et al. 2012). All treatment was given in the morning on an empty stomach. Participants in the intervention group (115) were given three different compounds at 40-min to 1-h intervals; compounds included 120 g of peeled, raw pumpkin seeds taken by chewing; 200 ml of areca nut extract, prepared as a soup (80 g of areca nut slices were placed in 600 ml water and decocted on medium heat until 200 ml liquid was left); and a magnesium sulfate solution at a dose of 0.5 g/kg body weight. After completion of treatment, each adult patient was asked to drink about 1.5 L water.

The study had two control groups: Control Group A consisted of 12 participants who received pumpkin seeds alone followed by a magnesium sulfate solution. Control Group B consisted of 11 participants who received areca nut extract alone, followed by a magnesium sulfate solution.

The fifth study in the systematic review was a combination of four studies, as all four reported the same outcomes (Srichaikul 2011, 2012a, b; Samappito 2012). Results of these studies should be interpreted with care, due to flaws in the research methodology as well as in the presentation of data. The methodology was very unclear, but it appears that there were 15 participants, aged between 16 and 65 years. The intervention group of five participants received 3–6 capsules (500 mg) of Thai traditional formula medicine (TTFM) daily before breakfast for three days. The TTFM is a mixture of seven plants and purified water, which is then dried and capsulated. The medicinal plants were: *Terminalia chebula* (Retz) (SamaoThai), *Terminalia Citrina* (Roxb.) (Samao Ted) (Arjun), *Curcuma zedoaria* (Berg) Rosco. (Kamin Aoi), *Terminalia Citriva* (Gaertn). Roxb. Flem (Samao Ngu), *Cuttle Bone* (Tricosan) (Lintalay), *Croton tiglium* Lin. (Purging croton) and *Diospyros mollis* (Griff.) (Ebony tree). Control Group A consisted of five participants who received 60–90 mg of *Areca catechu* Linn (Betel Nut) seeds, in dried powder capsules; this was mixed with syrupy water and taken once daily for three days. Control Group B also comprised five participants, and they were given 500 mg Mebendazole for three days.

## 15.7 Results of the Meta-Analysis

Four studies reported on the presence of helminths in faeces after treatment, and showed that the use of plant medicine was just as effective as routine treatment. A total of 59.6% of the samples showed the presence of worms after treatment with plant medicine, whilst 57.8% of the stool samples from the routine control group still revealed the presence of worms. This difference was not statistically significant, with a relative risk ratio of 0.80 (CI 0.37–1.70) (Fig. 15.2).

Five studies reported on adverse events. Slightly more adverse events were reported with TMPs (19.1% vs. 16.8%), but again the difference was not statistically significant, with an RR of 1.43 (CI 0.21–9.65) (Fig. 15.3). The most common adverse effects were diarrhoea, nausea, and vomiting.

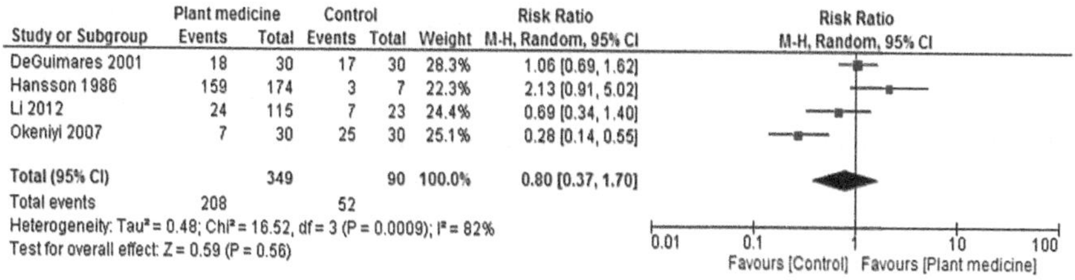


**Fig. 15.2** Number of participants who had worms present after treatment

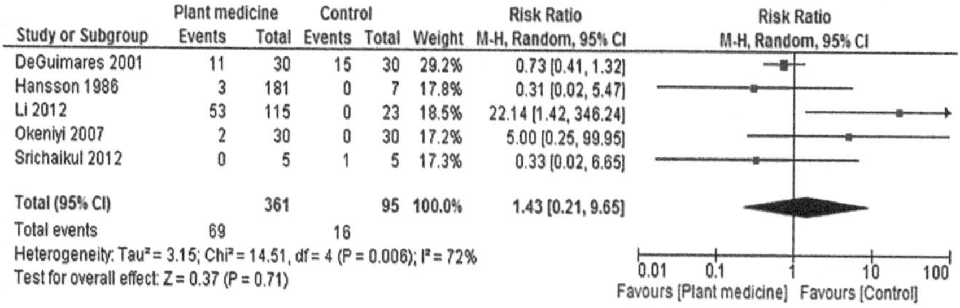


**Fig. 15.3** Adverse events

## 15.8 Discussion

The main research question which motivated this systematic review was: How efficacious are antihelmintic plants? This review showed that the predominant outcome arising from the included studies was the parasitic reduction rate, alternatively interpreted as the cure rate. The meta-analysis demonstrated that there was no statistically significant difference in the eradication of STH between pharmaceutical medicine and TMPs. Slightly more adverse events were reported amongst participants who received TMPs, but the differences were also not statistically significant.

This review also highlighted serious methodological flaws, particularly with the small sample sizes of most of the included studies. Such underpowered trials could influence the results of our meta-analysis, and we therefore identified a need for more professionally executed studies with larger samples: results from such superior quality trials could provide sufficient, conclusive evidence to influence policy regarding the integration of plant-based antihelmintic therapy into primary healthcare programs, especially in areas where there was very limited access to healthcare amenities. Another flaw was the possible bias emanating from the diagnostic method used: all the studies included in our review relied on a single stool sample examination for the presence of parasites, before and after treatment with experimental medicinal plants. The diagnostic method used in these studies may have limited sensitivity, particularly when the infection intensity was low (Keiser and Utzinger 2008).

Our review showed that very few RCT studies evaluating the efficacy of anti-helmintic plants, have been published. It could be that editors do not wish to publish studies showing non-significant differences; or simply that very few researchers have bothered to investigate this topic, despite evidence to prove that the majority of people, especially on the African continent, depend on TMPs to cure ailments such as STH.

This systematic review highlights the absence of high quality RCTs that are so critical to evidence-based healthcare. We therefore recommend that public health experts and clinicians conduct large-scale RCTs, evaluating how efficacious medicinal plants are against intestinal worms. Considering the risk of building resistance to the few existing drugs, there is a need to evaluate the use of TMPs, which may potentially offer new, safe, and readily available remedies.

## 15.9 Conclusions and Recommendations

We conclude that plant-based, antihelmintic treatment has the potential to treat intestinal helminthic infections effectively in humans, as was shown by the efficacy of *Dysphania ambrosioides*, *Carica papaya,* and pumpkin seeds. These results have important policy implications for public health experts when considering the integration of TMPs into primary healthcare programs. However, it must be acknowledged that not one of the studies was a true, high quality RCT, and hence, we also conclude that there is a need for research clinicians to conduct high quality, large-scale multicentre RCTs to investigate the efficacy and safety of antihelmintic TMPs in treating helminthic infections.

**Acknowledgements** We are grateful to the Southern African Young Scientist Summer Programme (SA-YSSP) in conjunction with the Department of Science and Technology (DST) and National Research Foundation (NRF) for funding. Sincere gratitude also goes to the University of Free State librarians, Hesma and Annamarie, for their assistance in searching for studies to be included in this review.

## References

Adams, V. J., Markus, M. B., Adams, J. F., Jordaan, E., Curtis, B., Dhansay, M. A., et al. (2005). Paradoxical helmenthiasis and giardiasis in Cape Town, South Africa: epidemiology and control. *African Health Sciences, 5*(3), 276–280.

Centers for Disease Control and Prevention [CDC]. (2013). *Parasites-soil transmitted helminths (sth's).* Global Health-Division of Parasitic Diseases and Malaria. Retrieved June 28, 2016, from http://www.cdc.gov/parasites/sth.

Chaudhary, A., & Singh, N. (2011). Contribution of world health organization in the global acceptance of ayurveda. *Journal of Ayurveda and Integrative Medicine, 2*(4), 179–186.

de Guimaraes, D. L., Llanos, R. S., & Acevedo, J. H. (2001). Ascariasis: comparison of the therapeutic efficacy between paico and albendazole in children from Huaraz. *Review Gastroenterology Peru, 21*(3), 1–6.

Hansson, A., Veliz, G., Naquira, C., Amren, M., Arroyo, M., & Arevalo, G. (1986). Preclinical and clinical studies with latex from Ficus Glabbrata HBK, a traditional intestinal anthelminthic in the Amazonian area. *Journal of Ethnopharmacology, 17,* 105–138.

Higgens, J. P., Altman, D. G., Gøtzsche, P. C., Jüni, P., Moher, D., Oxman, A. D., Sterne, J. A. et al. (2011). The Cochrane Collaboration's Tool for Assesing Risk of Bias in Randomised Trials. *British Medical Journal, 343*. http://dx.doi.org/10.1136/bmj.d5928.

Keiser, J., & Utzinger, J. (2008). Efficacy of current drugs against soil-transmitted helminth infections. *JAMA, 299*(16), 1937–1948. https://doi.org/10.1001/jama.299.16.1937.

Leyel, C. F. (2013). *Herbal delights—Tisanes, syrups, confections, electuaries, robs, juleps, vingars and conserves*. Worcestershire: Read Books LTD.

Li, T., Ito, A., Chen, X., Long, C., Okamoto, M., Raoul, F., et al. (2012). Usefulness of pumpkin seeds combined with areca nut extract in China. *Acta Tropica, 124,* 152–157.

Mali, R. G., & Metha, A. A. (2008). A review on anthelmintic plants. *Natural Product Radience*, 466–475.

Mireku, M. O., Boivin, M. J., Davidson, L. L., Ouédraogo, S., Koura, G. K., Koura, M. J., Bodeau-Livinec, F. et al. (2015). Impact of helminth infection during pregnancy on cognitive and motor functions of one-year-old children. *PLoS Neglegted Tropical Diseases, 9*(3). https://doi.org/10.1371/journal.pntd.0003463.

Mofid, L. S., Casapia, M., Montresor, A., Rahme, E., Fraser, W. D., Marquis, G. S., et al. (2015). Maternal deworming research study (MADRES) protocol: A double blind, placebo-controlled randomised trial to determine the effectiveness of deworming in the immediate post partum period. *BMJ*. https://doi.org/10.1136/bmjopen-2015-008560.

Nakazowa, R. A. (1996). Traditional medicine in the treatment of enteroparatisitosis. *Peru Journal of Gastroenterology, 16*(3), 1–7.

Nxasana, N., Baba, K., Bhat, V. G., & Vasaikar, S. D. (2013). Prevalence of intestinal parasites in primary school children of Mthatha, Eastern Cape Province, South Africa. *Annals of Medical and Health Sciences Research, 3*(4), 511–516. https://doi.org/10.4103/2141-9248.122064.

Okeniyi, J. A., Ogunlesi, T. A., Oyelami, O. A., & Adeyemi, L. A. (2007). Effectiveness of dried Carica papaya seeds against human intestinal parasitosis: A pilot study. *Journal of Medicinal Food, 10*(1), 194–196.

Pan, S., Zhou, S., Gao, S., Yu, Z., Zhang, S., Tang, M., Ko, K. et al. (2013). New perspectives on how to discover drugs from herbal medicines: CAM's outstanding contribution to modern therapeutics. *Evidence-Based Complementary and Alternative Medicine*, 1–25.

Peter, K. V., & Babu, K. N. (2012). Introduction. In K. V. Peter (Ed.), *Handbook of herbs and spices* (Vol. 2, p. 145). Boca Raton: Woodhead Publishing Ltd.

Pullan, R. L., Smith, J. L., Jasrasaria, R., & Brooker, S. J. (2014). Global numbers of infection and disease burden of soil transmitted helminth infections in 2010. *Parasites & Vectors, 7*(37), 1–19. Retrieved June 14, 2014, from https://parasitesandvectors.biomedcentral.com/articles/10.1186/1756-3305-7-37.

Sá, R. D., Santana, A. S., Silva, F. C., Soares, L. A., & Randau, K. P. (2016). Anatomical and histochemical analysis of Dysphania ambrosioides supported by light and electron microscopy. *Revista Brasileira de Farmacognosia, 26,* 533–543.

SABIN vaccine institute. (2014). *Global Network Neglected Tropical Diseases.* Retrieved June 9, 2014, from http://www.globalnetwork.org/neglected-tropical-diseases/fact-sheets.

Saboya, M. I., Catala, L., Nicholls, R. S., & Ault, S. K. (2013). Update on the Mapping of Prevalence and Intensity of Infection for Soil-Transmitted Helminth Infections in Latin America and the Caribbean: A Call for Action. *PLoS Neglected Tropical Diseasis, 7*(9). https://doi.org/10.1371/journal.pntd.0002419.

Samappito, S., Srichaikul, B., Viroj, J., & Bakker, G. (2012). Comparative study of double blind clinical trial in side-effects among Areca catechu l., Thai traditional herbal formula and mebendazole. *Trop Parasitololgy, 2*(2), 116–118. https://doi.org/10.4103/2229-5070.105176.

Saslis-Lagoudakis, C. H., Savolainen, V., Williamson, E. M., Forest, F., Wagstaff, S. J., Baral, S. R., Hawkins, J. A. et al. (2012). Phylogenies reveal predictive power of traditional medicine in bioprospecting. *Early Edition Journal*, 1–6. https://doi.org/10.1073/pnas.1202242109.

Shaw, J. G., & Friedman, J. F. (2011). Iron Deficiency Anemia: Focus on Infectious Diseases in Lesser Developed Countries. *Anemia*. https://doi.org/10.1155/2011/260380.

Soukhathammavong, P. A., Sayasone, S., Phongluxa, K., Xayaseng, V., Utzinger, J., Vounatsou, P., Odermat, P. et al. (2012). Low efficacy of single-dose Albendazole and Mebendazole against hookworm and effect on concomitant helminth infection in Lao PDR. *PLoS Neglegted Tropical Diseases, 6*(1). https://doi.org/10.1371/journal.pntd.0001417.

Srichaikul, B., Samappito, S., Bakker, G., Seubsoh, W., & Boonsong, K. (2012a). Comparative Double Blind Trial in Anthelmintic Efficacy Between Mebendzole and Areca catechu L. *Research Journal of Medicinal Sciences*, 124–128.

Srichaikul, B., Samappito, S., Nitiketkoon, C., Wongyai, S., & Bakker, G. (2012b). The Comparative Double Blind Clinical Trial of Anthelmintic Efficacy Among Mebendazole, Thai Traditional Herbal Formulae and Areca Catechu L. *Advances in Natural Science*, 51–55. https://doi.org/10.3968/j.ans.1715787020120502.1339.

Srichaikul, B., Samappito, S., Wongyai, S., & Bakker, G. (2011). The Comparative Double Blind Clinical Study in Anthelmintic Efficacy between Thai Traditional Herbal Formulae and Mebendazole. *Research Journal of Biological Sciences*, 552–557.

Sumner, J. (2000). *The natural history of medicinal plants*. Portland: Timber Press.

Tandon, V., Yadav, A. K., Roy, B., & Das, B. (2011). Phytochemicals as cure of worm infections in traditional medicine systems. *Emerging trends in Zoology*, 351–378.

Tapfumaneyi, K. D., & Rupande, G. (2013). Making libraries more relevant to communities, the inclusion of indigenous knowledge in library information services - the potential benefits and challenges: An Afrocentric librarian's perspective. *International Journal of Advanced Research, 1*(5), 579–586.

Taylor-Robinson, D. C., Maayen, N., Soares-Weiser, K., Dongegan, S., & Garner, P. (2015). Deworming drugs for soil-transmitted intestinal worms in children: effects on nutritional indicators, haemoglobin, and school performance. *Cochrane Database Systematic Review, 7*, 1–157. https://doi.org/10.1002/14651858.CD000371.pub6.

Ullah, I., Sarwar, G., Aziz, S. & Khan, M. H. (2009). Intestinal worm infestation in primary school children in rural Peshawar. *Gomal Journal of Medical Sciences 7*(2), 132–136.

Ukwubile, C. A. (2012). Anti-Helminthic Properties of Some Nigerian Medicinal Plants on Selected Intestinal Worms in Children (Age 5–13) in Ogurugu, South East Nigeria. *Bacteriology & Parasitology, 3*(9). http://dx.doi.org/10.4172/2155-9597.1000159.

Uneke, C., Eze, K., Oyibo, P., Azu, N., & Ali, E. (2006). Soil-transmitted helminth infection in school children in South-Eastern Nigeria: The public health implication. *The Internet Journal of Third World Medicine, 4*(1).

Vyagusa, D. B., Mubyazi, G. M., & Masatu, M. (2013). Involving traditional birth attendants in emergency obstetric care in Tanzania: policy implications of a study of their knowledge and practices in Kigoma Rural District. *International Journal for Equity in Health, 12*(83).

World Health Assembly [WHA]. (2001). *WHA54.19Schistosomiasis and soil-transmitted helminth infections*. Geneve: World Health Assembly. Retrieved from www.who.int/neglected_diseases/mediacentre/WHA_54.19_Eng.pdf.

World Health Organization [WHO]. (2017). *Soil-transmitted helminth infections. Retrieved January 2017, from*. Retrieved July 2017, from Media centre: http://www.who.int/mediacentre/factsheets/fs366/en/.

Yadav, N., Vasudeva, N., Singh, S., & Sharma, S. K. (2007). Medicinal properties of genus Chenopodium Linn. *Natural Product Radiance, 6*(2), 131–134.

Yaniv, Z. (2014). *Introduction: Medicinal plants in ancient traditions. In: Medicinal aromatic plants of the Middle-East*. Dorderecht, Nederlands: Springer.

# Chapter 16
# Evaluating Outcomes of the Antiretroviral Intervention in South Africa: A Systems Thinking Research Framework

**Johanna Ledwaba and Kambidima Wotela**

**Abstract** *Background and Significance of the topic:* HIV has become one of the most serious health and social development challenges in the world. Sub-Saharan Africa accounts for two-thirds of global HIV infections and almost half of these are in Southern Africa. Within the region, South Africa has the largest number (7.1 million) of people infected with HIV. The antiretroviral treatment intervention aims to reduce HIV transmission, improve the quality of life of people living with HIV, and to prolong life expectancy. To date, 3.4 million HIV-infected people receive antiretroviral treatment in South Africa. Notable literature is available on the outcomes of this intervention at country level and key urban areas. However, there is little, if any, knowledge on the outcomes in rural areas. Ultimately, this research intends to assess the outcomes of the antiretroviral treatment intervention in a South African rural setting called Mankweng in Limpopo Province using data abstracted from medical records to track viral load trends and to identify factors that affect virus suppression. The main focus of this paper is to propose a conceptual framework that should guide such an assessment. *Methodology:* We apply systems thinking approach to derive the theoretical framework and more importantly the conceptual framework in preparation for an empirical assessment of the antiretroviral intervention (ART) in a South African rural setting. Further, we then rely on summative content analysis to interrogate the academic and non-academic literature. *Application/Relevance to systems analysis:* The system thinking provides for a general approach that emphasises the product (theoretical and conceptual framework) rather than the process (literature review). More specifically, it allows us to integrate more meaningfully both the physical and academic context to deriving the theoretical and conceptual framework. *Policy and/or practice implications:* In general, development interventions established elsewhere must be contextualised

J. Ledwaba (✉) · K. Wotela
WITS Graduate School of Governance, University of the Witwatersrand (WITS), Johannesburg, South Africa
e-mail: johanna.ledwaba@gmail.com



P. Mensah et al. (eds.), *Systems Analysis Approach for Complex Global Challenges*,
https://doi.org/10.1007/978-3-319-71486-8_16

for the African continent and its sub-regions in order to achieve the best outcomes. *Discussion and conclusion:* Prominent in our interrogation of literature on outcomes of HIV interventions is the lack of a theory of change when undertaking summative evaluations. More specifically, the outcomes of the antiretroviral treatment interventions in rural areas remain unreported.

## 16.1 Introduction

According to the Joint United Nations Programme on HIV/AIDS (UNAIDS) (2017), HIV continues to be a severe global public health concern. By 2016, an estimated 36.7 million people were living with HIV globally. Sub-Saharan Africa remains the epicentre of the HIV epidemic with an estimated 25.5 million people infected with HIV. 70% of global HIV infections are localised in Sub-Saharan Africa, with Southern Africa accounting for almost two-thirds of all people living with HIV and AIDS worldwide. UNAIDS (2017) point out that within the Southern Africa region, South Africa has the largest number (7.1 million, 12.9%) of people infected with HIV. In 2012, Kwazulu-Natal province had the highest HIV prevalence (16.9%), while Western Cape (5.0%) had the lowest, followed by Northern Cape (7.4%) then Limpopo (9.2%) (Shisana et al. 2014).

The spread of HIV is driven by a number of factors. Some of the main risk factors for HIV infection in Sub-Saharan Africa include poverty, famine, low status of women in society, corruption, naive risk taking perception, resistance to sexual behaviour change, high prevalence of sexually transmitted infections (STI), internal conflicts and refugee status, antiquated beliefs, lack of recreational facilities, ignorance of individual's HIV status, child and adult prostitution, uncertainty of safety of blood intended for transfusion, widow inheritance, circumcision, illiteracy and female genital cutting and polygamy (Nyindo 2005).

In 2015, South Africa's poverty levels were 55.5% and a Gini Coefficient of 0.64 (StatsSA 2017). Poverty and inequality impede development, in addition to certain cultural and social norms hindering change in the HIV landscape in South Africa. To counter the scourge, South Africa has put in place a number of programmes and interventions to treat and mitigate the spread of HIV. In particular, the antiretroviral (ARV) treatment intervention aims to reduce HIV transmission, improve the quality of life of people living with HIV, and prolong life expectancy. To date, 3.7 million HIV-infected people receive antiretroviral treatment in South Africa. Studies have been conducted to assess outcomes and impact of the antiretroviral treatment intervention at country level and key urban areas (Fox et al. 2012b; Johnson et al. 2013; Sherman et al. 2014). These show that antiretroviral therapy has made a significantly positive difference to people living with HIV by *inter alia* reducing the mother-to-child transmission rate to 2.4% in 2012 and increasing life expectancy. However, existing gaps in knowledge about the outcomes of antiretroviral drugs in rural areas need to be addressed. Ultimately, this research intends to assess the outcomes of the antiretroviral intervention (ART) intervention in a South African rural setting called Mankweng in Limpopo Province using data abstracted from medical records to track viral load trends and to identify factors that influence virus suppression. This paper proposes a conceptual framework that should guide an assessment of outcome of the ART intervention prior to conducting the research. A conceptual framework will provide a roadmap to collecting, processing, and analysing data and information to assess and, thereafter, formulate a theoretical framework that allows interpretation of empirical findings. Rather than presenting and discussing empirical research results, the product is an interpretive framework and, consequently, a conceptual framework resulting from a systematic interrogation of academic and non-academic literature as described in Wotela (2017).

## 16.2 The Approach

Invariably, this research is an outcome evaluation—outcomes evaluation measures progress towards the impact or goal of the intervention by focussing on the short-term results of the intervention (Rossi et al. 2003). It is appropriated from a summative evaluation which assesses whether the effected change is as a result of the intervention (Patton 2008). To enable measurement of outcomes a results chain is used to understand the achievement of desired outcomes of an intervention. Further, the results chain links the results achieved to the theory of change that underpins the intervention. Therefore, this research will apply the results-chain of South Africa's ART programme and it's accompanying theory of change to interpret the empirical results that emanate from this evaluation. A conceptual framework that details steps to be followed while evaluating outcomes of the ART will be derived prior to conducting the evaluation. Before then, in this paper we derive a conceptual framework for evaluating outcomes of the antiretroviral intervention in a South African rural setting.

Technically, the main outcome of a literature review is a conceptual framework —defined implicitly by Kumar (2014) as an advanced outline of how a research should proceed after we have interrogated key literature on the research of interest. This implies that this outline resulting from a detailed understanding (and justifying) of the research problem, identifying the knowledge gap, and then developing an explanatory or theoretical framework for interpreting the research findings (Wotela 2017). Inferring from this, Wotela (2016) has applied systems thinking, described in Gharajedaghi (2006), to develop a conceptual framework or a literature review process that focuses on the following seven sub-components:

1. **Research setting or context analysis** to understand the physical research setting or context.
2. **Research problem analysis** to understand, streamline, and justify the research problem or opportunity or question that one intends to pursue.
3. **Research knowledge gap analysis** to establish the knowledge gap through reviewing past and current attempts on a similar research.
4. **Establishing and systematically discussing the academic field of study in which the research is located** to give the research an academic context for purposes of contextualising the attributes and variables of interest as well as the interpretative framework.
5. **Establishing and discussing the key research attributes and/or variables** to understand the information or data that one needs to answer the research questions or test the research hypotheses or prove the research propositions.
6. **Establishing and developing an interpretive (theory, model, perspective, or construct) framework** to apply when interpreting the empirical research findings.
7. **Summarising Steps 1 to 6 (above) into a conceptual framework**—that is, an outline of how the research or evaluation should proceed based on discussions and decisions emanating from the interrogated literature.

In each of these seven areas of focus, we apply a thematic summative content analysis (Hsieh and Shannon 2005) when synthesising the literature and reporting. The approach is thematic because we have devised key questions against which we interrogate and synthesise literature by theme. It is summative because we have derived these themes prior to commencing the literature review but we also allow and include those that arise during literature review. Lastly, it is 'content' because when reviewing literature we focus on explicit and implied subjective interpretation of the discussions.

We, therefore, apply this proposed outcomes-based literature review to develop a framework for interpreting our anticipated empirical research results, and more importantly, a conceptual framework to prepare for assessing the outcomes of ART intervention at Mankweng. We begin with an understanding of the context (Mankweng Hospital, Limpopo Province South Africa), followed by the ART intervention. Second, to establish the knowledge gap, we interrogate research approaches, designs, procedures and methods applied, as well as findings and conclusions realised by past and current studies on evaluations of ART. Apart from establishing the knowledge gap, we also use this interrogation to consider methodological options that we can employ for our assessment. Third, we propose and detail a framework that will facilitate the interpretation of the empirical research findings on the outcomes of the antiretroviral therapy in a South African rural setting. Worth mentioning, is the linkage between this interpretive framework and our key variables whose information we will collect during the research. Lastly, for now, we derive a conceptual framework that will guide the research when collecting, processing, analysing, and interpreting empirical results emanating from evaluating outcomes of the antiretroviral intervention in a South African rural setting.

## 16.3 Physical Research Setting/Context Analysis: Mankweng Hospital in Context

Mankweng Hospital was built in 1988 under the apartheid regime and commissioned by the then Lebowa government. It is a tertiary hospital found in the Capricorn District of the Limpopo Province under Polokwane local municipality. It is located in Mankweng Township, 30 km outside Polokwane city. Mankweng has an estimated population of 33,738 of whom 31.3% had no income in 2011 (Statistics South Africa (census) 2011). It provides healthcare services to Mankweng Township and seven surrounding villages as well as tertiary healthcare to all of Limpopo Province. The Portfolio Committee on Health (2011) reported that Mankweng hospital had only 509 beds and yet it was able to provide healthcare services to over 10,000 people from the surrounding communities. Without exception, the hospital also faces challenges of dilapidated infrastructure, shortage of medical professionals, procurement issues and disorganisation and poor managerial skills (Angoria medical technology hub 2014).

The physical research setting is located in the second poorest province in South Africa. It is located in the outskirts of the city centre and provides healthcare to the surrounding disadvantaged communities. It also serves as a tertiary hospital by providing a vast array of services including HIV/AIDS care and treatment. Further, the hospital also experiences challenges common to government healthcare, in particular decaying infrastructure, inadequately medical staff and retention of healthcare workers, and procurement.

## 16.4 The History and Description of the HIV Epidemic and Its Interventions

The first case of HIV was discovered in 1981 in the USA with South Africa identifying its first case a year later (Ras et al. 1983). The Joint United Nations Programme on HIV/AIDS (2017) estimated that in 2016 alone, 36.7 million people were infected with HIV and 1 million died from AIDS, globally. In view of the alarming HIV and AIDS statistics, there is evidence of disparities in the number of HIV infections in the world, in particular Africa. Sub-Saharan Africa bears the brunt of the HIV and AIDS with 79% of global infections concentrated in this region. The disease hit Southern Africa hard, with South Africa incurring 7.1 million HIV infections globally (UNAIDS 2017). Many of these infections occur among females aged 30–34 years (Shisana et al. 2014). In response to the epidemic, HIV interventions that focused on prevention, diagnosis, and treatment were initiated as early as 1990 and mid-2000, respectively (Drimie 2002). These were labelled as priority interventions by the World Health Organisation as they offered a comprehensive response to the HIV epidemic (World Health Organisation 2010).

The prevention intervention sought to avert HIV transmission through behavioural change in people who are vulnerable to HIV infection. This was done using the ABC approach—Abstinence, Be faithful and Condomise (Bonell and Imrie 2001). However, due to the complex nature of the HIV epidemic it became evident that effective HIV prevention needed a holistic approach that took into account other contextual factors (social, cultural and political). In the mid-2000, the combination of interventions adopted to reduce HIV transmission were *inter alia* ART and voluntary medical male circumcision (Joint United Nations Programme on HIV/AIDS 2010). A study by Marks and associates (Marks et al. 2006) highlights the importance of early HIV diagnosis to reinforce prevention interventions. Prevalence can be reduced by increasing the number of people who are aware of their HIV status, which further reduces risky behaviour. Moreover, literature shows that early HIV diagnosis presents an opportunity for people to start ART, leading to improved health outcomes and quality of life (Palella et al. 2003). Lastly, there is consensus that the ART programme can essentially control the HIV epidemic (Cohen et al. 2011).

Following a long debate on AIDS policy between the government and AIDS activists, the South African government finally introduced the HIV and AIDS Comprehensive Care, Management, and Treatment (CCMT) Plan in 2004, with a "goal to provide comprehensive care and treatment for people living with HIV and AIDS and to facilitate the strengthening of the national health system in South Africa" (DoH 2003, p. 24). The adoption of HIV and AIDS Care, management and treatment plan led to the development of the National Strategic Plan 2007–2011 that promoted access to ART and reduction of HIV infections by half (Nattrass 2008). The National Strategic Plan was developed to fast track implementation of the AIDS policy. The established CCMT plan was accompanied by a vertical monitoring and evaluation system to monitor the programme (Kawonga et al. 2012; DoH 2004).

UNAIDS (2017) estimated that 3.7 million of South Africans were accessing ART by end of 2016, which makes South Africa's ART intervention the largest in the world. Recently, the HIV continuum of care has become a key component of monitoring and evaluation of the ART intervention, with emphasis on evaluation of linkage into care, retention of patients throughout the continuum and achievement of viral load suppression as outcomes of the intervention (Cheever 2007). Viral load suppression can also be used as a measuring tool to improve performance of the intervention (McMahon et al. 2013).

A growing body of evidence suggests that HIV is a behavioural disease subject to environmental influence by factors that potentially drive the rapid spread of HIV in the region (Drimie 2002). It is evident from literature that poverty and inequality authenticate the challenges facing HIV interventions. For example, the 'sugar daddy' phenomena, commercial sex, migration and multiple sexual partners were reported as factors perpetuating the increase in HIV infections (Drimie 2002; Gregson et al. 2002; Shisana et al. 2014; Hanson and Hanson 2008). Another critical driver of HIV infections is gender imbalance, which leaves women vulnerable to HIV because they fail to negotiate safe sex if men refuse to use condoms (Langen 2007).

Studies have also shown that poverty and inequality are the main root causes of vulnerability to HIV infections (Mbirimtengerenji 2007; Parkhurst 2010). A number of authors have demonstrated that poverty can increase susceptibility to HIV infection through commercial sex, inferior health care, increased labour migration and the associated risk of having multiple partners (Drimie 2002; Mbirimtengerenji 2007; Parkhurst 2010). Drimie (2002) further argues that poor people living with HIV are more susceptible to illnesses than wealthy people and experience early mortality due to malnutrition, poor health and lack of access to health care and medication. Moreover, scholars appear to agree that, perhaps as a consequence of poverty from their own households, young women engage in sexual activities with older men in exchange for basic necessities (Drimie 2002; Mbirimtengerenji 2007; Parkhurst 2010).

On the other hand, Ghanotakis and co-workers (Ghanotakis et al. 2012) suggest that gender inequality influence vulnerability to HIV infection in women and girls. This was further explained by Ansari (2012) who observes that gender inequality manifests in violence, stigma and discrimination against women, where women have no power over men to negotiate safe sex. The author further argued that stigma is another risk factor that potentially influences personal decision regarding HIV testing and disclosure. Another scholar, Magadi (2011) has found that socioeconomically disadvantaged groups are most affected by HIV/AIDS. Similarly, a study conducted in the United States found that residential segregation, education and immigration are among the structural conditions that explain HIV/AIDS disparities among black communities (Denning and DiNenno 2010). The main consequence of HIV/AIDS in Africa is the slowing down and to some extent reversal of social and economic development (Taraphdar et al. 2011). Therefore, poverty reduction and elimination of structural inequalities are key targets to HIV interventions.

The question, however is, 'whether current ART programmes are comprehensive enough to address the underlying challenges of HIV infection and AIDS? Do the Theories of Change of these programmes allow for a comprehensive measurement of the outcomes (in addition to empirical measures)?'

Existing literature have shown challenges associated with HIV interventions, however, it is still important that HIV interventions move beyond exclusively targeting the so-called risky behaviours of the key populations to also address the root causes of the factors that makes HIV interventions difficult (Drimie 2002).

An impact evaluation by Drimie (2002) indicates that HIV and AIDS-related mortality have disproportionately affected women and children due to gender inequality and mother-to-child transmissions, respectively. One important indicator of the impact of HIV/AIDS is the increased dependency ratio among children and the elders (>60 years). Another factor that drives the social and economic effects of HIV/AIDS is the increased number of orphans; owing to the increased number of deaths among young adults that gives rise to child headed households (Drimie 2002). Furthermore, as a result of the epidemic, the public sector health systems are significantly overwhelmed due to increased demand for health services (Drimie 2002; Kaul et al. 2000). According to Drimie (2002), the effects of HIV/AIDS towards the economic development include shortage of labour force and lower productivity due to absenteeism and illness. Economic consequences of HIV suggest that the economic growth will decline because funds will be shifted away from savings to cover the costs of illness.

Outcomes evaluations are conducted to assess the efficacy of HIV interventions and form the core knowledge base that lead to effective interventions (Mantell et al. 1997). The importance of evaluations has increased in sub-Saharan Africa due to demand for accountability and to measure performance of HIV interventions. According to the World Health Organisation (2015, p. 24), evaluations provide strategic information that is used "(1) to understand the HIV epidemic—the extent of change results from interventions, (2) to track and gauge the health sector's response to HIV, particularly the health system inputs, coverage, quality services, outcomes and impact, (3) to inform programme, improvement, assuring quality and maximal return on resources invested and exposing bottlenecks and opportunities".

In summary, South Africa has the highest HIV infections worldwide. Poverty and inequality are hampering the efforts to cope with the impact of the epidemic. The underlying symptoms such as young girls sexually engaging with older men, sex violence, sex work, multiple sexual partners and gender imbalance increase vulnerability to HIV infections. Currently, HIV prevention and the ART programme are key interventions implanted to curb the scourge of HIV. Ultimately, Africa has to develop interventions that will recognise cultural norms and traditions and also address the predictors of poverty so as to maximally fight the scourge of HIV. Evaluations are necessary to provide feedback in order to improve HIV interventions.

## 16.5 Research Gap Analysis: Methods, Data, Findings, and Conclusions of Studies on and Evaluations of the ART Intervention

Having had a detailed understanding of the ART intervention, the following sections will review similar studies and evaluations. In doing so, the present study will identify research approaches, designs, procedures and methods applied as well as findings and conclusions reached by past and current studies that included evaluations of, ART intervention. Furthermore, the current study will uncover knowledge gaps in this area, especially as it relates to Mankweng Hospital in the Limpopo Province of South Africa. Lastly, with this knowledge, we begin to establish some methodological options that can be utilised for this research as part of our conceptual framework.

The review of current and previous studies is reported based on three key outcomes evaluation approaches among people living with HIV who are receiving ART, mostly from urban areas. These approaches are: (1) retention in care (Nguyen et al. 2013; Doshi et al. 2014; Saka et al. 2014), (2) viral load suppression (Cescon et al. 2011; Fielding et al. 2008; Nachega et al. 2009) and (3) mortality (Fatti et al. 2010; Fox et al. 2012b; Wandeler et al. 2012; Budgell et al. 2015).

Examination of these studies revealed that most outcome evaluations are carried out using a quantitative strategy to measure tangible progress towards the intervention's intended outcomes. In quantitative analysis, logistic regression models and Cox models were good candidates for identifying factors associated with each outcome while hypothesis testing are conducted using Chi-squared test (Fatti et al. 2010; Fielding et al. 2008; Fox et al. 2012a; Nachega et al. 2009; Saka et al. 2014; Wandeler et al. 2012). The most common denominator amongst outcomes evaluation studies was the use of a cross-sectional design, in which data was collected at one time point (Bryman 2012).

Notably, viral load suppression remains the key outcome for the programme, as it measures both the patient health outcome and the performance of the programme (Das et al. 2010). There is substantial agreement that lack of adherence to treatment is a key influencer of suboptimal viral load suppression (Fielding et al. 2008; Jobanputra et al. 2015). However, Mugavero and colleagues (Mugavero et al. 2012) have posed a hypothesis that suboptimal early retention in care (lost-to-follow up) represents a formidable obstacle to achieving viral load suppression, thus, early missed HIV care visits are associated with delayed viral load suppression. Similarly, Wandeler and co-workers (Wandeler et al. 2012) have suggested that the difficulty of retaining patients on ART in care results in poor long term outcomes especially in rural settings, subsequently posing a major threat to the sustainability and the success of the ART programme in Southern Africa. The high number of premature deaths that occur in people receiving treatment has led to low retention rates in rural settings. Rural settings experience inequitable health care services compared to urban areas, mainly due to limited access to healthcare facilities and treatment as well as shortage of medical health workers.

Other researchers have reported low retention rates of patients who started ART in recent years than those who initiated treatment in earlier years (Cornell et al. 2010; Nglazi et al. 2011; Fox et al. 2012b). However, some scholars attributed this to bias (Johnson et al. 2014). Should this be true, it simply means that either earlier treatment guidelines were associated with greater loss of patients from care or the expansion of the ART programme resulted in limited capacity to support patients in the long-term (Budgell et al. 2015).

In contrast, Fox and colleagues highlighted improvements and strong gains in long-term care after ART initiation, exposing the need for intensified interventions that address treatment retention in the first 12 months of ART administration to reduce overall attrition. Most literature showed that high rates of attrition and mortality occur during the first year of treatment (Fox et al. 2012b; Wandeler et al. 2012) and that most deaths occur in patients lost to follow up than in care (Saka et al. 2014; Budgell et al. 2015).

In conclusion, these studies have demonstrated that failure to achieve viral load suppression is largely attributed to a lack of adherence to long term HIV treatment. In addition, low rates of viral load suppression were observed in patients who were not in care. Patient retention was high for a period of one year and then decreased, while high mortality rates were more common in patients lost to follow up. Low retention rates pose a threat to sustainability of ART in Sub-Saharan Africa, especially in rural areas. Missing data was a common limitation identified in all the studies. The studies were largely conducted in urban areas in South Africa, while a few took place in the rural settings within the country and the continent. Rates of viral load suppression were equivalent between rural and urban settings, although patients from rural settings who sought care had advanced HIV disease. Furthermore, there were high mortality and attrition rates at rural settings than urban settings. There are limited studies that have evaluated outcomes of the ART intervention in rural areas, especially Limpopo—this study seeks to understand factors driving viral load suppression or failure to achieve viral load suppression in rural settings. In addition, none of these studies have explicitly detailed frameworks that have been used to interpret empirical evaluation findings.

## 16.6 Establishing a Framework for Interpreting Anticipated Empirical Evaluation Findings

### *16.6.1 An Introduction to Monitoring and Evaluation in Context of an ART Intervention*

There is increasing evidence that globally, monitoring and evaluation plays a significant role in development. It is a tool adopted by governments and organisations to improve accountability, effectiveness and efficiency (Kusek and Rist 2004). In development, monitoring and evaluation is essentially utilised to monitor activities of interventions while promoting learning, consequently providing improved performance and achievement of results of the intervention (Görgens-Albino and Kusek 2009; Holvoet and Inberg 2014). In addition, Guerra-López and Hicks (2015) point out that monitoring and evaluation influences policy and planning of interventions. Put differently, it highlights where the change is happening or lacking, what direction it is taking, and the progress.

According to Patton (2008, p. 5), to evaluate something means "determining its merit, worth, value or significance". However, the overarching goal is to utilise the data generated to inform decision-making that results in improved performance

(Guerra-López 2012). There are three main goals for evaluation, that is, to measure, understand and learn (Berriet-Solliec et al. 2014). Evaluation is also a tool meant to improve the design and operations of interventions in order to promote their continuation or cessation or justify their existence (Stoltzfus and Pillai 2002). In essence, evaluations measure effectiveness, appropriateness, acceptability and efficiency of interventions (Stoller and Rutherford 1989). To ensure that interventions deliver effective and efficient outputs and outcomes, evaluation is important. Stoller and Rutherford (1989) have suggested that a thorough evaluation entails assessment of events at various stages of programme development and implementation.

The use of evaluation results in decision making continues to generate heated debate in the evaluation sphere, with Patton (2008) asserting that there is a significant gap between the knowledge collected and knowledge applied. Recent debates are fixated on the credibility of evaluation results. The debate is largely focused on how to measure impact, whether or not experimental design in the form of randomised controlled trials (RCT) are the most optimal way to provide credible evidence on the past performance of an evaluation, rather than improving future performance evaluations using diverse methodologies (Gargani and Donaldson 2011), further, whether it should be quantitative or qualitative (White 2006). The proponents of RCT purport experimental design used as a benchmark or gold-standard evaluations (Bhana and Govender 2010), however, not all evaluation practitioners supports the use of RCTs for evaluations.

Another difference is failure to articulate the theory of change that underpins each intervention (Edgington 2010). A theory-based evaluation is highly recognised amongst theorists and evaluation practitioners; however, there are still challenges on how to apply this approach in practice and how to make valid deductions about contributions of certain interventions to observed changes (Nakrošis 2014). According to White (2006), a theory-based approach focuses on efficiency (how an intervention is working), not just how it operates, and this requires both quantitative and qualitative data. Furthermore, a theory-based approach provides details on the results chain from inputs to impacts, where some of the evaluations assess final outcomes. This is done by linking resources (inputs) to change in a community.

In summary, monitoring and evaluation is used for learning by defining what works and what doesn't work in an intervention. Understanding how and why interventions achieve intended or unintended outcomes provide an evidence base for decision making and future programme planning. Most importantly, it offers accountability to donors and also demonstrates results of interventions initiated and funded by them. It is sub-divided into 12 components, however, for the purposes of this research, the chapter focused on the five key relevant components: monitoring, formative evaluation, process evaluation, outcomes evaluation and impact evaluation. The literature presented highlighted impeccable results as a consequence of embracing monitoring and evaluation culture, however there are still ongoing differences among scholars regarding monitoring and evaluation. Such issues and discussions include the purpose of monitoring and evaluation in learning and advocacy, how to measure change, utilisation of evaluation results, and attribution of change to intervention.

### 16.6.2 Key Evaluation Variables in the Context of an ART Intervention

A key focus of this study is the use of a monitoring and evaluation approach (Kusek and Rist 2004; Tsui et al. 2014) in a results-chain of an ART programme, depicting a link of its inputs to its outcomes. Figure 16.1 presents the anticipated outcomes for a typical viral load suppression and mortality intervention. The outcome variable is linked to three other elements that contribute to attainment of the intended outcome; these include inputs, activities, and outputs, which address the monitoring effectiveness of the intervention.

For the ART programme to attain its goal and objectives, inputs such as money, staff, equipment, infrastructure must result in increased number of people on ART, increased number of viral load tests performed and increased number of adherence counselling performed by trained staff. If these activities and outputs are well implemented and reach the targeted population, the programme is likely to achieve its intended outcomes of increased viral load suppression among people taking ART, retain them in care and decreased mortality among people living with HIV. These encouraging short-term results (outcomes) should lead to changes in the

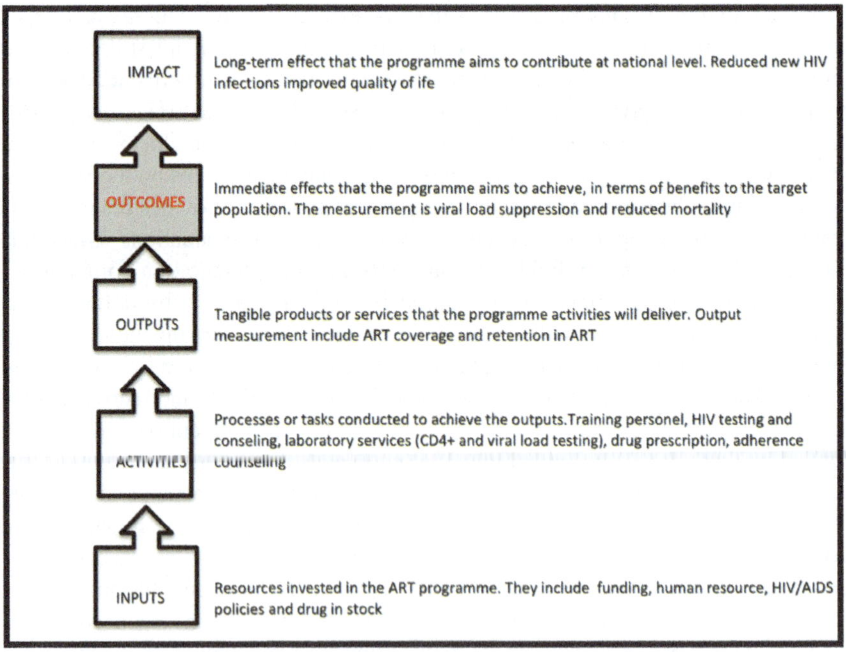


**Fig. 16.1** The results-chain of an ART programme, depicting a link of its inputs to its outcomes. *Source* Authors

long-term impact of the intervention, measured as increased life expectancy (survival) and reduced HIV transmissions.

To facilitate measurement of outcomes of the ART intervention, the results chain is used to understand the achievement of desired outcomes of an intervention. The results chain is comprised of the following elements, inputs, activities, outputs, outcomes and impact and each result level has an indicator that is utilised to collect data by tracking change over time, identifying problem areas as well as determining the effectiveness of the programme or monitoring programme goals (World Health Organisation 2015). In the context of this research, the results chain above depicts a series of events to achieve the intended outcomes under the following assumptions; (1) linking HIV diagnosed individuals to care in order to achieve increased ART coverage; (2) individuals on ART remain in care; (3) adherence to treatment yields sustained viral load suppression and survival; and (4) continued awareness of responsible sexual behaviour to reduce HIV transmissions.

Linked to each indicator is a means for collecting that information—the means of verification describes how information will be collected, and the frequency with which information must be collected (Kusek and Rist 2004). In addition, an indicator should provide a target of quantity, quality and timing of expected results (Kusek and Rist 2004; World Health Organisation 2005).

In conclusion, outcomes evaluation is focused on the outcomes found in the results chain. This outcome is linked to three other elements—inputs, activities and outputs—that demonstrate how the outcome will be achieved. Each outcome also has a set of indicators that measures its progress and achievement.

### *16.6.3 Documented Frameworks that Are Used to Interpret Empirical Findings of Summative Evaluations*

There is substantial evidence supporting the use of theory to guide the development of interventions for health promotion. Those applied in treatment adherence include the health belief model, trans-theoretical model, theories of reasoned action and planned behaviour, self-efficacy theory, and the social cognitive theory (Munro et al. 2007; Kinsey-Steele 2012). Collectively, these theories can explain (1) individual beliefs about health and the need to improve health, (2) key factors influencing behaviour and the short-term consequences that shape health behaviour, (3) the change process that defines the need for an intervention, and (4) determinants of health behaviour and methods of promoting change. Kinsey-Steele (2012) is of the opinion that the self-efficacy theory explains medication adherence behavioural change. However, a literature review undertaken by Munro and colleagues (Munro et al. 2007) suggests that none of these are useful in adherence research related to HIV interventions. Furthermore, none of them can explain the long-term effects of an intervention (Nutbeam et al. 2010).

In spite of it's limitations, the theory of change is used to explain outcomes of an intervention (Weiss 1997; Funnell and Rogers 2011). When using a theory of change approach "... activities A1, A2, and A3, if properly implemented (and with the ongoing presence of contextual factors X1, X2, and X3), should lead to outcomes O1, O2 and O3; and, if these activities, contextual supports, and all outcomes occur more or less as expected, the outcomes will be attributable to the intervention" (Connell and Kubisch 1998, p. 2). With this intention, Funnell and Rogers (2011, p. 387) argue that a "results chain represents the programme theory or theory of change as a linear process with inputs and activities at the front and long-term outcomes at the end", as it lays a pathway of how the intended change will be achieved. Indeed, the results chain is used to measure the results by clarifying the logic of the programme. It also defines the cause-effect relationship between inputs, output, outcomes and impact (Margoluis et al. 2013), and how they work together to affect a desired patient health outcome.

In the context of this research, the theory of change for the ART programme is 'to provide universal access to ART in order to improve survival rate and further reduce HIV transmissions'. Drawing from programme theory, we propose four assumptions that describe the sequential link between inputs, activities, outputs, outcomes, and impact; as described previously.

The importance of the results chain is embedded in its principal role of explaining 'if and when' logic, assuming that every level from the inputs way up to the impacts will occur (United Nations Development Programme 2011). To this end, the ideal theory of change, results chain and its accompanying framework play a major role in assessing success, measuring and interpreting the results of the ART programme. Like other theories, the theory of change has its own limitations. A weak and implausible theory of change can result in a failed intervention. In addition, a theory of change that fails to take external factors into account will not achieve the intended results.

## 16.7 Evaluating Outcomes of the ART Programme in Mankweng Hospital, a Conceptual Framework

Figure 16.2 depicts a proposed conceptual framework for evaluating the outcomes of the ART intervention in Mankweng Hospital. It shows a conclusive summary of how the evaluation question links in with the literature that explicitly details the procedure and techniques for resolving the question. The physical research setting is semi-rural, located in the second poorest province in South Africa. A notable proportion of its 33,000 inhabitants are instilled in their traditional and cultural norms. Further, they are disadvantaged, impoverished, and unemployed. Provision and access to health services is a challenge because of decaying infrastructure, lack of skilled personnel, failure to retain medical staff, and a lack of consistent essential

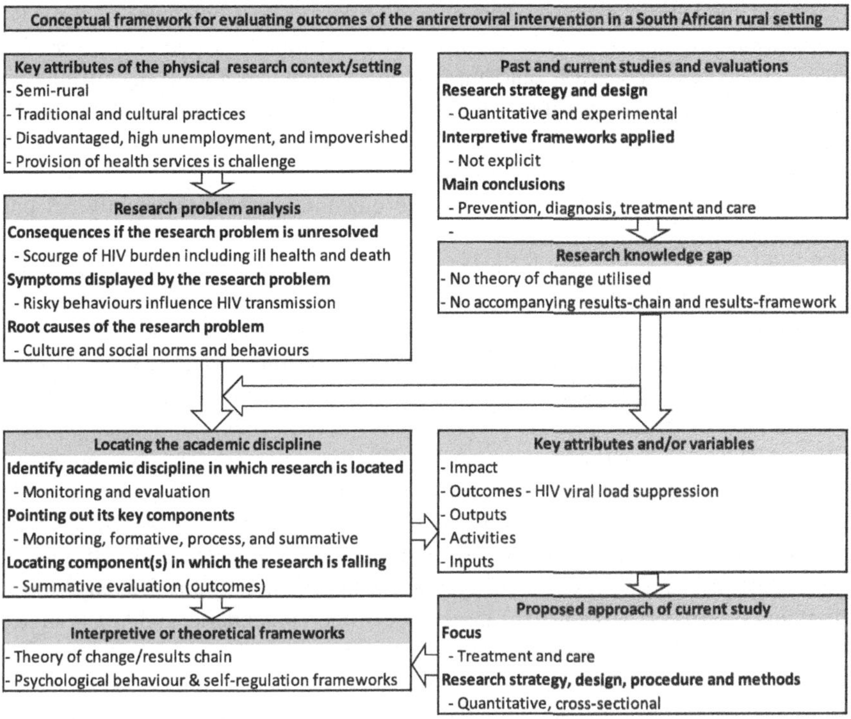


**Fig. 16.2** A proposed conceptual framework for evaluation of outcomes of the ART programme at Mankweng Hospital. *Source* Authors

medical supply. The government-funded Mankweng Hospital ART site is the largest HIV treatment site in the Capricorn district with over 2000 patients on ART. In addition, the facility also provides TB treatment and care. Patients visit this facility to receive treatment care, unlike other rural treatment sites in South Africa that provide treatment and care through community healthcare workers. Like the rest of South Africa, if not the continent, this area is exposed to the scourge of HIV/AIDS burden including ill health and ultimately death. This is due to risky sexual health behaviours exacerbated by traditional culture and social norms such as multiple sexual partners, commercial sex, and teenagers engaging in sex with older men. The risky sexual health behaviours do not only influence HIV transmission but present obstacles for HIV interventions, and hamper efforts to cope with the impact of the epidemic. Therefore, we need contextualised studies and evaluation to help develop interventions that take on board traditional culture and norms as well as poverty and inequality to fight the scourge of HIV.

Interrogating documentation on the current ART as well as past and current studies on and evaluation of similar interventions suggests that retention in care, viral load suppression and reduced mortality should comprise the main outcomes of HIV interventions. Most studies employed a quantitative research strategy and regression models for analysis. The studies reveal positive outcomes of the ART interventions conducted mostly in urban areas of South Africa. However, few studies have assessed ART outcomes in most rural areas and specifically the Limpopo Province. In addition, previous studies and evaluation have hardly employed an explicit theory of change and its accompanying results chain to interpret their respective findings. A theory of change logically stipulates how an intervention intends to achieve its intended outcomes and, consequently, its predicted impact (Funnell and Rogers 2011).

It is this understanding that led us to advance six questions that the empirical part of this study should address. First, '*how does socio-cultural conditions influence the intervention?*' Here we seek to understand whether there are other factors that could impede the full implementation of the intervention. Second, '*what was the retention in care?*' This question determines the efficiency of the ART intervention. Third, '*what is the proportion of adults who achieve viral load suppression among people taking ART for 12 months in rural areas?*' We downplay the other two outputs of the ART interventions for two reasons. The literature has always reported activities and outputs of the intervention. However, it has been more than a decade since the introduction of ART and, therefore, we wish to focus on the outcomes and possible impacts of the intervention. Furthermore, with the new Joint United Nations Programme on HIV/AIDS (2016) target to end the epidemic in 2030, it is imperative to start evaluating outcomes in order to obtain information that helps to fast track the intervention. Lastly, the current focus must shift towards evaluations that measure tangible progress towards the impact. Fourth, '*what are the predictors associated with viral load suppression among people on ART for 12 months in rural areas?*' and Fifth, '*what are the risk factors associated with failure to achieve viral load suppression among people on ART for 12 months in rural areas?*' Here we opt to understand factors that may influence or lead to the intended outcome. Sixth, '*what were the intended and unintended outcomes?*' Here we probe fidelity to the theory of change underpinning the intervention.

## 16.8 Conclusion

This research is focused on reviewing and developing a monitoring and evaluation framework of ART in order to manage and improve results of the intervention. It provides information on progress towards achieving results and evidence on 'what is working' and 'what appears not to be working'. The paper focused on the five key and relevant components of this field: monitoring, formative evaluation, process evaluation, outcomes evaluation and impact evaluation. Impact evaluation can further separate into *ex ante and ex post* impact evaluation. This study is invariably an outcomes evaluation because it intends to assess the effectiveness of the intervention (i.e. ART in a rural setting).

The key variables of this evaluation are impact, outcome, outputs, activities, and inputs. More specifically, the outcome was measured using HIV viral load value—

defined as virus level below 50 which can be obtained from laboratory results. This outcome provides an insight into viral load suppression rates in a community in order to prevent or reduce HIV transmissions.

Though the focus of HIV interventions includes prevention, diagnosis, treatment and care, the empirical study should deal with treatment and care since these are the core focus of ART measured using viral load suppression. For such a study, as evidenced by similar past studies, the most appropriate strategy is quantitative and the design is cross-sectional. First of all, for the evaluation of HIV treatment intervention, the follow-up study will use a qualitative strategy to supplement quantitative data in order to explore perceptions of patients on adherence and seeking HIV care early. Therefore, these choices become the hallmark of the empirical evaluation procedure and methods. The target population will be individuals on the Mankweng Hospital ART programme. We shall then randomly sample the medical records of these patients who have been on treatment for 12 months and transfer their information onto our full-structured data collection instrument or questionnaire. The questionnaire has 23 variables and will measure the following five components: (1) Patient profile or demographics, (2) clinical characteristics, (3) treatment history, (4) viral load monitoring, and (5) adherence characteristics. For analysis, we shall employ the logistic regression model to help us determine risk factors underlying failure to suppress viral load and a binomial regression model to probe predictors and risk factors underlying viral load suppression and the failure to achieve viral load suppression, respectively. We shall employ the last-value-carried-forward rule to avoid some limitations associated with missing data.

After data analysis, we propose to employ the theory of change and its accompanying results chain to interpret the empirical findings so that we explain the performance of the intervention and the workability of the causal link of inputs, activities, outputs, and intended outcomes. The theory of change assumes that viral load suppression will be achieved if people are adherent to treatment.

In conclusion, this paper provides a systems thinking approach towards the development of a conceptual framework rather than focusing on the literature review. It provides a roadmap that explicitly details steps on evaluating outcomes of the ART programme.

## References

Angoria (2014, August). Public hospitals in Limpopo discover hicts medical consulting expertise. Retrieved December 30, 2015, from http://old.agoria.be/www.wsc/old_agoriav3/en/Public-hospitals-in-Limpopo-discover-hict-s-medical-consulting-expertise.

Ansari, R. (2012). *Applications of public health education and health promotion interventions*. Bloomington: Trafford Publishing.

Berriet-Solliec, M., Labarthe, P., & Laurent, C. (2014). Goals of evaluation and types of evidence. *Evaluation, 20*(2), 195–213. https://doi.org/10.1177/1356389014529836.

Bhana, A., & Govender, A. (2010). *Evaluating interventions* (p. 60). Mental Health: Promoting.

Bonell, C., & Imrie, J. (2001). Behavioural interventions to prevent HIV infection: Rapid evolution, increasing rigour, moderate success. *British Medical Bulletin, 58*(1), 155–170.

Bryman, A. (2012). *Social research methods*. Oxford: Oxford University Press.

Budgell, E. P., Maskew, M., Long, L., Sanne, I., & Fox, M. P. (2015). Brief report: Does most mortality in patients on ART occur in care or after lost to follow-up? Evidence from the Themba Lethu Clinic, South Africa. *Journal of Acquired Immune Deficiency Syndromes 70*(3), 323.

Cescon, A. M., Cooper, C., Chan, K., Palmer, A. K., Klein, M. B., Machouf, N., et al. (2011). Factors associated with virological suppression among HIV-positive individuals on highly active antiretroviral therapy in a multi-site Canadian cohort. *HIV Medicine, 12*(6), 352–360.

Cheever, L. W. (2007). Engaging HIV-infected patients in care: their lives depend on it. *Clinical Infectious Diseases, 44*(11), 1500–1502.

Cohen, M. S., Chen, Y. Q., McCauley, M., Gamble, T., Hosseinipour, M. C., Kumarasamy, N., et al. (2011). Prevention of HIV-1 infection with early antiretroviral therapy. *New England Journal of Medicine, 365*(6), 493–505.

Connell, J. P., & Kubisch, A. C. (1998). Applying a theory of change approach to the evaluation of comprehensive community initiatives: progress, prospects, and problems. *New Approaches to Evaluating Community Initiatives, 2*(15–44). Retrieved from http://citeseerx.ist.psu.edu/viewdoc/download?doi=10.1.1.458.2459&rep=rep1&type=pdf.

Cornell, M., Grimsrud, A., Fairall, L., Fox, M. P., van Cutsem, G., Giddy, J., et al. (2010). Temporal changes in programme outcomes among adult patients initiating antiretroviral therapy across South Africa, 2002–2007. *AIDS (London, England), 24*(14), 2263.

Das, M., Chu, P. L., Santos, G.-M., Scheer, S., Vittinghoff, E., McFarland, W., et al. (2010). Decreases in community viral load are accompanied by reductions in new HIV infections in San Francisco. *PLoS ONE, 5*(6), e11068.

Denning, P., & DiNenno, E. (2010). Communities in crisis: Is there a generalized HIV epidemic in impoverished urban areas of the United States. Presented at the XVIII international AIDS conference.

Department of Health. (2003). Operational plan for comprehensive HIV and AIDS care, management and treatment for South Africa. Retrieved May 15, 2015 from https://www.gov.za/sites/www.gov.za/files/aidsoperationalplan1_0.pdf.

Department of Health. (2004). Monitoring and Evaluation Framework for Comprehensive HIV and AIDS Care, Management and Treatment Plan for South Africa. Retrieved March 30, 2016 from https://www.westerncape.gov.za/text/2004/12/monitorevaluation.pdf.

Doshi, R. K., Milberg, J., Isenberg, D., Matthews, T., Malitz, F., Matosky, M., Cheever, L. W. et al. (2014). High rates of retention and viral suppression in United States HIV safety net system: HIV care continuum in the Ryan White HIV/AIDS Program, 2011. *Clinical Infectious Diseases*, ciu722.

Drimie, S. (2002). The impact of HIV/AIDS on rural households and land issues in Southern and Eastern Africa. *A Background Paper Prepared for the Food and Agricultural Organization Sub-Regional Office for Southern and Eastern Africa. Pretoria, South Africa: Human Sciences Research Council.*

Edgington, N. (2010). Let's take a step back in the outcomes debate. Retrieved January 10, 2016 from http://www.socialvelocity.net/2010/01/lets-take-a-step-back-in-the-outcomes-debate.

Fatti, G., Grimwood, A., & Bock, P. (2010). Better antiretroviral therapy outcomes at primary healthcare facilities: an evaluation of three tiers of ART services in four South African provinces. *PLoS ONE, 5*(9), e12888.

Fielding, K. L., Charalambous, S., Stenson, A. L., Pemba, L. F., Martin, D. J., Wood, R., et al. (2008). Risk factors for poor virological outcome at 12 months in a workplace-based antiretroviral therapy programme in South Africa: a cohort study. *BMC Infectious Diseases, 8*(1), 93.

Fox, M. P., Shearer, K., Maskew, M., Macleod, W., Majuba, P., Macphail, P., et al. (2012a). Treatment Outcomes after Seven Years of Public-sector HIV treatment at the Themba Lethu Clinic in Johannesburg, South Africa. *AIDS (London, England), 26*(14), 1823.

Fox, M. P., Van Cutsem, G., Van Cutsem, J., Maskew, M., Keiser, O., Prozesky, H., et al. (2012b). Rates and predictors of failure of first-line antiretroviral therapy and switch to second-line ART in South Africa. *Journal of Acquired Immune Deficiency Syndromes (1999), 60*(4), 428.

Funnell, S. C., & Rogers, P. J. (2011). *Purposeful program theory: Effective use of theories of change and logic models (Vol. 31)*. New York: Wiley.

Gargani, J., & Donaldson, S. I. (2011). What works for whom, where, why, for what, and when? Using evaluation evidence to take action in local contexts. *New Directions for Evaluation, 2011*(130), 17–30.

Ghanotakis, E., Peacock, D., & Wilcher, R. (2012). The importance of addressing gender inequality in efforts to end vertical transmission of HIV. *Journal of the International AIDS Society, 15*(Suppl 2).

Gharajedaghi, J. (2006). Systems thinking: Managing chaos and complexity: A platform for designing business architecture. Amsterdam: Elsevier Inc.

Görgens-Albino, M., & Kusek, J. Z. (2009). *Making monitoring and evaluation systems work: a capacity development toolkit*. World Bank Publications.

Gregson, S., Nyamukapa, C. A., Garnett, G. P., Mason, P. R., Zhuwau, T., Caraël, M., et al. (2002). Sexual mixing patterns and sex-differentials in teenage exposure to HIV infection in rural Zimbabwe. *The Lancet, 359*(9321), 1896–1903.

Guerra-López, I. (2012). The monitoring and impact evaluation process: A Systemic approach to improving performance and impact. *International Journal of Environmental Science and Engineering Research, 3*(3), 80–85.

Guerra-López, I., & Hicks, K. (2015). The participatory design of a performance oriented monitoring and evaluation system in an international development environment. *Evaluation and Program Planning, 48*, 21–30.

Hanson, S., & Hanson, C. (2008). HIV control in low-income countries in sub-Saharan Africa: are the right things done? *Global Health Action*, 1.

Holvoet, N., & Inberg, L. (2014). Taking stock of monitoring and evaluation systems in the health sector: findings from Rwanda and Uganda. *Health Policy and Planning, 29*(4), 506–516.

Hsieh, H., & Shannon, S. (2005). Three approaches to qualitative content analysis. *Qualitative health Research, 15*(9),1277–88.

Jobanputra, K., Parker, L. A., Azih, C., Okello, V., Maphalala, G., Kershberger, B., et al. (2015). Factors Associated with Virological Failure and Suppression after Enhanced Adherence Counselling, in Children, Adolescents and Adults on Antiretroviral Therapy for HIV in Swaziland. *PLoS ONE, 10*(2). Retrieved from http://dx.plos.org/10.1371/journal.pone.0116144.

Johnson, L. F., Estill, J., Keiser, O., Cornell, M., Moolla, H., Schomaker, M., Boulle, A. (2014). Do increasing rates of loss to follow-up in antiretroviral treatment programs imply deteriorating patient retention? *American Journal of Epidemiology*, kwu295.

Johnson, L. F., Mossong, J., Dorrington, R. E., Schomaker, M., Hoffmann, C. J., Keiser, O., et al. (2013). Life expectancies of South African adults starting antiretroviral treatment: collaborative analysis of cohort studies. *PLoS Medicine, 10*(4), e1001418.

Joint United Nations Programme on HIV/AIDS (UNAIDS). (2010). *Combination HIV prevention: tailoring and coordinating biomedical, behavioural and structural strategies to reduce new HIV infections*. Geneva: UNAIDS.

Joint United Nations Programme on HIV/AIDS. (2016). Global AIDS Update 2016. Retrieved March 16, 2017, from http://www.unaids.org/sites/default/files/media_asset/global-AIDS-update-2016_en.pdf.

Joint United Nations Programme on HIV/AIDS (UNAIDS). (2017). UNAIDS DATA 2017. Retrieved September 06, 2017, from http://www.unaids.org/sites/default/files/media_asset/20170720_Data_book_2017_en.pdf.

Kaul, R., Makadzange, T., & Rowland-Jones, S. (2000). AIDS in Africa: a disaster no longer waiting to happen. *Nature Immunology, 1*(4), 267–270.

Kawonga, M., Blaauw, D., & Fonn, S. (2012). Aligning vertical interventions to health systems: A case study of the HIV monitoring and evaluation system in South Africa. *Health Res Policy Syst, 10*(2), 10–1186.

Kinsey-Steele, D. (2012). Effects of Psychosocial Issues on Medication Adherence Among HIV/AIDS Patients. *FEDERAL PRACTITIONER*.

Kumar, R. (2014). Research Methodology: A Step-by-Step Guide for Beginners. Los Angeles: Sage Publications.

Kusek, J. Z., & Rist, R. C. (2004). *Ten steps to a results-based monitoring and evaluation system: a handbook for development practitioners*. World Bank Publications. Retrieved from https://openknowledge.worldbank.org/handle/10986/14926.

Langen, T. T. (2007). Gender power imbalance on women\'s capacity to negotiate self-protection against HIV/AIDS in Botswana and South Africa. *African Health Sciences, 5*(3), 188–197.

Magadi, M. A. (2011). Understanding the gender disparity in HIV infection across countries in sub-Saharan Africa: Evidence from the demographic and health surveys. *Sociology of Health & Illness, 33*(4), 522–539.

Mantell, J. E., Di Vittis, A. T., & Auerbach, M. I. (1997). *Evaluating HIV prevention interventions*. Berlin: Springer Science & Business Media.

Margoluis, R., Stem, C., Swaminathan, V., Brown, M., Johnson, A., Placci, G., et al. (2013). Results chains: A tool for conservation action design, management, and evaluation. *Ecology and Society, 18*(3), 22.

Marks, G., Crepaz, N., & Janssen, R. S. (2006). Estimating sexual transmission of HIV from persons aware and unaware that they are infected with the virus in the USA. *Aids, 20*(10), 1447–1450.

Mbirimtengerenji, N. D. (2007). Is HIV/AIDS epidemic outcome of poverty in Sub-Saharan Africa? *Croatian Medical Journal, 48*(5), 605.

McMahon, J. H., Elliott, J. H., Bertagnolio, S., Kubiak, R., & Jordan, M. R. (2013). Viral suppression after 12 months of antiretroviral therapy in low-and middle-income countries: a systematic review. *Bulletin of the World Health Organization, 91*(5), 377–385.

Mugavero, M. J., Amico, K. R., Westfall, A. O., Crane, H. M., Zinski, A., Willig, J. H., et al. (2012). Early retention in HIV care and viral load suppression: implications for a test and treat approach to HIV prevention. *Journal of Acquired Immune Deficiency Syndromes, 59*(1), 86.

Munro, S., Lewin, S., Swart, T., & Volmink, J. (2007). A review of health behaviour theories: how useful are these for developing interventions to promote long-term medication adherence for TB and HIV/AIDS? *BMC Public Health, 7*(1), 104.

Nachega, J. B., Nguyen, H., Dowdy, D. W., Chaisson, R. E., Regensberg, L., Maartens, G., et al. (2009). Antiretroviral therapy adherence, virologic and immunologic outcomes in adolescents compared with adults in southern Africa. *Journal of Acquired Immune Deficiency Syndromes (1999), 51*(1), 65.

Nakrošis, V. (2014). Theory-based evaluation of capacity-building interventions. *Evaluation, 20*(1), 134–150.

Nattrass, N. (2008). AIDS and the scientific governance of medicine in post-apartheid South Africa. *African Affairs, 107*(427), 157–176.

Nglazi, M. D., Lawn, S. D., Kaplan, R., Kranzer, K., Orrell, C., Wood, R., et al. (2011). Changes in programmatic outcomes during 7 years of scale-up at a community-based antiretroviral treatment service in South Africa. *Journal of Acquired Immune Deficiency Syndromes, 56*(1), e1.

Nguyen, D. B., Do, N. T., Shiraishi, R. W., Le, Y. N., Tran, Q. H., Nguyen, H. H., et al. (2013). Outcomes of antiretroviral therapy in Vietnam: Results from a national evaluation. *PloS ONE, 8*(2), e55750.

Nutbeam, D., Harris, E., & Wise, W. (2010). *Theory in a nutshell: A practical guide to health promotion theories*. New York: McGraw-Hill.

Nyindo, M. (2005). Complementary factors contributing to the rapid spread of HIV-I in Sub-Saharan Africa: A review. *East African Medical Journal, 82*(1), 40–46.

Palella, F. J., Deloria-Knoll, M., Chmiel, J. S., Moorman, A. C., Wood, K. C., Greenberg, A. E., et al. (2003). Survival benefit of initiating antiretroviral therapy in HIV-infected persons in different CD4+ cell strata. *Annals of Internal Medicine, 138*(8), 620–626.

Parkhurst, J. O. (2010). Understanding the correlations between wealth, poverty and human immunodeficiency virus infection in African countries. *Bulletin of the World Health Organization, 88*(7), 519–526.

Patton, M. Q. (2008). *Utilization-focused evaluation*. Thousand Oaks: SAGE.

Portfolio Committee for Health. (2011). Report of the Portfolio Committee on Health on the oversight visit to Mankweng, Louis Trichardt Hospitals and Madombidza Clinic in the Limpopo Province from 10–12 August 2011. Retrieved September 28, 2015, from http://pmg-assets.s3-website-eu-west-1.amazonaws.com/doc/2012/comreports/120319pchealthreport.htm.

Ras, G., Simson, I., Anderson, R., Prozesky, O., & Hamersma, T. (1983). Acquired immunodeficiency syndrome. *South African Medical Journal, 64,* 140–142.

Rossi, P. H., Lipsey, M. W., & Freeman, H. E. (2003). *Evaluation: A systematic approach*. Thousand Oaks: SAGE.

Saka, B., Landoh, D. E., Patassi, A., d'Almeida, S., Singo, A., Gessner, B. D., & Pitché, P. V. (2014). Loss of HIV-infected patients on potent antiretroviral therapy programs in Togo: risk factors and the fate of these patients. *Pan African Medical Journal, 15*(1).

Sherman, G., Lilian, R., Bhardwaj, S., Candy, S., & Barron, P. (2014). Laboratory information system data demonstrate successful implementation of the prevention of mother-to-child transmission programme in South Africa. *SAMJ. South African Medical Journal, 104*(3), 235–238.

Shisana, O., Rehle, T., Simbayi, L., Zuma, K., Jooste, S., Zungu, N., Ramlagan, S. (2014). South African national HIV prevalence, incidence and behaviour survey, 2012. Cape Town.

Statistics of South Africa. (2011). Mankweng. Retrieved January 12, 2016, from http://www.statssa.gov.za/?page_id=4286&id=13123.

Statistics of South Africa. (2017). Poverty trends in South Africa. An examination of absolute poverty between 2012 and 2015. Retrieved Sempember 08, 2017, from http://www.statssa.gov.za/publications/Report-03-10-06/Report-03-10-062015.pdf.

Stoller, E. J., & Rutherford, G. W. (1989). Evaluation of AIDS prevention and control programs. *AIDS, 3*(1), S289–S296.

Stoltzfus, R. J., & Pillai, G. (2002). Measuring performance: A strategy to improve programs. *The Journal of Nutrition, 132*(4), 845S–848S.

Taraphdar, P., Guha, R. T., Haldar, D., Chatterjee, A., Dasgupta, A., Saha, B., et al. (2011). Socioeconomic consequences of HIV/AIDS in the family system. *Nigerian Medical Journal: Journal of the Nigeria Medical Association, 52*(4), 250.

Tsui, J., Hearn, S., & Young, J. (2014). *Monitoring and evaluation of policy influence and advocacy*. Working Paper 395, London: Overseas Development Institute.

United Nations Development Programme. (2011). Outcome level evaluation: A companion guide to the handbook of planning monitoring and evaluating for development for development results for programme units and evaluators. Retrieved on 20 December 2015 from http://web.undp.org/evaluation/documents/guidance/UNDP_Guidance_on_Outcome-Level%20_Evaluation_2011.pdf.

Wandeler, G., Keiser, O., Pfeiffer, K., Pestilli, S., Fritz, C., Labhardt, N. D., et al. (2012). Outcomes of antiretroviral treatment programs in rural Southern Africa. *Journal of Acquired Immune Deficiency Syndromes, 59*(2), e9.

Weiss, C. H. (1997). Theory-based evaluation: Past, present, and future. *New Directions for Evaluation,* (76), 41–55.

White, H. (2006). Impact evaluation: the experience of the Independent Evaluation Group of the World Bank.

World Health Organisation. (2005). National AIDS Programmes: A guide to indicators for monitoring and evaluating National AIDS Programmes. Retrieved December 24, 2015, from www.who.int/hiv/pub/me/en/naparv.pdf.

World Health Organisation. (2010). Priority interventions. Retrieved December 20, 2015 from http://www.who.int/hiv/pub/priority_interventions_web.pdf.

World Health Organisation. (2015). Consolidated strategic information guidelines for HIV in the health sector. Retrieved December 30, 2015, from http://apps.who.int/iris/bitstream/10665/164716/1/9789241508759_eng.pdf?ua=1&ua=1.

Wotela, K. (2016). Towards a systematic approach to reviewing literature for interpreting public and business management research results. *Electronic Journal of Business Research Methods, 14*(2), 83–97.

Wotela, K. (2017). Conceptualising conceptual frameworks in public and business management research: Paper presented at the 16th European Conference on Research Methodology for Business and Management Studies. Dublin Institute of Technology. June 22–23, 2017. Reading: Academic Conferences and Publishing International Limited.

# Chapter 17
# New Approaches to Measuring Ageing in South Africa

**Mercy Shoko**

**Abstract** *Background and Significance of the topic:* Health expectancy is generally expressed as life expectancy free of disability and is a useful tool for assessing the interaction between health, ill-health and mortality with age. The present study explores ageing trajectories using health expectancy for the four population subgroups in South Africa, namely Black African, Indian/Asian, Coloured (of mixed descent) and White. It provides a fresh look at ageing in South Africa with important implications for evidence-based long-term plans and policies to address the current and future needs of older persons. *Methodology*: Estimated health expectancy was determined using the Sullivan method which requires the use of data on both morbidity and mortality. *Application/Relevance to systems analysis*: Research on health variation in older persons across population groups is central to demographic systems, and helps to reveal the socioeconomic vulnerability of the older adult population. *Policy and/or practice implications*: The research informs potential areas for policy change relating to older adults and highlights the planning required to enable the provision of age-appropriate services. *Discussion and conclusion*: The present study is important because it showed population group heterogeneity which characteristically gets masked at national level. The data indicated that in addition to age and sex variations, there is a population group hierarchy in health expectancy. The findings also supported the "health-survival paradox" in disability free life expectancy and the general differences in health expectancy suggested in the study has implications for retirement ages, which is currently 60. The study has clear policy implications, one of them being the need for age-appropriate planning for health and social services.

---

M. Shoko (✉)
Statistics South Africa and School of Public Health,
University of the Witwatersrand, Johannesburg, South Africa
e-mail: Mercysh@statssa.gov.za



P. Mensah et al. (eds.), *Systems Analysis Approach for Complex Global Challenges*,
https://doi.org/10.1007/978-3-319-71486-8_17

## 17.1 Introduction

Demographic ageing is a shift of the age structure towards the older population. Morbidity and mortality levels and patterns are expected to change with growing age and are influenced by health-related factors, but, also by factors other than health such as social determinants of health (SDH), which have become increasingly important in explaining health differentials among subpopulations. The World Health Organisation (WHO) acknowledges that non-health related factors influence health outcomes and defines social determinants of health as "…the circumstances in which people are born, grow up, live, work and age, and the systems put in place to deal with illness. These circumstances are in turn shaped by a wider set of forces: economics, social policies, and politics" (WHO 2016). A related concept is the "social gradient in health which runs from top to bottom of the socioeconomic spectrum' and explains why the worst health outcomes are observed among the most disadvantaged in society and the best health outcomes occur among the more affluent (ibid).

Globally, falling fertility and mortality have given rise to an ageing population (Kinsella and Velkoff 2001). Africa currently has a young population, however due to a combination of decreasing fertility, improved health and longevity, it is projected that the numbers and proportions of the older population will increase four-fold by 2050 (Golaz et al. 2012). In South Africa, research on ageing is conducted against a backdrop of evidence suggesting that the country has one of the fastest ageing population in sub-Saharan Africa, and along with it the need to estimate the levels, patterns and trends of ageing, the health and socioeconomic vulnerability of the older adult population, and the ensuing policy implications (Phaswana-Mafuya et al. 2013; Udjo 2011; Ralston et al. 2016; Goodrick and Pelser 2014; Kinsella and Ferreira 1997). However, current research on ageing does not adequately reflect the diversity of the South African population, as it presents older adults as a homogenous group.

A recent study by Chirinda and Chen (2017) used the Sullivan method to compare the pattern of health expectancy in the context of ageing in low- and middle-income countries. The present study extends the discourse on ageing to include population group, which given South Africa's historical segregation legacy, is an indicator for socioeconomic inequalities and provides a possible explanation for the health variations in the country. Exploring health expectancy variation would therefore be critical for planning, based on the survival and the health of older adults, and reveal gaps between population groups that need to be addressed.

## 17.2 Background to South Africa

South Africa has four population groups, namely, Black African, Indian/Asian, Coloured (of mixed descent) and White. This population group classification is a legacy of apartheid created specifically with the Population Registration Act of 1950 which required classification and registration of all individuals in accordance with their racial characteristics and created demographic and socioeconomic cleavage along racial lines. South Africa has a long history and relatively high coverage of vital registration but censuses and surveys remain the principal sources of demographic information. The repeal of the Population Registration Act in 1991 saw the drop of the population group as one of the individual attributes in the National population register (NPR). There is debate on the relevance of population group classification following the demise of the apartheid system, however it continues to be used by the democratic government as a means for monitoring and tracking progress in redressing the iniquities of apartheid social and economic policies (Moultrie and Dorrington 2011; Christopher 2002; Bah 1999).

According to the 2011 Census, South Africa has a total population of approximately 52 million, of which 51% are females and the remaining 49% are males. Demographic and socioeconomic outcomes in the country differ in important ways for the four population groups. This is illustrated using replacement level fertility, prevalence of HIV and income inequality. The population group share of Black African, Indian/Asian, Coloured and White is 79.2, 2.5, 8.9 and 8.9% respectively (Statistics South Africa 2012a). Replacement level fertility is the total fertility rate at which a population exactly replaces itself from one generation to the next, without migration (Craig 1994). It is typically represented as the average number of children born per woman and set around 2.1. All population groups in South Africa have experienced fertility decline that dates back to the mid-1960s, although at different rates (Moultrie and Timaeus 2002). Fertility among Whites is below replacement level (1.9), Indian/Asians are at replacement level (2.1), for Coloureds and Black Africans, it is 2.6 and 2.8 respectively (Udjo 2014).

The South African population is disproportionately affected by HIV/AIDS, which accounts for 31% of the total disability-adjusted life years (Coovadia et al. 2009). According to the 2012 National HIV Prevalence, Incidence and Behaviour Survey, an estimated 6.4 million people are living with HIV, but the HIV prevalence varies greatly across population groups, with 15% among Black Africans, 3% among Coloureds and under 1% each for Whites and Indian/Asians (Shisana et al. 2014).

South Africa is one of Africa's largest economies, but with a Gini coefficient of 0.64, it has one of the most unequal societies in the world (UNDP 2013). The Gini coefficient is the measure of income inequality, ranging from 0 to 1. 0, where 0 is a perfectly equal society and a value of 1 represents a perfectly unequal society. The Wold Bank definition states that it measures the extent to which the distribution of income or consumption expenditure, among individuals or households within an economy, deviates from a perfectly equal distribution (The World Bank 2017).

Inequality in South Africa is not population group neutral. The average annual household income is estimated at R61 000 for households with a Black African head, R112 000 for Coloureds, R250 000 for India/Asians and R365 000 for Whites (Statistics South Africa 2012a). The convergence of declining fertility, increased adult mortality due to HIV/AIDS and a high Gini coefficient all influence the vulnerability of older persons. This is on top of apartheid era policies that discriminated against the majority of current elderly persons by limiting opportunities for obtaining formal education during their younger years and thus their socioeconomic status in their old age (Statistics South Africa 2014a).

Legislations relating to older adults in South Africa date back to the Aged Persons Act of 1967 and its successor the Older Persons Act 13 of 2006 (Malherbe 2007). Initially, the threshold for aged males and females was set at 65 and 60 years respectively (ibid). This was revised to 60 years for both genders in the provision of a social pension, conditional on a means test threshold. The social pension has reportedly reduced the country's overall poverty gap by 21% and by 54% for households with older adults. The pension also led to a 98% reduction of the poverty gap for households with only older adults (Samson and Kaniki 2008).

The present study explored ageing trajectories using health expectancy for the four demographic groups in South Africa, namely Black African, Indian/Asian, Coloured and White.

## 17.3 Methodology

Data for this study was obtained from Statistics South Africa, a government department mandated with the production of official statistics in the country. The 2011 census and 2011 mortality and cause of death datasets were publicly available, however the latter did not include population group variable. A special request was made for national data on mortality and cause of death that include the variable population group for the deceased.

It is important to note that the Births and Deaths Registration Act No. 51 of 1992 (as amended) makes it mandatory for all deaths that occur in the country to be registered. The registration process is administered by the Department of Home Affairs (DHA) using the Notification/Register of death/stillbirth form (either the new DHA-1663 form introduced in 2012 or the BI 1663 being phased out). The death certificate issued by the DHA is then used to update the national population register for South Africans and permanent residents who were on the population register. Thereafter, the forms are obtained by Statistics South Africa for processing, with the addition of forms for South Africans who are not on the national population register and for non-South Africans (Statistics South Africa 2014b). Like other sub-Saharan African countries, under-reporting of deaths undermines the mortality data, although studies indicate that the coverage for adults is very high at approximately 93 or 94% (Dorrington et al. 2014; Statistics South Africa 2014b).

In summary the present study uses as its basis births, population, deaths, health and functioning data, collected from the census, in conjunction with mortality and causes of death data. Questions related to the variables used in the study were asked of everybody enumerated during census. The exceptions were questions on health and functioning that were asked of individuals in a household setting and not of persons in institutions such as frail care centres and retirement villages. It must be noted however that the total non-household population was relatively small at just above 1%, so their exclusion was not expected to have a significant effect on the estimates.

With respect to health and functioning, the 2011 census included a set of questions recommended by the United Nations' commissioned Washington Group on Disability Statistics for use to measure disability (Madans et al. 2011). The questions included six functional domains: seeing, hearing, walking, communication, cognition and self-care. Every member of the household was asked about the degree of difficulty in all the functional domains at the time of the census with responses on an ordinal scale: no difficulty, some difficulty, a lot of difficulty, cannot do at all, do not know (because of proxy responses) and cannot yet be determined (for children) (Statistics South Africa 2014c).

The population group in the census data was self-reported, and the allocation of the population group of the deceased in the mortality and cause of death data was at the discretion of the health practitioner certifying the death. In the BI 1663, effective at the reference period, the health practitioner was asked "do you consider the deceased to be African, White, Indian, Coloured, Other (specify)" (Statistics South Africa 2012b). Race was unspecified for approximately 20% of registered deaths (left blank, classified as other or unknown for population group in the mortality and cause of death dataset). Census data included less than 1% of the population that selected the "Other" category without any additional specifics provided. Research has shown potential overlap or "misclassification" in Black African and Coloured population groups in South Africa (Udjo and van Aardt 2012; Republic of South Africa 1968 cited in Khalfani et al. 2005).

The analysis commenced with an estimation of the completeness of reporting of deaths. Deceased individuals whose age or sex were unspecified constituted less than 1% each, and were distributed proportionally among the specified data, assuming that the distribution of the known is similar to that for the unknown. Deaths with unspecified population group were proportionally distributed across the four population groups for ages five years and older, and were included with the Black African deaths for children under five years. The population classified as Other in the census data was used to boost the Coloured population group, so was the proportion of the disabled. Black African deaths for children under five years were adjusted for underreporting using the expert assessments of completeness of deaths, available by age (Dorrington et al. 2014). These completeness estimates are also used to adjust the deaths for ages five years and older, and the fraction used to adjust deaths proportionally distributed across the four population groups.

This study focused on health expectancy which includes life expectancy (LE) and disability-free life expectancy (DFLE). Life expectancy is obtained from a constructed life table as an estimate of the remaining number of years expected to be lived at a given age, if the prevailing mortality conditions persisted. Disability-free life expectancy is the life expectancy adjusted for prevalence of a given health dimension at the given ages. The Sullivan method for computing health expectancy (Sullivan 1971) was chosen because of its robustness, its feasibility with simple and cross sectional data and ease of interpretation (Imai and Soneji 2007). Additionally, the method is independent of the age structure allowing for comparison between two population groups (ibid).

As noted by Sagner and associates (Sagner et al. 2002), different age markers have been used in gerontological studies conducted in Africa to indicate the beginning of old age and it ranges from 50 to 65 years. That age range was therefore selected for the determination of life expectancy and disability-free life expectancy years.

Adjusted births, population and death data disaggregated by single ages, sex and population group were input data for the life tables. The disaggregation by sex was informed by the differential force of mortality and health by sex, which is well established in the literature. Age specific disability (ill-health) proportions derived from the health and functionality data, were also disaggregated by single ages, by sex and population group and formed part of the input data for deriving health expectancy. Ill-health was defined as "a lot of difficulty", or "unable to do" in any one of the six functional domains consistent with the Washington Group for disability statistics, or "some difficulty" in at least two consistent with that of Statistics South Africa (2014c). The Sullivan Method was applied using the input data described above for the computation of both life expectancy and disability-free life expectancy (Sullivan 1971; Jagger et al. 2014). The formula is explained below:

$$e_x = \frac{1}{l_x} \sum_{x}^{\infty} {}_nL_x$$

and

$$e'_x = \frac{1}{lx} \sum (1 - nD)_n^* L_x$$

where

$e_x$ is life expectancy at time t
$e'_x$ is disability-free life expectancy at time t
$l_x$ is survivorship at time t
$D_x$ the prevalence of disability in the age interval
${}_nL_x$ Person Years lived in the age group $x, x+n$

## 17.4 Results

The results presented in Tables 17.1 and 17.2 are for the life expectancy and health expectancy (or disability-free life expectancy) in brackets computed by the author for males and females. Life expectancy measures the number of remaining years to be lived at a particular age, considering the current mortality level of the country. Taking into account both mortality and ill-health at particular ages makes it possible to separate the remaining number of years into years spent in good and bad health.

The results showed that life and health expectancy differed by population group, age and sex. Life expectancy and disability-free life years in general, decreased with increasing age, while health expectancy was lower than the life expectancy at each age. Compared to their male counterparts and in line with global trends, South African females had a survival advantage as shown by a higher life expectancy at birth (Mathers et al. 2001; Udjo 2014). Females continued to have a survival advantage with consistently higher life expectancy values beyond age 50. There was also a distinct population group hierarchy at birth, with Black Africans having the lowest life expectancy (56.4 years for males and 62.3 years for females), followed by Coloureds (64.8 years for males and 71.0 years for females respectively), Indian/Asian (67.8 years for males and 73.9 years for females, respectively), and then Whites (83.8 and 88.8 years for males and females respectively), (results not shown). Around age 50, all the population groups converged, with the exception of Whites (both male and female), who continue to have a survival advantage.

**Table 17.1** Life and (health expectancy in brackets) for males calculated by the author using the South Africa 2011 census and mortality and cause of death data

| Age | Black African | Coloured | Indian/Asian | White | South Africa |
|---|---|---|---|---|---|
| 50 | 19.7 (15.4) | 18.3 (15.5) | 23.0 (19.1) | 46.6 (33.9) | 21.2 (16.9) |
| 51 | 19.2 (14.9) | 17.5 (14.8) | 22.2 (18.4) | 46.1 (33.3) | 20.6 (16.4) |
| 52 | 18.7 (14.4) | 16.7 (14.0) | 21.5 (17.7) | 45.7 (32.8) | 20.0 (15.8) |
| 53 | 18.2 (13.9) | 15.9 (13.3) | 20.9 (17.0) | 45.2 (32.2) | 19.5 (15.3) |
| 54 | 17.7 (13.4) | 15.2 (12.6) | 20.2 (16.4) | 44.7 (31.7) | 19.0 (14.8) |
| 55 | 17.2 (12.9) | 14.4 (11.9) | 19.4 (15.7) | 44.3 (31.2) | 18.4 (14.2) |
| 56 | 16.7 (12.4) | 13.7 (11.2) | 18.7 (15.0) | 43.9 (30.7) | 17.9 (13.7) |
| 57 | 16.2 (12.0) | 13.0 (10.6) | 18.1 (14.4) | 43.4 (30.1) | 17.4 (13.2) |
| 58 | 15.8 (11.5) | 12.3 (9.9) | 17.5 (13.8) | 43.0 (29.6) | 16.9 (12.7) |
| 59 | 15.3 (11.1) | 11.6 (9.3) | 16.9 (13.2) | 42.6 (29.1) | 16.4 (12.2) |
| 60 | 14.8 (10.6) | 10.9 (8.7) | 16.2 (12.6) | 42.2 (28.6) | 15.8 (11.7) |
| 61 | 14.4 (10.2) | 10.3 (8.1) | 15.6 (12.0) | 41.8 (28.1) | 15.3 (11.3) |
| 62 | 14.0 (9.8) | 9.7 (7.6) | 15.0 (11.5) | 41.4 (27.6) | 14.9 (10.8) |
| 63 | 13.6 (9.4) | 9.1 (7.1) | 14.4 (10.9) | 41.0 (27.1) | 14.4 (10.4) |
| 64 | 13.2 (9.0) | 8.5 (6.6) | 13.9 (10.4) | 40.6 (26.7) | 13.9 (9.9) |
| 65 | 12.8 (8.6) | 8.0 (6.1) | 13.4 (10.0) | 40.2 (26.1) | 13.5 (9.5) |

**Table 17.2** Life and (health expectancy in brackets) for females calculated by the author using the South Africa 2011 census and mortality and cause of death data

| Age | Black African | Coloured | Indian/Asian | White | South Africa |
|---|---|---|---|---|---|
| 50 | 25.5 (17.9) | 23.1 (18.8) | 26.8 (21.2) | 48.5 (35.0) | 26.4 (19.4) |
| 51 | 24.9 (17.3) | 22.2 (18.0) | 25.9 (20.4) | 47.8 (34.3) | 25.7 (18.7) |
| 52 | 24.2 (16.7) | 21.4 (17.2) | 25.1 (19.6) | 47.1 (33.6) | 25.0 (18.1) |
| 53 | 23.6 (16.1) | 20.5 (16.4) | 24.2 (18.8) | 46.5 (32.9) | 24.3 (17.5) |
| 54 | 23.0 (15.5) | 19.6 (15.6) | 23.4 (18.0) | 45.9 (32.2) | 23.7 (16.8) |
| 55 | 22.4 (15.0) | 18.8 (14.9) | 22.6 (17.3) | 45.3 (31.5) | 23.0 (16.2) |
| 56 | 21.7 (14.4) | 18.0 (14.1) | 21.8 (16.6) | 44.6 (30.8) | 22.3 (15.6) |
| 57 | 21.1 (13.8) | 17.1 (13.4) | 21.0 (15.8) | 44.0 (30.2) | 21.7 (15.0) |
| 58 | 20.5 (13.3) | 16.3 (12.6) | 20.2 (15.1) | 43.4 (29.5) | 21.0 (14.4) |
| 59 | 19.9 (12.8) | 15.5 (11.9) | 19.4 (14.3) | 42.8 (28.9) | 20.4 (13.8) |
| 60 | 19.3 (12.2) | 14.7 (11.2) | 18.6 (13.6) | 42.2 (28.2) | 19.7 (13.2) |
| 61 | 18.7 (11.7) | 13.9 (10.5) | 17.8 (13.0) | 41.6 (27.6) | 19.1 (12.7) |
| 62 | 18.1 (11.2) | 13.2 (9.9) | 17.1 (12.3) | 41.1 (27.0) | 18.4 (12.1) |
| 63 | 17.6 (10.7) | 12.5 (9.3) | 16.4 (11.6) | 40.6 (26.4) | 17.8 (11.6) |
| 64 | 17.0 (10.2) | 11.8 (8.6) | 15.6 (11.0) | 39.9 (25.7) | 17.2 (11.0) |
| 65 | 16.4 (9.7) | 11.1 (8.1) | 15.1 (10.4) | 39.3 (25.1) | 16.6 (10.5) |

A plot of health expectancy against disability free life expectancy for men and women is presented in Figs. 17.1 and 17.2 respectively. They reveal that females have a lower health expectancy, at all the ages and across the four population groups.

The life expectancy of Whites of both genders was almost double or higher than that of the other three population groups from age 50. White males and females who are 50–65 years old, could be expected to live a further 40 years. The disability-free life expectancy was approximately 35 years and 25 years for both sexes, at age 50 and 65 respectively.

Both life expectancy and disability-free life years were the lowest for the Coloured population group. At age 50, life expectancy was estimated to be 18 years for males and 23 for females, with a drop to 8 years and 11 years, respectively, by age 65. The disability-free life years at age 50, was 15 and 18 years for males and females respectively. The fall in both life expectancy and disability-free life years is also steepest in the Coloured population group. The population group life and health expectancy for Black Africans mirrored that of South Africa in general.

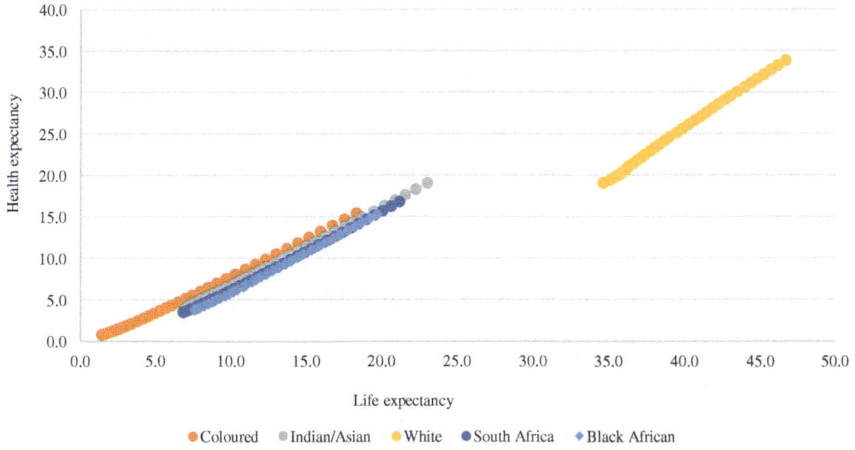


**Fig. 17.1** Life and health expectancy by age and population group for males calculated by the author using the South Africa 2011 census and mortality and cause of death data

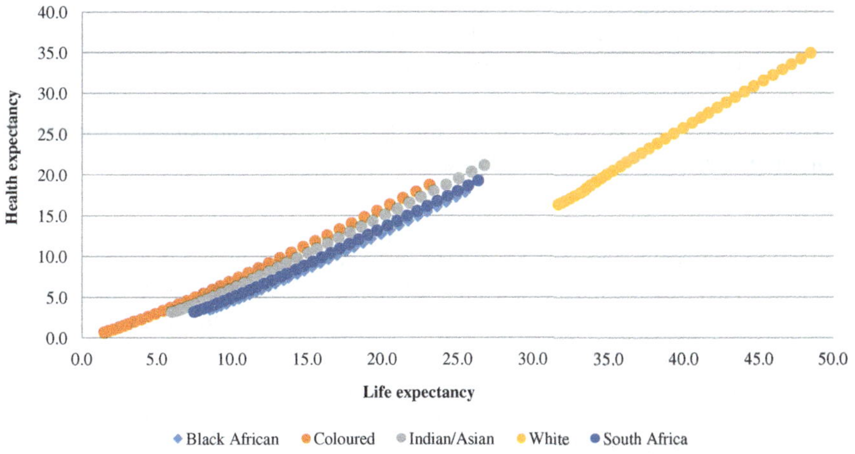


**Fig. 17.2** Life and health expectancy by age and population group for females calculated by the author using the South Africa 2011 census and mortality and cause of death data

## 17.5 Discussion

In the past, data on life expectancy, infant mortality and the causes of death used to be seen as a sufficient basis for assessing population health status and determining public health priorities. As mortality rates continue to decline and life expectancy increases, however, more and more questions arose about the quality of the years lived. The typical indicators remain indispensable as there are still major inequalities in mortality between countries and between groups within countries.

Nevertheless, changes during the last 20 years have demonstrated the need for indicators of a new type, namely disability-free life expectancy, healthy life expectancy or active life expectancy. These provide information on the functional state and vitality of populations as well as on their quality of life (Robine et al. 1999).

To recap, life expectancy measures the number of remaining years to be lived at a particular age, considering the current mortality level of the country. Health expectancy, or disability-free life expectancy, shows how life expectancy is divided into years spent with and without disability at each age. The focus of this study was the application of the Sullivan method to determine life expectancy and disability-free life expectancy (DFLE) variation in the four population groups in South Africa. The results confirmed the expectations that life expectancy and disability-free life expectancy vary by population group (Statistics South Africa 2014a). Overall, health expectancy fell with increasing age and more rapidly than life expectancy.

The variation between life and health expectancy for males and females was consistent with the "health-survival paradox" (Oksuzyan et al. 2008), whereby females live longer than males on average but their longer life expectancy is accompanied by more years with disability, both in absolute terms and as a proportion of remaining life.

The life and disability-free life expectancy variation suggested that Coloureds were the least advantaged, followed by Black African, then Indian/Asian (although with very similar patterns), and then lastly White. The expectation at the commencement of the study was that patterns observed at the beginning of life would persist at older ages, and given the fact that Indians/Asians are similar to Whites in many important ways, they would demonstrate similar trends for life expectancy and health expectancy, while comparable trends would be observed between the Black African and Coloured population groups (City of Cape Town 2010). Instead, similar life and health expectancy was observed between Coloureds, Indians/Asians and Black Africans, while a discernible difference existed between Indians/Asians and Whites, although the Coloured population group were found to be the most disadvantaged overall. One explanation for the observed pattern is the notion that persons who survive to older ages after having been subjected to disproportionate mortality conditions in early life, are the healthiest of their fraction and may be healthier or not importantly different compared to their counterparts who had a survival advantage from the beginning (Anderson et al. 2004). Life expectancy and disability-free life years for Black Africans mirrored the South African average because they form the majority of the population.

The findings of this study are subject to some limitations. The life expectancy estimates were largely dependent on the quality of the data from which they were derived, and results may reflect the quality of the input data, other than the health variations across population groups. Although the quality of the vital registration and vital statistics systems data is affected by incompleteness of both birth and death registrations, it was deemed a credible data source. Additional discussion of these data have been published elsewhere (Dorrington 2013; Dorrington et al. 2014; Udjo 2017; Shoko et al. 2016). Self-assessment of functional limitations among the elderly showed that the burden of ill health and disabilities increases with age, creating an increased demand for health care and other services as the proportion of elderly persons continues to grow. The results confirmed that disability is gendered and racially skewed. Women reported poorer health than men, and black Africans, Coloureds and Indians reported poorer health than whites (Statistics South Africa 2014c).

Previous studies have demonstrated that while eliminating fatal diseases such as cancer through improved health care leads to an increase in health expectancy, it may also cause an increase in life expectancy with disability, thus increasing the burden of disability on society. In contrast, the removal of non-lethal diseases, such as arthritis, would strongly increase disability-free life expectancy without changing total life expectancy, and would therefore reduce life expectancy with disability to a marked extent. Between these two extremes, the removal of both lethal and disabling diseases, such as heart diseases, would increase life expectancy and disability-free life expectancy in various proportions (Robine et al. 1999). Therefore, it would be possible to determine priorities in health policies through the calculation of potential gains in health expectancy deriving from the elimination of different causes of morbidity. Such calculations distinguish between gains in mortality and gains in morbidity or disability, and thus show whether the elimination of one or another pathology can be expected to reduce or increase overall morbidity (Ibid).

The onset of old age presents physical and cognitive limitations and the use of assistive devices and chronic medication plays a critical role in ensuring an independent life among elderly persons. The unmet need for assistive devices among particular groups of elderly persons requires attention. This will translate into the realisation of improved health and overall well-being for this particular vulnerable group (Statistics South Africa 2014a). Improvements in infrastructure relating to access to health care centres and opportunities that promote healthy living, will also have the net effect of a decline in the disease burden among the old, and an increase in productivity in old age (Bloom and Canning 2004).

Given the injustices and discriminatory laws propagated by the apartheid government that led to unequal access to education, services and income, it is inevitable that the majority of current elderly persons in South Africa had limited opportunities for obtaining formal education during their younger years. Lack of skills and education during their productive years has had consequences on the current socioeconomic status of many elderly people. This includes the inability to create an affordable retirement funding plan. Consequently, the lower level of education and socioeconomic status in turn has negative implications on the health and social well-being of elderly persons (Ramashala 2002; Makiwane and Kwizera 2006).

One way to address old-age poverty is through social pensions. They are non-contributory cash benefits paid to older people, either universally or sometimes subject to a means test (Samson and Kaniki 2008). Historically, social security in South Africa was one way of meeting the needs of the white minority. Since the onset of democracy in 1994, however, the social security system has undergone reform and provides support to men and women over the age of 60 via the State Old Age Pension (Samson et al. 2006). The effectiveness of South Africa's social security system in improving the welfare of beneficiaries has been widely recognised as evidenced by a reduction in hunger and extreme poverty, while improving health care, education and gender equality (Samson and Kaniki 2008). It must be acknowledged however that retirement from active employment marks the onset of 'dependence' on the state and significant others to provide for the needs of the elderly. Such transition has implications on the immediate elderly person, the family, the community and the state, as they provide for their social, health and financial needs. A longer working life is beneficial not only to individuals in the form of improved health and well-being, but also to the state through tax revenue contribution, and therefore this needs to be encouraged through progressive ageing-related policies (Statistics South Africa 2014a).

Recommending a retirement age however is a complex task. This is particularly so if population group consideration is thrown into the mix; with the health variation across population groups provided in this paper. The age dilemma is also relevant to social pensions. For example, some population groups may be "eligible" for the grant way before the prescribed age of 60 years and others long after. Potential challenges are *inter alia* cost implications, and feasibility of equitable policies, particularly by population group. One way of tackling the life and health expectancy divide would be to address the socioeconomic environment that contributes to the differential health outcomes observed at both the levels of population group and gender.

The deterioration in the health of older adults shown by the declining health expectancy is a cause for concern, particularly for Black Africans who have a long history of playing caregiver roles to their grandchildren. These households also known as skip generation households, where only children and older adults are found, are potentially vulnerable (Statistics South Africa 2014a).

## 17.6 Conclusion and Recommendations

The use of health expectancy demonstrated important ageing disparities across population groups in South Africa. With age, similar life and health expectancy was observed among Coloureds, Indians/Asians and Black Africans, while a discernible difference existed among Indians/Asians and Whites, although the Coloured population group were found to be the most disadvantaged. Life expectancy and disability-free life years for Black Africans mirrored the South African average because they form the majority of the population. The findings supported the "health-survival paradox" in disability-free life expectancy and the general differences in health expectancy suggested in the study has implications for retirement ages, which is currently 60. The State Old Age Pension is an important mechanism that contributes to reduction in hunger and extreme poverty, while improving health care, education and gender equality.

This study is important because it showed population group heterogeneity which characteristically gets masked at national level, while at the same time presenting policy challenges. The study has clear policy implications, one of them being the need for age-appropriate planning for health and social services. Further research is required to explore the rapid fall in life expectancy, including the causes of death to establish whether deaths in the various population groups are premature deaths or conditions associated with ageing.

**Acknowledgements** This is a revised version of a paper written at the 2013/14 Southern African Young Scientists Summer Programme (SA-YSSP). The author therefore acknowledges the International Institute of Applied Systems Analysis, University of the Free State, South Africa's National Research Foundation and Department of Science and Technology for co-ordinating and funding the SA-YSSP and Statistics South Africa for allowing access to the datasets used for the study. Many thanks also go to Warren C. Sanderson, Sergei Scherbov and Nancy Phaswana-Mafuya, who introduced the concept of new ways of measuring ageing during the programme and helped to conceptualise and develop the research, and to anonymous individuals for invaluable comments on earlier drafts of the paper. The views expressed in this paper are however, those of the author and do not represent any organisations or individuals mentioned here.

## References

Anderson, N. B., Bulatao, R. A., Cohen, B., & National Research Council. (U.S.). (2004). *Critical perspectives on racial and ethnic differences in health in late life.* National Academies Press.

Bah, S. (1999). The making and unmaking of a national but stratified vital statistics system in the republic of South Africa and the new making of a more comprehensive vital statistics system. *South African Journal of Demography, 2001,* 45–50.

Bloom, D. E. & Canning, D. (2004). *Proceedings—Economic policy symposium.* http://www.nber.org/papers/w10817. Accessed February 2017.

Chirinda, W., & Chen, H. (2017). Comparative study of disability-free life expectancy across six low- and middle-income countries. *Geriatr Gerontol Int, 17*(4), 637–644.

Christopher, A. J. (2002). 'To define the indefinable': Population classification and the census in South Africa. *Area, 34*(4), 401–408.

City of Cape Town. (2010). Discussion paper: Demographics Scenario.

Coovadia, H., Jewkes, R., Barron, P., Sanders, D., & McIntyre, D. (2009). The health and health system of South Africa: Historical roots of current public health challenges. *The Lancet, 374* (9692), 817–834.

Craig, J. (1994). Replacement level fertility and future population growth. *Population Trends, Winter* (78), 20–22.

Dorrington, R. (2013). Alternative South African midyear estimates, 2013. Centre for Actuarial Research Monograph 13, University of Cape Town.

Dorrington, R., Bradshaw, D., & Laubscher, R. (2014). *Rapid mortality surveillance report 2012.* Cape Town: South African Medical Research Council.

Golaz, V., Nowik, L., & Sajoux, M. (2012). Africa, a Young but Ageing Continent. *Population & Societies, 491.*

Goodrick, W., & Pelser, A. (2014). The Greying of a rainbow nation: Policy responses to the implications of population ageing in South Africa. *African Population Studies: Supplement on Population Issues in South Africa, 28*(1).

Imai, K., & Soneji, S. (2007). On the estimation of disability-free life expectancy. *Journal of the American Statistical Association, 102*(480), 1199–1211. https://www.ncbi.nlm.nih.gov/pubmed/26279593.

Jagger, C., Oyen, H. V., & Robine, J-M. (2014). *Health expectancy calculation by the Sullivan method: A practical guide*. Newcastle University Institute of Ageing.

Khalfani, A. K., Zuberi, T., Bah, S., & Lehohla, P. J. (2005). Population statistics. In T. Zuberi, A. Sibanda, & E. Udjo (Eds.), *The demography of South Africa* (Vol.). New York: M.E. Sharpe.

Kinsella, K., & Ferreira, M. (1997). Aging trends: South Africa. *International Brief*: U.S. Bureau of the Census.

Kinsella, K., & Velkoff, V. (2001). *An aging world: 2001 P95/01-1*. Washington: US Government Printing Office: US Census Bureau.

Madans, J., Loeb, M., & Altman, B. (2011). Measuring disability and monitoring the UN convention on the rights of persons with disabilities: The work of the Washington group on disability statistics. *BMC Public Health, 11*(4), 1–8.

Makiwane, M., & Kwizera, S. A. (2006). An investigation of quality of life of the elderly in South Africa, with specific reference to Mpumalanga. *Applied Research in Quality of life, 1*(3–4), 297–313.

Malherbe, K. (2007). Older persons act: Out with the old and in the with the older? *Law, Democracy & Development, 11*(2001).

Mathers, C. D., Sadana, R., Salomon, J. A., Murray, C. J. L., & Lopez, A. D. (2001). Healthy life expectancy in 191 countries, 1999. *The Lancet, 357*(9269), 1685–1691.

Moultrie, T., & Dorrington, R. (2011). Used for Ill; used for good: A century of collecting data on race in South Africa. *Ethnic and Racial Studies, 35*(8), 1447–1465.

Moultrie, T., & Timaeus, I. (2002). *Trends in South African fertility between 1970 and 1998: An analysis of the 1996 census and the 1998 demographic and health survey technical report*. Cape Town: Burden of Disease Research Unit: Medical Research Council.

Oksuzyan A., Juel K., Vaupel J. W., & Christensen, K. (2008). Men: good health and high mortality. Sex differences in health and aging. *Aging Clinical and Experimental Research, 20* (2), 91–102.

Phaswana-Mafuya, N., Peltzer, K., Chirinda, W., Kose, Z., Hoosain, E., Ramlagan, S., et al. (2013). Self-rated health and associated factors among older South Africans: Evidence from the study on global ageing and adult health. *Global Health Action, 6*, 10.

Ralston, M., Schatz, E., Menken, J., Gómez-Olivé, F. X., & Tollman, S. (2016). Who benefits—or does not—from South Africa's old age pension? Evidence from characteristics of rural pensioners and non-pensioners. *International Journal of Environmental Research and Public Health, 13*(2001), 85. https://doi.org/10.3390/ijerph13010085.

Ramashala, M. F. (2002). *Living arrangements, poverty and the health of older persons in Africa*. Department of Economic and Social Affairs. Population Division. United Nations. http://www.un.org/esa/population/publications/bulletin42_43/ramashala.pdf. Accessed February 2017.

Robine, J.-M., Romieu, I., & Cambois, E. (1999). Health expectancy indicators. *Bulletin of the World Health Organization, 77*(2), 181–185.

Sagner, A., Dowd, E. J., & Lowal, P. (2002). *Defining "Old Age". Markers of old age in Sub-Saharan Africa and the implications for crosscultural research*. Technical report Commissioned by WHO Minimum Data Set (MDS) Project. WHO.

Samson, M., & Kaniki, S. (2008). Social pensions as developmental social security for Africa. In D. Hailu, & F. V. Soares (Eds.), *Cash transfers in Africa and Latin America: An overview*. Brazil: International Poverty Centre.

Samson, M., MacQuene, K., & Van Niekerk, I. (2006). Social grants, South Africa (Policy brief 1). https://www.odi.org/sites/odi.org.uk/files/odi-assets/publications-opinion-files/1688.pdf. Accessed February 2017.

Shisana, O., Rehle, T., Simbayi, L., Zuma, K., Jooste, S., N, Z., et al. (2014). *South African national Hiv prevalence, incidence and behaviour survey, 2012*. Cape Town: South Africa: HSRC Press.

Shoko, M., Collinson, M. A, Lefakane, L., Kahn, K., & Tollman, S. M. (2016). What can we learn about South African households by comparing the national Census 2011 with the Agincourt Health and Demographic surveillance system data in rural northeast Mpumalanga? *African Population Studies 30*(2).

Statistics South Africa. (2012a). *Statisticial release (revised): Census 2011*. Pretoria: South Africa.

Statistics South Africa. (2012b). *Cause of Death Certification: A Guide for Completing the Notice of Death/Stillbirth (DHA-1663)*. Pretoria: South Africa.

Statistics South Africa. (2014a). *Census 2011: Profile of older persons in South Africa*. Pretoria: South Africa.

Statistics South Africa. (2014b). *Mortality and causes of death in South Africa, 2011: Findings from Death Notification*. Pretoria: South Africa.

Statistics South Africa. (2014c). *Census 2011: Profile of persons with disabilities in South Africa*. Pretoria: South Africa.

Sullivan, D. F. (1971). A single index of mortality and morbidity. *HSMHA Health Reports, 86*(4), 347–354.

The World Bank (2017). GINI index (World Bank estimate). http://databank.worldbank.org/data/Views/Metadata/MetadataWidget.aspx?Name=GINI%20index%20(World%20Bank%20estimate)&Code=SI.POV.GINI&Type=S&ReqType=Metadata&ddlSelectedValue=SAU&ReportID=43276&ReportType=Table. Accessed October 2017.

Udjo, E. (2011). Workforce and retirement age: Modelling the impact of mandatory retirement age on the size of the workforce in tertiary sector organisations in South Africa. *South African Journal of Labour Relations, 35*(2).

Udjo, E. (2014). Estimating the demographic parameters from the 2011 South African Population Census. *Supplement on Population Issues in South Africa, 28*(1).

Udjo, E. (2017). Can estimating completeness of death registration be used as evidence of inaccuracy of population size estimates from a census? The Case of the 2011 South African Population Census. *African Population Studies 31*(1).

Udjo, E., & van Aardt, C. (2012). Evaluating the demographic, economic and socioeconomic aspects of the 2011 South African Census. UNISA Bureau of Market Research.

UNDP. (2013). *Human development report 2013: The rise of the south: Human progress in a diverse world*. New York.

WHO. (2016). *Social determinants of health: Key concepts*. http://www.who.int/social_determinants/thecommission/finalreport/key_concepts/en/. Accessed September 19, 2016.

# Erratum to: Cities as Forces for Good in the Environment: A Systems Approach

**M. Bruce Beck, Dillip K. Das, Michael Thompson, Innocent Chirisa, Stephen Eromobor, Serge Kubanza, Tejas Rewal and Everardt Burger**

## Erratum to:
## Chapter 2 in: P. Mensah et al. (eds.), *Systems Analysis Approach for Complex Global Challenges*, https://doi.org/10.1007/978-3-319-71486-8_2

In the original version of the book, the following corrections have to be incorporated in Chapter 2:

Corresponding author name "Bruce M. Beck" has to be changed to read as "M. Bruce Beck".

The inadvertently deleted term "Kinshasa" in the abstract has to be included again.

Chapter author name "Tegas Rewal" has to be changed to read as "Tejas Rewal".

The erratum chapter and the book have been updated with the changes.

---

The updated online version of this chapter can be found at
https://doi.org/10.1007/978-3-319-71486-8_2



P. Mensah et al. (eds.), *Systems Analysis Approach for Complex Global Challenges*,
https://doi.org/10.1007/978-3-319-71486-8_18

# Index





P. Mensah et al. (eds.), *Systems Analysis Approach for Complex Global Challenges*,
https://doi.org/10.1007/978-3-319-71486-8

The manufacturer's authorised representative in the EU is Springer
Nature Customer Service Centre GmbH, Europaplatz 3, 69115 Heidelberg,
Germany. If you have any concerns regarding our products, please
contact ProductSafety@springernature.com

Printed and bound by CPI Group (UK) Ltd, Croydon, CR0 4YY
30/09/2026
02225662-0012